# Dunes of the World

**Series Editors**

Sand dune systems of Quaternary age occur on all continents and at all latitudes and comprise coastal dunes and inland or continental dunes. Coastal dunes are best developed along windward coasts, where sand sized sediment is abundant. Inland dunes occur primarily in low- and mid-latitude arid and semi-arid regions, although there are many examples of cold climate dune systems. Inland dune systems are sensitive to the effects of anthropogenic disturbance (grazing, agriculture, off-road vehicles), as well as climate change and variability (drought cycles). Coastal dunes are impacted by coastal development, storms, and sea level change.

**Aims & Scope**

This series of volumes is intended to provide students and professionals in earth and environmental sciences with an overview of major coastal and inland dune fields. Information will facilitate decision-making and environmental management. The volumes will be regionally-based and will provide up to date information and reviews of dune field characteristics (morphology, vegetation, sediments), sediment sources, dune field history and response to climate and sea level change past, present and future. Volumes may also provide information on dune (field) processes; relations between geomorphology and ecosystem processes (e.g. dune vegetation and its effects on sediment transport and erosion and deposition patterns); dune flora and fauna; habitat restoration etc.

More information about this series at http://www.springer.com/series/15468

Nicholas Lancaster • Patrick Hesp
Editors

# Inland Dunes of North America

Springer

*Editors*
Nicholas Lancaster
Desert Research Institute
Reno, NV, USA

Patrick Hesp
College of Science and Engineering
Flinders University
Adelaide, SA, Australia

ISSN 2509-7806 ISSN 2509-7814 (electronic)
Dunes of the World
ISBN 978-3-030-40497-0 ISBN 978-3-030-40498-7 (eBook)
https://doi.org/10.1007/978-3-030-40498-7



This Springer imprint is published by the registered company Springer Nature Switzerland AG.
The registered company address is: Gewerbestrasse 11, 6330 Cham, Switzerland

# Contents

# Editors and Contributors

## About the Editors

**Nicholas Lancaster** Desert Research Institute, Reno, NV, USA

**Patrick Hesp** College of Science and Engineering, Flinders University, Adelaide, SA, Australia

## Contributors

**Brian Bodenbender** Geological and Environmental Sciences Department, Hope College, Holland, MI, USA

**Robin Davidson-Arnott** Department of Geography, University of Guelph, Guelph, ON, Canada

**Suzanne DeVries-Zimmerman** Geological and Environmental Sciences Department, Hope College, Holland, MI, USA

**Ryan C. Ewing** Department of Geology and Geophysics, Texas A&M University, College Station, TX, USA

**Alan F. Halfen** Department of Geography, University of Wisconsin-Milwaukee, Milwaukee, WI, USA

**Edward Hansen** Geological and Environmental Sciences Department, Hope College, Holland, MI, USA

**Paul R. Hanson** CSD, School of Natural Resources, University of Nebraska, Lincoln, NE, USA

**William C. Johnson** Department of Geography and Atmospheric Sciences, University of Kansas, Lawrence, KS, USA

**Zoran Kilibarda** Department of Geosciences, Indiana University Northwest, Gary, IN, USA

**Aaron N. Koop** Department of Geography and Atmospheric Sciences, University of Kansas, Lawrence, KS, USA

**David B. Loope** University of Nebraska-Lincoln, Lincoln, NE, USA

**Joseph A. Mason** University of Wisconsin, Madison, WI, USA

**Margaret H. Redsteer** School of Interdisciplinary Arts & Sciences, University of Washington Bothell, Bothell, WA, USA

**Christopher S. Swezey** U.S. Geological Survey, Reston, VA, USA

**James B. Swinehart** University of Nebraska-Lincoln, Lincoln, NE, USA

**Todd Thompson** Indiana Geological Survey, Indiana University, Bloomington, IN, USA

**Andrew Valdez** Great Sand Dunes National Park and Preserve, National Park Service, Mosca, CO, USA

**Deanna van Dijk** Geology, Geography and Environmental Studies Department, Calvin College, Grand Rapids, MI, USA

**Brian Yurk** Mathematics Department, Hope College, Holland, MI, USA

**James R. Zimbelman** Center for Earth and Planetary Studies, National Air and Space Museum, Smithsonian Institution, Washington, DC, USA

# About the Editors

**Patrick Hesp** (PhD; DSc) is a Strategic Professor of Coastal Studies, College of Science and Engineering at Flinders University, Australia. He has held academic positions in NSW, Western Australia, Singapore, USA, and NZ; non-academic positions in the WA State Department of Agriculture, Geomarine P/L, and the Rottnest Island Authority; held visiting professorships and fellowships in South Africa, Namibia, Israel, Holland, China, Brazil, Italy, Malaysia, Thailand, and France; and has worked on beaches and coastal and desert dunes all over the world. He is an expert on coastal dune geomorphology and has published over 290 articles in his career to date.

**Nicholas Lancaster** is an Emeritus Research Professor at the Desert Research Institute, Nevada, USA. His decades of research on sand dunes has taken him to deserts in Africa (Namib, Kalahari, northern and western Sahara), Arabia, Antarctica, and the western United States (Mojave and Sonoran Deserts). His work has resulted in more than 150 scientific papers and several books and has been recognized by awards from the Geological Society of America, the Association of America Geographers, the International Society for Aeolian Research, the International Quaternary Association, and the Nevada System of Higher Education.

# Chapter 1
# Introduction to Inland Dunes of North America

**Nicholas Lancaster and Patrick Hesp**

**Abstract** This chapter provides an introduction to the volume and summarizes the occurrence of inland dunes in North America, the history of dune studies, and aspects of dune chronology.

**Keywords** Dune fields · USA · Canada · Mexico · Luminescence chronology · Sediment supply

## 1.1 Introduction

Inland sand dunes are widespread in North America and are found from the North Slope of Alaska to the Sonoran Desert in northern Mexico and from the Delmarva Peninsula in the east to Southern California in the west (Fig. 1.1). They cover an area of approximately 459,165 km$^2$ of the United States and 42,000 km$^2$ of Canada (Wolfe et al. 2009). Many of these dune fields are small and isolated, and are now stabilized by vegetation and inactive or degraded in current conditions of climate and sand supply. In combination with luminescence and radiocarbon dating of periods of aeolian accumulation or stability, these dune systems provide information on past environmental conditions, including past wind regimes and periods of drought. Active (vegetation-free or sparsely vegetated) dunes are mostly restricted to parts of the southern Great Plains and the deserts of the Southwestern USA and Northern Mexico, although small areas of active dunes do occur in boreal locations, e.g. Great Kobuk Sand Dunes, Alaska (Mann et al. 2002).

N. Lancaster (✉)
Desert Research Institute, Reno, NV, USA
e-mail: nick.lancaster@dri.edu

P. Hesp (✉)
College of Science and Engineering, Flinders University, Adelaide, SA, Australia
e-mail: patrick.hesp@flinders.edu.au



N. Lancaster, P. Hesp (eds.), *Inland Dunes of North America*,
Dunes of the World, https://doi.org/10.1007/978-3-030-40498-7_1

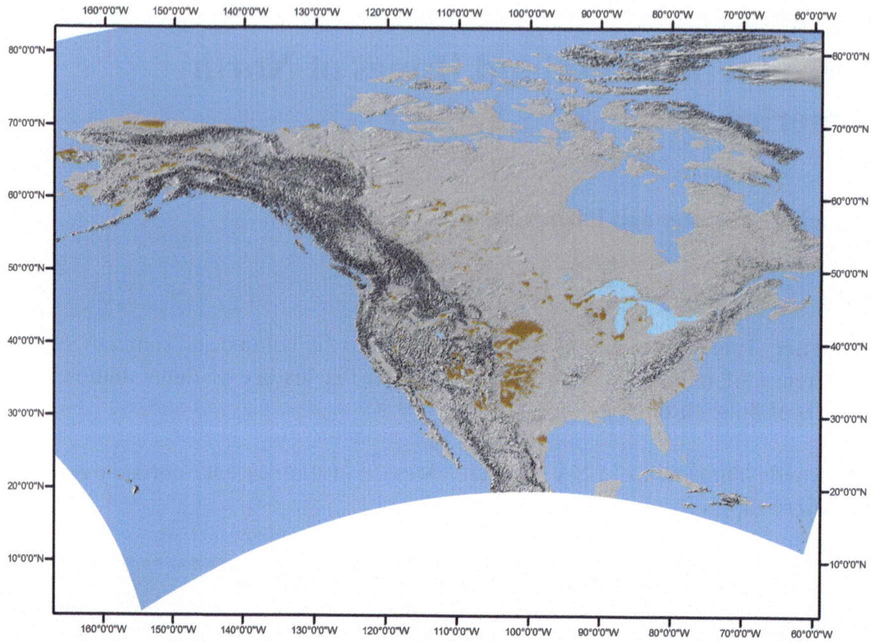


**Fig. 1.1** Inland dune systems of North America. Dunefield extent from Wolfe et al. (2009) and Soller et al. (2009), supplemented by Lancaster mapping

In this volume, we provide an overview of and highlight recent research on areas of inland dunes in North America that span a range from those that are actively accumulating in current conditions of climate and sediment supply to those that were formed in past conditions and are now degraded relict systems. The contributions include detailed analyses of individual active dune systems at White Sands, New Mexico; Great Sand Dunes, Colorado; and the Laurentian Great Lakes; as well as the vegetation-stabilized dunes of the Nebraska Sand Hills and the Colorado Plateau. Additional chapters discuss the widespread partially vegetated dune systems of the central and southern Great Plains; the relict dunes of the Atlantic Coastal Plain of the eastern USA; and active and stabilized dunes of the Colorado Plateau and the southwestern deserts of the USA and northern Mexico.

## 1.2 Inland Dune Studies in North America

There is a long history of observations and studies of inland dunes in North America. European travelers and survey parties noted and, in some cases, mapped the occurrence of dunes (often referred to as "sand hills"). Their observations provide a valuable source of information on the state of dune fields on the Great Plains in the nineteenth and early twentieth century, as discussed by Muhs and Holliday (1995).

Many of these early observers also commented on the scenic beauty of the dunes. For example Russell (1885) in his studies of the Lake Lahontan basin noted

"The sand here is of a light creamy-yellow color, and forms beautifully curved ridges and waves that are covered with fret-work of wind-ripples, and frequently marked in the most curious manner by the foot-prints of animals thus forming strange hieroglyphics that are sometimes difficult to translate". Zebulon Pike happened on the Great Sand Dunes of Colorado in January 1807 and observed that the dunes appeared "exactly that of a sea in a storm (except as to color) and not the least sign of vegetation".

Mapping of soils and Quaternary deposits in the late nineteenth and early twentieth centuries provided important information on the nature and extent of dunes in the Midwest and Northeastern states (see references in Cooper (1935)), and in southern California (Thompson 1929). The availability of aerial photographs in the 1920s and 1930s prompted more systematic investigations. One of the first to provide a comprehensive and detailed classification of dunes and to assess geomorphic and age relations between different generations of dunes was the work of Melton (1940), in the southern High Plains. Melton also suggested that dune-forming wind regimes had changed over time from northwesterly to southerly, a change confirmed by more recent studies (see Halfen and Johnson (2013) and Sridhar et al. (2006)). Working at the same time, Hack (1941) mapped dunes in NE Arizona and provided a seminal classification of dune type in relation to vegetation cover, sand supply, and wind energy.

The first compilation of the extent of dune areas in the USA and parts of Canada was undertaken by Thorp and Smith (1952) who published a map of sand and loess deposits, based on state-by-state soil mapping. More detailed regional surveys of dune occurrence and characteristics include those by Eymann (1953) and Dean (1978) for deserts in southern California. H.T.U Smith and his son Roger (R.S.U.) Smith compiled major surveys of dunes for the central Great Plains (Smith 1965) and the North American deserts (Smith 1982). H.T.U. Smith was, in addition, one of the first to recognize the importance of past wind action in shaping the dune systems of the Mojave Desert (Smith 1967).

Despite the widespread nature of dune areas in North America, major reviews of Quaternary landforms and deposits such as Wright and Frey (1965) and Schultz and Frye (1965) focused on the extensive loess deposits of North America. It was not until the work of Busacca et al. (2003) and Muhs and Zárate (2001) that comprehensive reviews of dune areas and their context were attempted. The mapping by Thorp and Smith (1952) was updated by GIS based mapping that covers all northern areas of North America (Wolfe et al. 2009), and dune and sand sheet areas in the conterminous USA are included in the USGS digital surficial deposit map compilation of Soller et al. (2009). Additional regional studies of dune distribution and chronology are provided by Halfen and Johnson (2013) for the central and southern Great Plains; Muhs and Wolfe (1999) and Wolfe et al. (2004) for the northern Great Plains; and Markewich et al. (2015) for the eastern USA; while dune distribution and characteristics in the deserts of the southwestern USA and northern Mexico are summarized by Lancaster (this volume).

Studies of dune fields in North America have provided understanding of many fundamental aspects of dune dynamics and history. Landmark investigations include studies of the internal sedimentary structure of dunes at White Sands, New Mexico (McKee 1966); and the pioneering investigations of the Algodones, Salton Sea, and Kelso Dunes in California (Norris 1966; Norris and Norris 1961; Sharp 1966), which provided the background for many subsequent investigations of dune dynamics and sediment sources. Although North American dunes were not the primary focus of the USGS Global Sand Seas project of the 1970s, the approaches inspired by this group led to many important advances, including work on cold climate dunes (Ahlbrandt and Andrews 1978), sand sheets (Fryberger et al. 1979), and the sedimentology of Great Sand Dunes, Colorado (Andrews 1981). The recognition of dunes on Mars provided a great incentive for terrestrial analogue studies of dunes, including those in the deserts of the southwestern USA (Breed 1977; Greeley 1986) and also resulted in studies of dune fields using remote sensing data sets (e.g. Blount et al. 1990; Paisley et al. 1991; Ramsey et al. 1999). Renewed interest in planetary dunes has come as a result of the data from Mars Science Laboratory Curiosity Rover, prompting new investigations of terrestrial analogues in North America (Ewing et al. 2015; Szynkiewicz et al. 2010).

Studies of modern dune sediments as a means to better interpret the characteristics of ancient aeolian sandstones of the Colorado Plateau and elsewhere has motivated multiple studies in the Desert Southwest, (e.g. Havholm and Kocurek 1988; Hunter 1977; Kocurek and Nielson 1986; Nielson and Kocurek 1986; Simpson and Loope 1985). The application of geochemical and mineralogical methods to understand dune sand provenance, especially in the Plains and Desert Southwest, was pioneered by Muhs and colleagues, and is summarized in Muhs (2017).

The creation of better instrumentation, an increased understanding of flow dynamics, computer modeling, and realization of the importance of climate and vegetation changes to dune activity has resulted in important investigations of winds and sediment transport on dunes based on field experiments in North America, (e.g. Barchyn and Hugenholtz 2012b; Frank and Kocurek 1994; Lancaster 1989; Lancaster et al. 1996; McKenna Neuman et al. 1997; Pelletier and Jerolmack 2014; Sweet and Kocurek 1990; Walker and Nickling 2003), with applications to both inland and coastal dune systems.

## 1.3 Dune History and Chronology

Understanding of dune field history may provide information on past periods of aridity and dune building, as exemplified by research into the history of dune accumulation on the Great Plains of the USA and Canada, where the response of these dune systems to episodes of severe drought and the possible effects of global warming has prompted many studies (Barchyn and Hugenholtz 2012a; Barchyn and Hugenholtz 2013; Miao et al. 2007; Muhs and Maat 1993; Wolfe et al. 2006).

Dune orientations separately, or in combination with data on loess thickness and particle size trends, provide information on past wind regimes, for the last glacial maximum period (Markewich et al. 2015; Mason et al. 2011), and for Holocene drought episodes (Schmeisser et al. 2010; Sridhar et al. 2006). Such data sets are valuable in making model-data comparisons and to validate paleo-climate models (Conroy et al. 2019).

Numerical chronologies for periods of dune accumulation and stability in North America were first developed using conventional $^{14}C$ ages of organic matter from palaeosols and peat layers (e.g. Filion 1987). Subsequently, chronologies were developed using accelerator mass spectrometry (AMS) $^{14}C$ dates (Ahlbrandt et al. 1983; Mason et al. 2004). These chronologies not only bracket periods of sand accumulation, but provide useful information on periods of stability, especially when the ages are from paleosols. They are, however, limited by the availability of organic horizons in dunes, which restricts their utility to dunes in more humid areas, or dunes associated with wetlands (Mehringer and Warren 1976).

With the development and widespread application of luminescence dating techniques that provide a direct age for periods of aeolian sand accumulation, luminescence-dated numerical chronologies have been developed, beginning with the work of Forman and Maat (1990) in Colorado and Edwards (1993) at Kelso Dunes, California. These investigations used TL (Thermoluminesence) and IRSL (Infra-red stimulated luminescence), respectively. Subsequent studies have mostly employed OSL (Optically stimulated luminescence) with SAR protocols, especially on the Great Plains, where quartz-rich sands provide consistent results. In the Great Basin and Mojave deserts, however, feldspar-rich dune sands favor use of IR stimulated luminescence protocols.

The available chronologic information was summarized for dune areas in Canada and the USA north of 38°N by Wolfe et al. (2009) and then comprised 163 luminescence and 880 radiocarbon dates. This database provided the basis for a global chronologic database – the INQUA Dunes Atlas database (Lancaster et al. 2016). Currently, there are 1286 luminescence dates in the database for North America (Canada, Mexico, and the USA). Their spatial distribution is shown in Fig. 1.2. A review and interpretation of these ages is provided by Halfen et al. (2015). It is clear from Fig. 1.2 that the coverage of dated sites is uneven. In particular, there are relatively fewer published ages from dunes in the southern Great Plains, the intermountain west, Mexico, and Alaska. The temporal distribution of ages for the region is complex: multiple periods of Holocene dune accumulation and reworking have occurred and indicate the sensitivity of dunes in many areas to climate change.

Given the widespread distribution of active and vegetation-stabilized dunes in North America, it might be expected that the boundary conditions of sediment supply, availability and mobility (Kocurek and Lancaster 1999) would be similarly diverse. However, this does not appear to be the case. In areas adjacent to the Laurentide Ice Sheet, deglaciation provided an abundant source of sand from glaciofluvial deposits, leading to the formation of dune fields throughout the northern Plains and the upper Midwest (Arbogast et al. 2015; Halfen et al. 2015). Elsewhere formation of dune fields in many areas is clearly linked to enhanced sediment supply from fluvial sources, as in the Great Plains (Halfen and Johnson 2013) and the

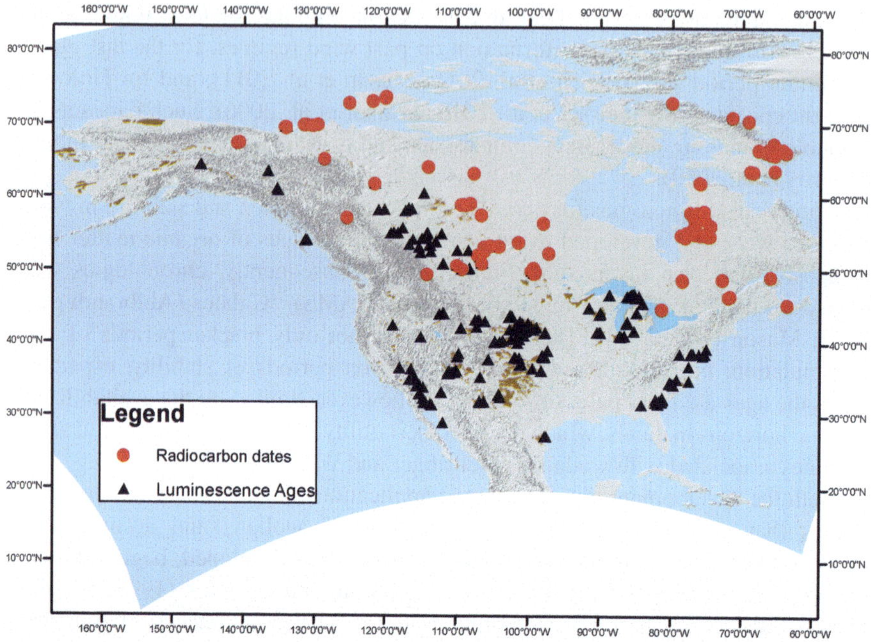


**Fig. 1.2** Luminescence and radiocarbon dated dunes sites in North America. Dune extent as Fig. 1.1. Sites from INQUA Dunes Atlas Chronologic Database, http://inquadunesatlas.dri.edu

southeast coastal plain (Swezey et al. 2016). The record is more complex in areas of the southwestern deserts, in part because of the lack of luminescence ages, but fluvial sources are clearly indicated for the Algodones and Parker dunes (Muhs et al. 2003).

## 1.4 Conclusions

The widespread occurrence of dune fields in North America is indicative of the importance of aeolian activity in many different landscapes, from the margins of the boreal forest to hot deserts. The occurrence of the dune fields and their history reflect a variety of boundary conditions, including increased sediment supply during the late Pleistocene and Pleistocene-Holocene transition; and mid- to late-Holocene drought periods. The variety of dune field environments has promoted a range of investigations, from modern dune dynamics to Quaternary history. These different approaches are well-exemplified in this volume of studies. They also indicate the areas in which further research is needed, including application of modern luminescence dating techniques to dunes in the desert southwest.

## References

Ahlbrandt TS, Andrews S (1978) Distinctive sedimentary features of cold-climate eolian deposits, North Park, Colorado. Palaeogeogr Palaeoclimatol Palaeoecol 25:327–351

Ahlbrandt TS, Swinehart JB, Maroney DG (1983) The dynamic Holocene dune fields of the Great Plains and Rocky Mountain Basins, U.S.A. In: Brookfield ME, Ahlbrandt TS (eds) Eolian sediments and processes. Developments in Sedimentology. Elsevier, Amsterdam/Oxford/New York/Tokyo, pp 379–406

Andrews S (1981) Sedimentology of Great Sand Dunes, Colorado. In: Ethridge FP, Flores RM (eds) Recent and ancient non marine depositional environments: models for exploration. The Society of Economic Paleontologists and Mineralogists, Tulsa, pp 279–291

Arbogast AF, Luehmann MD, Miller BA, Wernette PA, Adams KM, Waha JD, O'Neil GA, Tang Y, Boothroyd JJ, Babcock CR, Hanson PR, Young AR (2015) Late-Pleistocene paleowinds and aeolian sand mobilization in north-central Lower Michigan. Aeolian Res 16(0):109–116

Barchyn TE, Hugenholtz CH (2012a) Aeolian dune field geomorphology modulates the stabilization rate imposed by climate. J Geophys Res 117(F2):F02035

Barchyn TE, Hugenholtz CH (2012b) Winter variability of aeolian sediment transport threshold on a cold-climate dune. Geomorphology 177: 38–50

Barchyn TE, Hugenholtz CH (2013) Dune field reactivation from blowouts: Sevier Desert, UT, USA. Aeolian Res 11(0):75–84

Blount G, Smith MO, Adams JB, Greeley R, Christensen PR (1990) Regional aeolian dynamics and sand mixing in the Gran Desierto: evidence from Landsat Thematic Mapper images. J Geophys Res 95(B10):15463–15482

Breed CS (1977) Terrestrial analogs of the Hellespontus dunes, Mars. Icarus 30:326–340

Busacca, A.J., Begét, J.E., Markewich, H.W., Muhs, D.R., Lancaster, N., Sweeney, M.R., 2003. Eolian sediments. In: A.R. Gillespie, S.C. Porter, B.F. Atwater (Eds.), Developments in Quaternary Sciences. Elsevier, Amsterdam pp. 275–309

Conroy JL, Karamperidou C, Grimley DA, Guenthner WR (2019) Surface winds across eastern and midcontinental North America during the Last Glacial Maximum: a new data-model assessment. Quat Sci Rev 220:14–29

Cooper WS (1935) The history of the Upper Mississippi River in late Wisconsin and postglacial time

Dean LE (1978) The California Desert Sand Dunes. University of California, Riverside

Edwards SR (1993) Luminescence dating of sand from the Kelso Dunes, California. In: Pye K (ed) Dynamics and environmental context of aeolian sedimentary systems. Geological Society of London, Special Publication, London, pp 59–68

Ewing RC, McDonald GD, Hayes AG (2015) Multi-spatial analysis of aeolian dune-field patterns. Geomorphology 240(0):44–53

Eymann JL (1953) A study of sand dunes in the Colorado and Mojave Deserts. M.S., University of Southern California, Los Angeles, 91 pp

Filion L (1987) Holocene development of parabolic dunes in the central St. Lawrence lowland, Quebec. Quat Res 28:196–209

Forman SL, Maat P (1990) Stratigraphic evidence for late quaternary dune activity near Hudson on the Piedmont of northern Colorado. Geology 18(8):745–748

Frank A, Kocurek G (1994) Effects of atmospheric conditions on wind profiles and aeolian sand transport with an example from White Sands National Monument. Earth Surf Process Landf 19(8):735–745

Fryberger S, Ahlbrandt T, Andrews S (1979) Origin, sedimentary features, and significance of low-angle eolian "sand sheet" deposits, Great Sand Dunes National Monument and vicinity, Colorado. J Sediment Petrol 49(3):733–746

Greeley R (1986) Aeolian landforms: laboratory simulations and field studies. In: Nickling WG (ed) Aeolian geomorphology. Allen and Unwin, Boston/London/Sydney, pp 195–211
Hack JT (1941) Dunes of the Western Navajo County. Geogr Rev 31(2):240–263
Halfen AF, Johnson WC (2013) A review of Great Plains dune field chronologies. Aeolian Res 10:135–160
Halfen AF, Lancaster N, Wolfe SA (2015) Interpretations and common challenges of aeolian records from North American dune fields. Quat Int 410 (Part B): 75–95
Havholm KG, Kocurek G (1988) A preliminary study of the dynamics of a modern draa, Algodones, southeastern California, USA. Sedimentology 35:649–669
Hunter RE (1977) Basic types of stratification in small eolian dunes. Sedimentology 24:361–388
Kocurek G, Lancaster N (1999) Aeolian system sediment state: theory and Mojave Desert Kelso dune field example. Sedimentology 46:505–515
Kocurek G, Nielson J (1986) Conditions favourable for the formation of warm-climate aeolian sand sheets. Sedimentology 33:795–816
Lancaster N (1989) The dynamics of star dunes: an example from the Gran Desierto, Mexico. Sedimentology 36:273–289
Lancaster N, Nickling WG, McKenna Neuman CK, Wyatt VE (1996) Sediment flux and airflow on the stoss slope of a barchan dune. Geomorphology 17(1-3):55–62
Lancaster N, Wolfe S, Thomas D, Bristow C, Bubenzer O, Burrough S, Duller G, Halfen A, Hesse P, Roskin J, Singhvi A, Tsoar H, Tripaldi A, Yang X, Zárate M (2016) The INQUA Dunes Atlas chronologic database. Quat Int 410(Part B):3–10
Mann DH, Heiser PA, Finney BP (2002) Holocene history of the Great Kobuk Sand Dunes, Northwestern Alaska. Quat Sci Rev 21(4–6):709–731
Markewich HW, Litwin RJ, Wysocki DA, Pavich MJ (2015) Synthesis on Quaternary aeolian research in the unglaciated eastern United States. Aeolian Res 17(0):139–191
Mason JA, Swinehart JB, Goble RJ, Loope DB (2004) Late-Holocene dune activity linked to hydrological drought, Nebraska Sand Hills, USA. Holocene 14(2):209–217
Mason JA, Swinehart JB, Hanson PR, Loope DB, Goble RJ, Miao X, Schmeisser RL (2011) Late Pleistocene dune activity in the central Great Plains, USA. Quat Sci Rev 30(27,Äì28):3858–3870
McKee ED (1966) Structures of dunes at White Sands National Monument, New Mexico (and a comparison with structures of dunes from other selected areas). Sedimentology 7(1):1–69
McKenna Neuman C, Lancaster N, Nickling WG (1997) Relations between dune morphology, air flow, and sediment flux on reversing dunes, Silver Peak, Nevada. Sedimentology 44(6):1103–1114
Mehringer PJ, Warren CN (1976) Marsh, dune, and archaeological chronology, Ash Meadows, Amargosa Desert, Nevada. In: Elston R (ed) Holocene environmental change in the Great Basin, Nevada Archaeological Survey Research Paper No. 6. University of Nevada, Reno
Melton FA (1940) A tentative classification of sand dunes. J Geol 48:113–174
Miao X, Mason JA, Swinehart JB, Loope DB, Hanson PR, Goble RJ, Liu X (2007) A 10,000 year record of dune activity, dust storms, and severe drought in the central Great Plains. Geology 35(2):119–122
Muhs DR (2017) Evaluation of simple geochemical indicators of aeolian sand provenance: late Quaternary dune fields of North America revisited. Quat Sci Rev 171:260–296
Muhs DR, Holliday VT (1995) Active dune sand on the Great Plains in the 19th century: evidence from accounts of early explorers. Quat Res 43:118–124
Muhs DR, Maat PB (1993) The potential response of eolian sands to greenhouse warming and precipitation reduction on the Great Plains of the United States. J Arid Environ 25:351–361
Muhs DR, Wolfe SA (1999) Sand dunes of the northern Great Plains of Canada and the United States. In: Lemmen DS, Vance RE (eds) Holocene climate and environmental change in the Palliser Triangle: a geoscientific context for evaluating the effects of climate change on the southern Canadian Prairies. Geological Survey of Canada, Ottawa, pp 183–197

Muhs DR, Zárate M (2001) Late Quaternary eolian records of the Americas and their paleoclimatic significance. In: Markgraf V (ed) Interhemispheric climate linkages. Academic, New York, pp 183–216

Muhs DR, Reynolds RR, Been J, Skipp G (2003) Eolian sand transport pathways in the southwestern United States: importance of the Colorado River and local sources. Quat Int 104:3–18

Nielson J, Kocurek G (1986) Climbing zibars of the Algodones. Sediment Geol 48:1–15

Norris RM (1966) Barchan dunes of Imperial Valley, California. J Geol 74:292–307

Norris RM, Norris KS (1961) Algodones dunes of southeastern California. Geol Soc Am Bull 72:605–620

Paisley ECI, Lancaster N, Gaddis L, Greeley R (1991) Discrimination of active and inactive sands by remote sensing: Kelso Dunes, Mojave Desert, California. Remote Sens Environ 37:153–166

Pelletier JD, Jerolmack DJ (2014) Multi-scale bed form interactions and their implications for the abruptness and stability of the downwind dune-field margin at White Sands, New Mexico, U.S.A. J Geophys Res Earth Surf 119:JF003210

Ramsey MS, Christensen PR, Lancaster N, Howard DA (1999) Identification of sand sources and transport pathways at the Kelso Dunes, California using thermal infrared remote sensing. Geol Soc Am Bull 111:646–662

Russell IC (1885) Geological history of Lake Lahontan, a Quaternary lake of northwestern Nevada. US Geol Surv Monogr 11:288

Schmeisser RL, Loope DB, Mason JA (2010) Modern and late Holocene wind regimes over the Great Plains (central U.S.A.). Quat Sci Rev 29(3,Äì4):554–566

Schultz CB, Frye JC (eds) (1965) Loess and related eolian deposits of the world. University of Nebraska Press, Lincoln

Sharp RP (1966) Kelso Dunes, Mohave Desert, California. Geol Soc Am Bull 77:1045–1074

Simpson E, Loope D (1985) Amalgamated interdune deposits, White Sands, New Mexico. J Sediement Petrology 55(3):0361–0365

Smith HTU (1965) Dune morphology and chronology in central and western Nebraska. J Geol 73:557–578

Smith HTU (1967) Past versus present wind action in the Mojave Desert region, California. AFCRL-67-0683, U.S. Army Cambridge Research Laboratory

Smith RSU (1982) Sand dunes in the North American deserts. In: Bender G (ed) Reference handbook of the deserts of North America. Greenwood Press, Westport, pp 481–554

Soller DR, Reheis MC, Garrity CP, Van Sistine DR (2009) Map database for surficial materials in the conterminous United States, U.S. Geological Survey Data Series 425. USGS, Washington, DC

Sridhar V, Loope DB, Swinehart JB, Mason JA, Oglesby RJ, Rowe CM (2006) Large wind shift on the Great Plains during the medieval warm period. Science 313:345–347

Sweet ML, Kocurek G (1990) An empirical model of aeolian dune lee-face airflow. Sedimentology 37(6):1023–1038

Swezey CS, Fitzwater BA, Whittecar GR, Mahan SA, Garrity CP, Alemán González WB, Dobbs KM (2016) The Carolina Sandhills: quaternary eolian sand sheets and dunes along the updip margin of the Atlantic Coastal Plain province, southeastern United States. Quat Res 86(3):271–286

Szynkiewicz A, Ewing RC, Moore CH, Glamoclija M, Bustos D, Pratt LM (2010) Origin of terrestrial gypsum dunes—implications for Martian gypsum-rich dunes of Olympia Undae. Geomorphology 121(1–2):69–83

Thompson DG (1929) The Mojave Desert region, California. United States Geological Survey, Water Supply Paper, 578, 759 pp.

Thorp J, Smith HSU (1952) Pleistocene eolian deposits of the United States, Alaska, and parts of Canada. Geological Society of America, New York

Walker IJ, Nickling WG (2003) Simulation and measurement of surface shear stress over isolated and closely spaced transverse dunes in a wind tunnel. Earth Surf Process Landf 28(10):1111–1124

Wolfe SA, Huntley DJ, Ollerhead J (2004) Relict late Wisconsinan dune fields of the Northern Great Plains, Canada. Géogaphie Physique et Quaternaire 58(2-3):323–336
Wolfe SA, Ollerhead J, Huntley DJ, Lian OB (2006) Holocene dune activity and environmental change in the prairie parkland and boreal forest, central Saskatchewan, Canada. Holocene 16(1):17–29
Wolfe SA, Gillis A, Robertson L (2009) Late Quaternary eolian deposits of northern North America: age and extent. Geological Survey of Canada, Ottawa
Wright HE, Frey DG (eds) (1965) The Quaternary of the United States. Princeton University Press, Princeton

# Chapter 2
# Quaternary Eolian Dunes and Sand Sheets in Inland Locations of the Atlantic Coastal Plain Province, USA

**Christopher S. Swezey**

**Abstract** Quaternary eolian dunes and sand sheets that are stabilized by vegetation are present throughout many inland locations of the Atlantic Coastal Plain province (USA). These locations include river valleys, the Carolina Sandhills region, adjacent to Carolina Bays, and upland areas of the northern coastal plain. The eolian dunes are primarily parabolic in river valleys and in upland areas of the northern coastal plain, linear in the Carolina Sandhills region, and arcuate adjacent to Carolina Bays. Optically stimulated luminescence (OSL) ages from the eolian sands range from circa (ca.) 92–5 ka, revealing that they are relict features that are not active today. These sands have been degraded by vegetation and pedogenic processes, and are stabilized under modern environmental conditions. Most of the OSL ages are approximately coincident with the last glacial maximum (LGM), when conditions were generally colder, drier, and windier. Various features associated with these eolian dunes and sand sheets suggest that the winds that mobilized the sand blew from the northwest in the coastal plain region of Maryland and Delaware, and from the west in the coastal plain region of North Carolina, South Carolina, and Georgia. Most of the eolian dunes and sand sheets are composed of fine to medium sand, although a substantial silt component is present in the northern coastal plain, and a substantial coarse sand component is present in the Carolina Sandhills region. Eolian sand mobilization would have been facilitated by conditions of stronger wind velocity (at least 4–6 m/s), lower air temperature, lower air humidity, and (or) reduced vegetation cover. Eolian sediment mobilization appears to have occurred episodically at any given site, although sites that are farther south have preserved a greater proportion of eolian sands yielding pre-LGM ages (indicating that the southern landscapes farther from the ice sheet have experienced less reworking).

**Keywords** Quaternary · Aeolian · Dune · Carolina Bay · Carolina Sandhills · U.S.A.

---

C. S. Swezey (✉)
U.S. Geological Survey, Reston, VA, USA
e-mail: cswezey@usgs.gov



N. Lancaster, P. Hesp (eds.), *Inland Dunes of North America*,
Dunes of the World, https://doi.org/10.1007/978-3-030-40498-7_2

## 2.1 Introduction

In the eastern United States (U.S.), the Atlantic Coastal Plain province (Fig. 2.1) extends from New York to Florida, and contains strata and sediments of Cretaceous to Quaternary age. Until recently, much of the Quaternary record in this province has been considered to be relatively sparse, consisting primarily of a few onshore lacustrine and paludal records, some beach and barrier island complexes, and some offshore sand and mud. However, with the advent of optically stimulated luminescence (OSL) dating techniques and high-resolution topographic information from Light Detection and Ranging (LiDAR) data, new studies have revealed that the Quaternary record of Atlantic Coastal Plain province is much more extensive and complex than had previously been perceived. Some of these new studies have focused on fluvial settings (e.g., Leigh 2006, 2008; Suther et al. 2011), whereas others have focused on modern coastal settings (e.g., Mallinson et al. 2008; Scott et al. 2010; Timmons et al. 2010; Parham et al. 2013; Seminack and Buynevich 2013; Peek et al. 2014). One of the more surprising revelations from these new studies is the recognition of widespread Quaternary eolian sand dunes and sand sheets of approximately synchronous age throughout many inland locations of the U.S. Atlantic Coastal Plain province (e.g., Ivester et al. 2001; Ivester and Leigh 2003; Markewich et al. 2009; Swezey et al. 2013, 2016a, b).

Inland locations of the U.S. Atlantic Coastal Plain province are not settings in which one would typically expect widespread eolian sands because the modern climate is not conducive to eolian sediment mobilization. Indeed, most of these inland Quaternary eolian sediments are stabilized by vegetation, and the dune and sand sheet morphologies have been degraded by erosion and pedogenic processes. In other words, these eolian sediments are relict features from times when conditions were different from the modern environment. Although future work will undoubtedly reveal additional locations and features, this publication provides a summary of Quaternary eolian sand dunes and sand sheets in the following four inland settings of the U.S. Atlantic Coastal Plain province: (1) river valleys; (2) the Carolina Sandhills region; (3) Carolina Bays; and (4) upland areas of the northern Atlantic Coastal Pain.

## 2.2 Modern Climate

From northern Delaware to northern Florida, the modern climate of the Atlantic Coastal Plain province is humid and mesothermal with little or no water deficiency during any season (climate classification of Thornthwaite 1931, 1948). During January the mean temperature varies from ~0 °C in northern Delaware to ~12 °C in northern Florida, whereas during July the mean temperature varies from ~12 °C in northern Delaware to ~30 °C in northern Florida (Fig. 2.2). Precipitation occurs throughout the year, and mean annual precipitation ranges from ~110 cm in

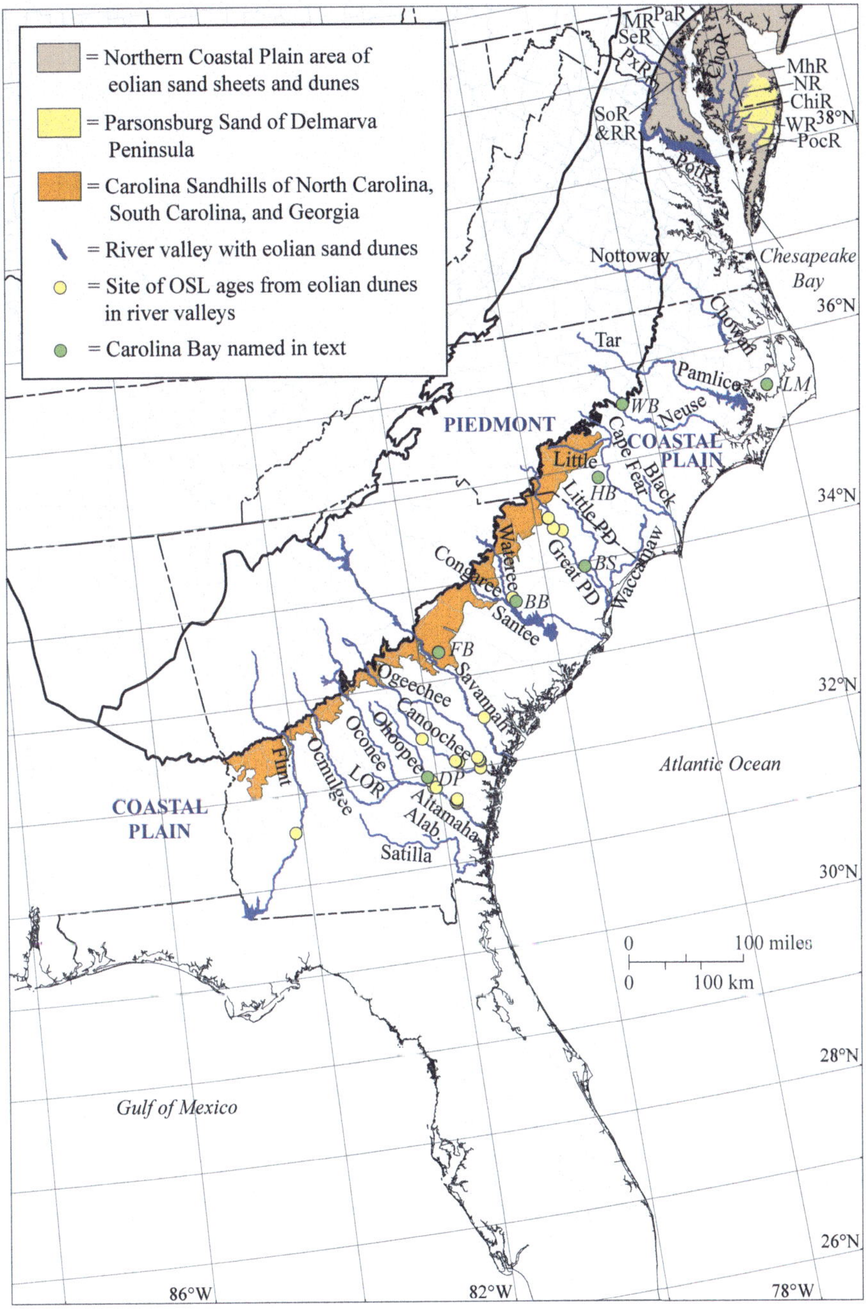


**Fig. 2.1** Coastal Plain location map. The location of the Carolina Sandhills is from Griffith et al. (2001, 2002). *Alab.* Alabaha River, *BB* Big Bay, *BS* Bear Swamp, *ChiR* Chicomacomico River, *ChoR* Choptank River, *DP* Dukes Pond, *FB* Flamingo Bay, *Great PD* Great Pee Dee River, *HB* Herndon Bay, *Little PD* Little Pee Dee River, *LM* Lake Mattamuskeet, *LOR* Little Ocmulgee River, *MhR* Marsheyhope Creek, *MR* Magothy River, *NR* Nanticoke River, *PaR* Patapsco River, *PxR* Patuxent River, *PocR* Pocomoke River, *PotR* Potomac River, *RR* Rhode River, *SeR* Severn River, *SoR* South River, *WB* Wilson's Bay, *WR* Wicomico River

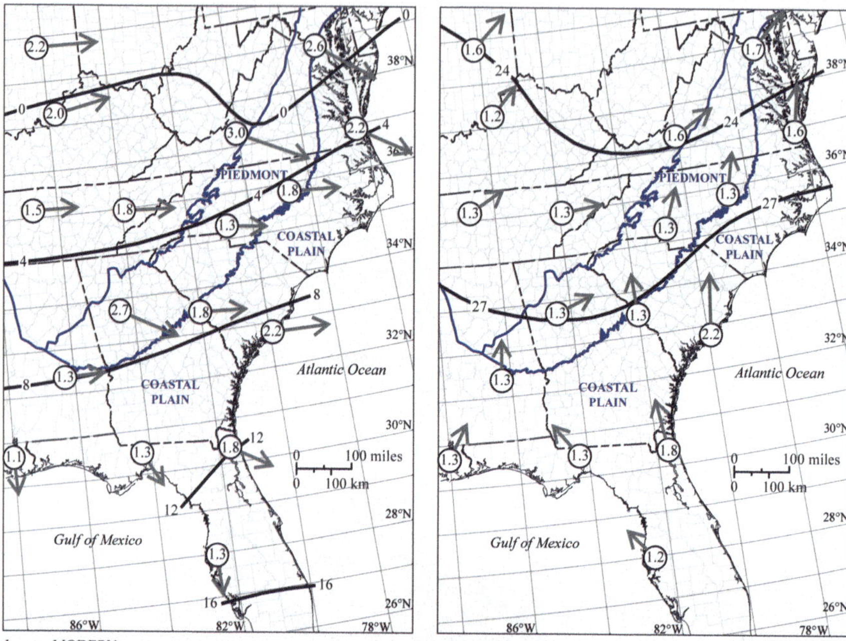


January MODERN JULY Modern

**Fig. 2.2** Modern climate data of the southeastern United States. Mean temperature data in degrees Celsius are from Webb et al. (1993), and mean resultant wind data are from Baldwin (1975). The mean resultant wind is the vectorial average of all surface wind velocities and wind directions on the basis of hourly observations at a given place during the specified months for 1951–1960. The velocity of surface wind in meters per second (m/s) is written inside each circle, and is proportional to the length of the gray arrows

northern Delaware to ~140 cm in northern Florida (Fig. 2.2). Average annual free water surface (FWS) evaporation values range from ~92 cm in northern Delaware to ~122 cm in northern Florida (Farnsworth et al. 1982). Average annual potential evapotranspiration values range from ~74 cm in northern Delaware to ~107 cm in northern Florida (Fig. 2.2). These values yield ratios of annual precipitation to potential evapotranspiration (P:PE) that vary from 1.49 in northern Delaware to 1.31 in northern Florida. For reference, a P:PE ratio between 0.50 and 0.75 denotes a "sub-humid" climate in the UNESCO (1979) classification of arid regions.

The directions of surface winds in the southeastern United States vary seasonally (Fig. 2.2) and are governed primarily by the following three variables: (1) the westerlies; (2) the polar front jet stream; and (3) the Bermuda High. During winter, the westerlies and the polar front jet stream are stronger, the polar front jet stream moves to lower latitudes, and the Bermuda High is weaker (Sahsamanoglou 1990; Harman 1991; Davis et al. 1997). As a result, during winter the surface winds over the Atlantic Coastal Plain province blow predominantly from the west and west-northwest. Most precipitation during winter is frontal in association with the polar front jet stream where cold and dry continental air from Canada is in contact with warm and humid maritime air from the Gulf of Mexico (Court 1974; Soulé 1998; Katz et al. 2003). In contrast, during summer the westerlies and the polar front jet stream are weaker, the polar front jet stream moves to higher latitudes, and the Bermuda High is stronger (Sahsamanoglou 1990; Harman 1991; Davis et al. 1997). As a result, during summer the surface winds over the U.S. Atlantic Coastal Plain province blow from the south via the Bermuda High, bringing moisture to the Atlantic Coastal Plain from the Gulf of Mexico and (or) the Atlantic Ocean (Court 1974; Soulé 1998; Katz et al. 2003). Most precipitation during summer is associated with convection rather than fronts.

The mean resultant velocity of surface winds in the U.S. Atlantic Coastal Plain province is <3 m/s during any given month (Fig. 2.2), but there is some variability ("gustiness") around the mean. For example, wind velocities of 6 m/s or greater occurred ~8% of the time per whole year during the interval of 1981–2010 according to hourly data from the Metropolitan Airport at the city of Columbia, South Carolina (www.ncdc.noaa.gov; accessed 18 August 2016). In relatively warm low-latitude regions, however, typical threshold wind velocities for sustained eolian mobilization of 0.25–0.50 mm diameter quartz sand are 4–6 m/s (e.g., Hsu 1974), and therefore modern surface winds in inland locations of the Atlantic Coastal Plain province are really not sufficient for much sustained eolian sand transport.

## 2.3 Age Data

The age data presented in this paper were obtained by radiocarbon techniques and (or) luminescence techniques. Unless otherwise stated, the radiocarbon ages are reported in radiocarbon years ($^{14}C$ yr) before present (BP), using the Libby half-life of 5568 years and with 0 $^{14}C$ year BP being equivalent to AD 1950. In contrast, the luminescence ages are reported in calibrated years (cal year) BP with 0 cal year BP being the year that a specific luminescence age was determined. The luminescence ages presented in this paper were compiled from different sources, and different authors used different statistical models to determine their best estimates of the ages. Where available, information on these different statistical models is given in Tables 2.1, 2.2, 2.3, and 2.4. For luminescence ages published for the first time in this paper (Table 2.2), the choice of statistical model that is thought to yield the most accurate age follows criteria discussed in Swezey et al. (2016b). In brief, if the dispersion was <25% (as determined by the R program radial plot, following Galbraith and Roberts 2012), then the preferred age was the age obtained by the weighted mean. If the dispersion was ≥25%, then the preferred age was the age obtained by the Minimum Age Model-3.

**Table 2.1** OSL data from eolian dunes in river valleys of the U.S. Atlantic Coastal Plain province

| Sample ID | References | State | River valley | LAT (North) | LONG (West) | ELEV (m) | DEPTH (cm) | AGE (ka) | Preferred age model |
|---|---|---|---|---|---|---|---|---|---|
| UGA-DL-3 | Leigh et al. (2004) | South Carolina | Great Pee Dee River | 34.53512 | −79.81282 | unspec. | 210–180 | 16.6 ± 1.8 | unspec. |
| UGA-DL-4 | Leigh et al. (2004) | South Carolina | Great Pee Dee River | 34.53259 | −79.81730 | unspec. | 190–160 | 20.2 ± 2.4 | unspec. |
| UGA-DL-2 | Leigh et al. (2004) | South Carolina | Great Pee Dee River | 34.41460 | −79.76242 | 26 | 210–180 | 15.0 ± 1.4 | unspec. |
| UGA-4-SC34-Br. | Leigh (2008) | South Carolina | Great Pee Dee River | 34.37303 | −79.64535 | unspec. | 210 | 19.5 ± 1.8 | unspec. |
| Wateree01 | Ivester et al. (2002) and Brooks et al. (2010) | South Carolina | Wateree and Congaree Rivers | 33.7900 | −80.4852 | 64 | unspec. | 74.3 ± 7.1 | unspec. |
| Wateree02 | Ivester et al. (2002) and Brooks et al. (2010) | South Carolina | Wateree and Congaree Rivers | 33.7866 | −80.4891 | 69 | unspec. | 29.6 ± 2.4 | unspec. |
| Wateree03 | Ivester et al. (2002) and Brooks et al. (2010) | South Carolina | Wateree and Congaree Rivers | 33.8008 | −80.4989 | 72 | unspec. | 33.2 ± 2.8 | unspec. |
| GAW 03a | Swezey et al. (2013) | South Carolina | Savannah River | 32.49416 | −81.19051 | 11 | 150 | 22.9 ± 1.7 | Weighted |
| GAW 03b | Swezey et al. (2013) | South Carolina | Savannah River | 32.49416 | −81.19051 | 11 | 150 | 24.0 ± 1.2 | Weighted |
| GAW 04a | Swezey et al. (2013) | South Carolina | Savannah River | 32.49416 | −81.19051 | 11 | 140 | 18.6 ± 0.1 | MAM |
| GAW 04b | Swezey et al. (2013) | South Carolina | Savannah River | 32.49416 | −81.19051 | 11 | 140 | 19.1 ± 0.1 | MAM |
| GAW05 | Swezey et al. (2013) | South Carolina | Savannah River | 32.54332 | −81.26128 | 12 | 152 | 32.7 ± 2.1 | Weighted |
| GAW06 | Swezey et al. (2013) | South Carolina | Savannah River | 32.54423 | 81.26348 | 12 | 148 | 32.9 ± 2.5 | Weighted |

| | | | | | | | | | |
|---|---|---|---|---|---|---|---|---|---|
| GAW07 | Swezey et al. (2013) | South Carolina | Savannah River | 32.54423 | −81.26348 | 12 | 20 | 24.1 ± 1.5 | Weighted |
| GAW08 | Swezey et al. (2013) | South Carolina | Savannah River | 32.52192 | −81.23143 | 12 | 96 | 30.8 ± 1.6 | Weighted |
| GAW09 | Swezey et al. (2013) | South Carolina | Savannah River | 32.49084 | −81.19570 | 9 | 78 | 21.5 ± 1.3 | Weighted |
| GAW10 | Swezey et al. (2013) | South Carolina | Savannah River | 32.48911 | −81.18024 | 8 | 40 | 31.3 ± 1.8 | Weighted |
| GAW11 | Swezey et al. (2013) | South Carolina | Savannah River | 32.48457 | −81.20333 | 9 | 60 | 10.2 ± 0.7 | MAM |
| GAW12 | Swezey et al. (2013) | South Carolina | Savannah River | 32.42373 | −81.15730 | 8 | 80 | 19.2 ± 1.2 | Weighted |
| GAW13 | Swezey et al. (2013) | South Carolina | Savannah River | 32.43740 | −81.14857 | 8 | 90 | 10.3 ± 0.7 | Weighted |
| GAW30 | Swezey et al. (2013) | South Carolina | Savannah River | 32.55597 | −81.27906 | 8 | 549 | 30.5 ± 2.2 | Weighted |
| GAW36 | Swezey et al. (2013) | South Carolina | Savannah River | 32.50996 | −81.22414 | 11 | 82 | 23.3 ± 1.9 | Weighted |
| Dalhousie-DL17 | Leigh (2008) | Georgia | Ogeechee River | 32.09682 | −81.37651 | unspec. | 210–185 | 45.3 ± 6.7 | unspec. |
| Dalhousie-DL15 | Leigh (2008) | Georgia | Ogeechee River | 32.06689 | −81.35625 | unspec. | 195 | 37.5 ± 2.8 | unspec. |
| Dalhousie-DL18 | Leigh (2008) | Georgia | Ogeechee River | 32.05596 | −81.34448 | unspec. | 210–185 | 31.4 ± 5.9 | unspec. |
| Dalhousie-AI-22 | Ivester et al. (2001) | Georgia | Canoochee River | 32.38440 | −82.12426 | unspec. | 350 | 24.3 ± 3.3 | unspec. |
| Dalhousie-AI-4 | Ivester et al. (2001) | Georgia | Canoochee River | 32.08542 | −81.68432 | 21 | 257–233 | 26.6 ± 2.3 | unspec. |
| Dalhousie-AI-5 | Ivester et al. (2001) | Georgia | Canoochee River | 32.08001 | −81.67687 | unspec. | 277–253 | 32.8 ± 3.3 | unspec. |
| Dalhousie-AI-3 | Ivester et al. (2001) | Georgia | Canoochee River | 32.06734 | −81.64418 | 19 | 200 | 25.8 ± 1.2 | unspec. |

(continued)

**Table 2.1** (continued)

| Sample ID | References | State | River valley | LAT (North) | LONG (West) | ELEV (m) | DEPTH (cm) | AGE (ka) | Preferred age model |
|---|---|---|---|---|---|---|---|---|---|
| Dalhousie-AI-2 | Ivester et al. (2001) | Georgia | Canoochee River | 32.06552 | −81.64565 | 20 | 201–183 | 29.9 ± 6.2 | unspec. |
| Dalhousie-DL10 | Leigh (2008) | Georgia | Canoochee River | 31.97694 | −81.35153 | unspec. | 200–175 | 34.0 ± 2.6 | unspec. |
| Dalhousie-AI-18 | Ivester et al. (2001) | Georgia | Ohoopee River | 31.95195 | −82.09706 | 52 | 177–153 | 77.4 ± 6.6 | unspec. |
| Dalhousie-AI-15 | Ivester et al. (2001) | Georgia | Ohoopee River | 31.94299 | −82.10378 | 32 | 127–103 | 23.6 ± 5.4 | unspec. |
| UGA-BH-2 | Leigh et al. (2004) | Georgia | Altamaha River | 31.86451 | −82.04860 | 25 | 210–180 | 18.6 ± 1.9 | unspec. |
| UGA-BH5.2 | Leigh et al. (2004) | Georgia | Altamaha River | 31.86882 | −82.08279 | 25 | 210–180 | 17.6 ± 2.6 | unspec. |
| Dalhousie-AI-9 | Ivester et al. (2001) | Georgia | Altamaha River | 31.70031 | −81.78545 | 16 | 227–203 | 45.0 ± 7.4 | unspec. |
| Dalhousie-AI-11 | Ivester et al. (2001) | Georgia | Altamaha River | 31.69402 | −81.79448 | 16 | 227–203 | 38.1 ± 5.0 | unspec. |
| Dalhousie-AI-10 | Ivester et al. (2001) | Georgia | Altamaha River | 31.67538 | −81.80657 | 16 | 235–225 | 4.9 ± 0.5 | unspec. |
| Dalhousie-AI-12 | Ivester et al. (2001) | Georgia | Altamaha River | 31.67307 | −81.80575 | 17 | 250 | 20.9 ± 1.2 | unspec. |
| Dalhousie-AI-13 | Ivester et al. (2001) | Georgia | Altamaha River | 31.65689 | −81.79436 | 14 | 250 | 16.2 ± 2.0 | unspec. |
| Dalhousie-AI-6 | Ivester et al. (2001) | Georgia | Flint River | 31.57104 | −84.13042 | 74 | 200 | 8.6 ± 0.9 | unspec. |
| Dalhousie-AI-7 | Ivester et al. (2001) | Georgia | Flint River | 31.57104 | −84.13042 | 74 | 325 | 15.8 ± 1.7 | unspec. |
| Dalhousie-AI-8 | Ivester et al. (2001) | Georgia | Flint River | 31.56852 | −84.13158 | 66 | 460 | 17.5 ± 1.7 | unspec. |

*DEPTH* Depth below land surface at which OSL sample was collected, *ELEV* elevation of land surface, *LAT* latitude, *LONG* longitude, *MAM* Minimum Age Model-3 (Galbraith and Laslett 1993; Galbraith et al. 1999), *Weighted* age in thousands of years (ka) ago using the weighted mean OSL value for equivalent dose (DE) determinations (similar to the Central Age Model of Galbraith et al. 1999), *unspec.* unspecified

**Table 2.2** OSL data from eolian sand in the Carolina Sandhills region of the U.S. Atlantic Coastal Plain province

| Sample ID | References | State | Site descript. | LAT (North) | LONG (West) | ELEV (m) | DEPTH (cm) | AGE (ka) | Preferred age model |
|---|---|---|---|---|---|---|---|---|---|
| UGA-DB-TU4-90 | Leigh (2008) | North Carolina | Bogwater, Fort Bragg | 35.12665 | −79.22776 | unspec. | 90 | 22.7 ± 5.9 | unspec. |
| UGA-DB-TU4-153 | Leigh (2008) | North Carolina | Bogwater, Fort Bragg | 35.12665 | −79.22776 | unspec. | 153 | 24.1 ± 6.1 | unspec. |
| USGS1585 | Swezey et al. (2016) | South Carolina | Site 3 | 34.56355 | −80.13365 | 113 | 47–42 | 10.2 ± 0.8 | MAM3 |
| USGS1586 | Swezey et al. (2016) | South Carolina | Site 3 | 34.56355 | −80.13365 | 113 | 70–65 | 24.1 ± 2.2 | MAM3 |
| USGS1588 | Swezey et al. (2016) | South Carolina | Site 5 | 34.56355 | −80.10430 | 125 | 42 | 9.4 ± 0.8 | Weighted |
| USGS1589 | Swezey et al. (2016) | South Carolina | Site 5 | 34.56355 | −80.10430 | 125 | 85 | 19.3 ± 1.6 | MAM3 |
| USGS1642 | Swezey et al. (2016) | South Carolina | Site 4 | 34.55686 | −80.12839 | 119 | 200 | 46.3 ± 3.1 | MAM3 |
| USGS1643 | Swezey et al. (2016) | South Carolina | Site 4 | 34.55686 | −80.12839 | 119 | 60 | 25.5 ± 2.3 | MAM3 |
| USGS1715 | Swezey et al. (2016) | South Carolina | Site 5 | 34.62072 | −80.05958 | 76 | 210 | 39.8 ± 2.7 | MAM3 |
| USGS1716 | Swezey et al. (2016) | South Carolina | Site 5 | 34.62072 | −80.05958 | 76 | 250 | 69.6 ± 5.6 | Weighted |
| USGS1717 | Swezey et al. (2016) | South Carolina | Site 5 | 34.62072 | −80.05958 | 76 | 250 | 53.7 ± 5.1 | MAM3 |
| USGS1718 | Swezey et al. (2016) | South Carolina | Site 2 | 34.52557 | −80.22427 | 122 | 40 | 7.0 ± 0.9 | MAM3 |
| USGS1719 | Swezey et al. (2016) | South Carolina | Site 2 | 34.52557 | −80.22427 | 122 | 145 | 48.9 ± 7.7 | Weighted |

(continued)

**Table 2.2** (continued)

| Sample ID | References | State | Site descript. | LAT (North) | LONG (West) | ELEV (m) | DEPTH (cm) | AGE (ka) | Preferred age model |
|---|---|---|---|---|---|---|---|---|---|
| USGS1720 | Swezey et al. (2016) | South Carolina | Site 1 | 34.62237 | −80.23235 | 125 | 90 | 49.8 ± 3.7 | Weighted |
| USGS1721 | Swezey et al. (2016) | South Carolina | Site 1 | 34.62237 | −80.23235 | 125 | 160 | 9.2 ± 0.6 | Weighted |
| USGS1722 | Swezey et al. (2016) | South Carolina | Site 1 | 34.62237 | −80.23235 | 125 | 210 | 10.9 ± 0.6 | MAM3 |
| USGS GAW-35 | Previously unpublished | South Carolina | White Pond Dune | 34.16364 | −80.77531 | 91 | 90 | 92.3 ± 5.2 | Weighted |

*DEPTH* Depth below land surface at which OSL sample was collected, *ELEV* elevation of land surface, *LAT* latitude, *LONG* longitude, *MAM* Minimum Age Model-3 (Galbraith and Laslett 1993; Galbraith et al. 1999), *Weighted* age in thousands of years (ka) ago using the weighted mean OSL value for equivalent dose (DE) determinations (similar to the Central Age Model of Galbraith et al. 1999), *unspec*. unspecified

**Table 2.3** OSL data from sand ridges of Carolina Bays in the U.S. Atlantic Coastal Plain province

| Sample ID | References | State | Carolina Bay Name | LAT (North) | LONG (West) | ELEV (m) | DEPTH (cm) | AGE (ka) | Preferred age model |
|---|---|---|---|---|---|---|---|---|---|
| UW2786 (core 1) | Moore et al. (2016) | North Carolina | Herndon Bay | 34.8603 | −78.9391 | 50.8 | 180–200 | 27.2 ± 2.8 | CAM |
| UW2787 (core 2) | Moore et al. (2016) | North Carolina | Herndon Bay | 34.8602 | −78.9377 | 52.3 | 300–330 | 29.6 ± 3.1 | CAM |
| UW2788 (core 4) | Moore et al. (2016) | North Carolina | Herndon Bay | 34.8599 | −78.9354 | 52.1 | 160–190 | 36.7 ± 4.1 | LC |
| unspec. | Ivester et al. (2003) and Brooks et al. (2010) | South Carolina | Big Bay | 33.7676 | −80.4590 | 58 | 60–75 | 2.1 ± 0.3 | unspec. |
| unspec. | Ivester et al. (2003) and Brooks et al. (2010) | South Carolina | Big Bay | 33.7661 | −80.4584 | 58 | 60–75 | 11.2 ± 0.9 | unspec. |
| unspec. | Ivester et al. (2003) and Brooks et al. (2010) | South Carolina | Big Bay | 33.7654 | −80.4647 | 58 | 60–75 | 25.2 ± 1.9 | unspec. |
| unspec. | Ivester et al. (2003) and Brooks et al. (2010) | South Carolina | Big Bay | 33.7709 | −80.4798 | 59 | 60–75 | 35.7 ± 2.6 | unspec. |
| unspec. | Ivester et al. (2003) and Brooks et al. (2010) | South Carolina | Unnamed bay near Big Bay | 33.7643 | −80.4683 | 59 | 60–75 | 20.4 ± 1.6 | unspec. |
| unspec. | Moore et al. (2012) | South Carolina | Flamingo Bay | unspec. | unspec. | unspec. | 35 | 5.0 ± 0.5 | MAM |
| unspec. | Moore et al. (2012) | South Carolina | Flamingo Bay | unspec. | unspec. | unspec. | 50 | 9.2 ± 1.0 | MAM |
| unspec. | Moore et al. (2012) | South Carolina | Flamingo Bay | unspec. | unspec. | unspec. | 65 | 11.5 ± 1.3 | MAM |
| unspec. | Moore et al. (2012) | South Carolina | Flamingo Bay | unspec. | unspec. | unspec. | 78 | 15.5 ± 1.8 | MAM |
| unspec. | Moore et al. (2012) | South Carolina | Flamingo Bay | unspec. | unspec. | unspec. | 95 | 13.1 ± 1.7 | MAM |

(continued)

**Table 2.2** (continued)

| Sample ID | References | State | Carolina Bay Name | LAT (North) | LONG (West) | ELEV (m) | DEPTH (cm) | AGE (ka) | Preferred age model |
|---|---|---|---|---|---|---|---|---|---|
| unspec. | Ivester et al. (2002) and Brooks et al. (2010) | South Carolina | Flamingo Bay | unspec. | unspec. | unspec. | unspec. | 108.7 ± 10.9 | unspec. |
| unspec. | Ivester et al. (2002) and Brooks et al. (2010) | South Carolina | Flamingo Bay | unspec. | unspec. | unspec. | unspec. | 40.3 ± 4.0 | unspec. |
| unspec. | Ivester et al. (2002) and Brooks et al. (2010) | South Carolina | Bay-40 (SRS 40) | unspec. | unspec. | unspec. | unspec. | 77.9 ± 7.6 | unspec. |

*CAM* Central Age Model, *DEPTH* Depth below land surface at which OSL sample was collected, *ELEV* elevation of land surface, *LAT* latitude, *LC* Largest Component, *LONG* longitude, *MAM* Minimum Age Model-3 (Galbraith and Laslett 1993; Galbraith et al. 1999), *unspec.* unspecified

**Table 2.4** OSL data from eolian dunes on upland areas of the northern Atlantic Coastal Plain province

| Sample ID | Reference | Location | Site descript. | LAT (North) | LONG (West) | ELEV (m) | DEPTH (cm) | AGE (ka) | Preferred age model |
|---|---|---|---|---|---|---|---|---|---|
| UIC2020BL | Lowery et al. (2010) | Miles Point, Maryland | Tilghman Soil | unspec. | unspec. | unspec. | 85 | 27.9 ± 1.6 | unspec. |
| UIC2019BL | Lowery et al. (2010) | Miles Point, Maryland | Tilghman Soil | unspec. | unspec. | unspec. | 90 | 29.5 ± 1.7 | unspec. |
| UIC2014BL | Lowery et al. (2010) | Miles Point, Maryland | Miles Point Loess | unspec. | unspec. | unspec. | 135 | 34.8 ± 2.0 | unspec. |
| UIC2013BL | Lowery et al. (2010) | Miles Point, Maryland | Miles Point Loess | unspec. | unspec.. | unspec. | 137 | 40.9 ± 2.4 | unspec. |
| UIC2013BL | Lowery et al. (2010) | Miles Point, Maryland | Pre-Loess Sequum | unspec. | unspec. | unspec. | 148 | 41.1 ± 2.4 | unspec. |
| UIC2013BL | Lowery et al. (2010) | Miles Point, Maryland | Pre-Loess Sequum | unspec. | unspec. | unspec. | 161 | 40.6 ± 2.7 | unspec. |
| UGA07 OSL-512 | Markewich et al. (2009) | Brandywine2, Maryland | unspec. | 38.66106 | −76.86119 | unspec. | 66 | 19.2 ± 2.7 | MEAN |
| UGA07 OSL-472 | Markewich et al. (2009) | Fenwick, Maryland | Dune | 38.64756 | −77.10817 | unspec. | unspec. | 26.7 ± 3.1 | MEAN |
| UGA07 OSL-479 | Markewich et al. (2009) | Goose Bay, Maryland | Dune | 38.51428 | −77.26185 | unspec. | 183 | 30.4 ± 4.0 | MEAN |
| UGA07 OSL-514 | Markewich et al. (2009) | Goose Bay, Maryland | Dune | 38.51428 | −77.26185 | unspec. | 305 | 24.4 ± 4.1 | MEAN |
| UGA07 OSL-478 | Markewich et al. (2009) | Chapman Landing, Maryland | unspec. | 38.62139 | −77.11917 | unspec. | unspec. | 27.0 ± 3.3 | MEAN |
| Pedon VA1 | Feldman et al. (2000) | Pedon VA1, Virginia | Loess | 38.68786 | −77.31222 | 97 | <51 | 13.8 ± 1.0 | TL |

*DEPTH* Depth below land surface at which OSL sample was collected, *ELEV* elevation of land surface, *LAT* latitude, *LONG* longitude, *MEAN* age in thousands of years (ka) ago using the mean OSL value for equivalent dose (DE) determinations, *TL* Thermoluminescence date, *unspec.* unspecified

## 2.4 Descriptions of Eolian Dunes and Sand Sheets

Vegetated (stabilized) eolian dunes and sand sheets of Quaternary age are found in many settings throughout inland locations of the U.S. Atlantic Coastal Plain province. These eolian sediments may be divided into the following four categories according to geographic location: (1) in river valleys; (2) in the Carolina Sandhills region; (3) adjacent to Carolina Bays; and (4) on upland areas of the northern Atlantic Coastal Plain.

### 2.4.1 *Eolian Dunes in River Valleys*

Vegetated (stabilized) eolian sand dunes are present within numerous river valleys in the U.S. Atlantic Coastal Plain province (Fig. 2.1). At the time of this writing, such dunes have been identified in river valleys of all of the eastern coastal states from Delaware to Georgia. On the Delmarva Peninsula (Delaware, Maryland, Virginia), eolian dunes are present in the valleys of the Choptank River, Chicamacomico River, Marsheyhope Creek, Nanticoke River, Wicomico River, and Pocomoke River (Denny and Owens 1979; Denny et al. 1979; Newell and DeJong 2011; Markewich et al. 2015), and on the east side of the Potomac River/Chesapeake Bay in Virginia (Mixon 1985). Elsewhere in Maryland, eolian dunes and (or) sand sheets are present in the valleys of the Patapsco River, Magothy River, Severn River, South River, Rhode River, Potomac River, and Patuxent River (Hack 1955; Markewich et al. 2009; Newell and DeJong 2011). On the mainland part of Virginia, eolian dunes are present in the valley of the Nottoway River (Powars et al. 2016). In North Carolina, eolian dunes are present in the valleys of the Chowan River, Tar River, Pamlico River, Neuse River, Black River, Little River, and Cape Fear River (Daniels et al. 1969; Thom 1970; Miller 1979; Soller 1988; Markewich and Markewich 1994; Markewich et al. 2015). In South Carolina, eolian dunes are present in the valleys of the Little Pee Dee River, Great Pee Dee River, Waccamaw River, Wateree River, Congaree River, Santee River, and Savannah River (Daniels et al. 1969; Thom 1970; Pickering and Jones 1974; Markewich and Markewich 1994; Brooks et al. 2010; Swezey et al. 2013). In Georgia, eolian dunes are present in the valleys of the Ogeechee River, Canoochee River, Ohoopee River, Little Ohoopee River, Oconee River, Ocmulgee River, Little Ocmulgee River, Altamaha River, Alabaha River, Satilla River, and Flint River (Thom 1970; Pickering and Jones 1974; Markewich and Markewich 1994; Ivester et al. 2001; Ivester and Leigh 2003).

In most publications, eolian dunes in river valleys of the Atlantic Coastal Plain are identified as sand of Quaternary age, but a formal stratigraphic formation name is not applied to the dunes. On the Delmarva Peninsula, however, both eolian sand dunes within river valleys and eolian sand dunes and sand sheets on upland terraces are mapped as the Quaternary Parsonsburg Sand (Denny and Owens 1979; Denny et al. 1979). Because the reference locality for the Parsonsburg Sand is located in an upland area of the Delmarva Peninsula (test hole Wi-Bg 11 at the north end of Parsonsburg Ridge, Wicomico County, Maryland; Rasmussen and Slaughter 1955), it is not appropriate to apply the Parsonsburg Sand nomenclature to the river valley sand dunes that have a discontinuous distribution from Delaware to Georgia. In other words, it is better to restrict the name Parsonsburg Sand to eolian sand in upland areas of the Delmarva Peninsula.

Despite their discontinuous distribution, eolian dunes in the river valleys show some consistent patterns with respect to geographic and stratigraphic location. For example, the eolian dunes are located on the east side of the modern river channels (Pickering and Jones 1974; Carver and Brook 1989; Markewich and Markewich 1994; Ivester and Leigh 2003; Swezey et al. 2013). An excellent example of such dunes is present in the valley of the Savannah River, Jasper County, South Carolina (Fig. 2.3). With respect to stratigraphic location, the eolian dunes in river valleys are located above unconformities on terraces within their respective valleys. Depending upon location, these terraces are composed of either (1) Quaternary sand, mud, and (or) peat of fluvial and paludal origin; or (2) pre-Quaternary sand and (or) mud upon which a paleosol has developed (Ivester and Leigh 2003; Swezey et al. 2013).

In many river valleys, the eolian dunes occur as groups of dunes or dune fields that are elongate parallel to the rivers. On the Delmarva Peninsula the linear ridges of eolian sand may be as much as 5 m high and up to 5 km long (Denny and Owens 1979), whereas in Georgia the linear ridges may be 2–14 m high and 6 to >100 km long (Ivester and Leigh 2003). At some locations, the eolian dunes in river valleys form multiple linear ridges that are parallel to the modern river channel. In river valleys where multiple eolian dune ridges are present, the western ridges typically have morphologies that are more distinct and regular (Ivester and Leigh 2003).

Many of the individual dunes in river valleys have parabolic shapes or infilled parabolic shapes, and the orientations of these dunes show a predictable geographic variability. In Maryland and Delaware, the tails of parabolic dunes point northwest (Denny and Owens 1979; Denny et al. 1979; Carver and Brook 1989; Markewich et al. 2009). In contrast, in North Carolina, South Carolina, and Georgia, the tails of parabolic dunes point west or southwest (Thom 1970; Carver and Brook 1989; Ivester and Leigh 2003; Swezey et al. 2013).

The height (sand thickness) of individual eolian dunes in river valleys generally increases towards the south. Specifically, dune height is typically 5 m or less in Virginia, Delaware, and Maryland, 2–10 m in North Carolina and South Carolina, and 2–14 m in Georgia. On the Delmarva Peninsula in Delaware and Maryland, eolian dunes in river valleys are either parabolic forms up to 1 m high or linear ridges up to 5 m high (Hack 1955; Denny et al. 1979; Denny and Owens 1979). On the Delmarva Peninsula in Virginia, however, some larger eolian dunes (6–15 m high) form linear and arcuate dune fields as much as 0.5 km wide and 3 km long parallel to the Chesapeake Bay shoreline (Mixon 1985). Farther south in North Carolina, eolian dune height is typically 1–7 m in the valley of the Cape Fear River (Soller 1988) and up to 5 m in the valley of the Neuse River (Daniels et al. 1969). In South Carolina, eolian dune height is up to 5 m in the valleys of the Great Pee Dee

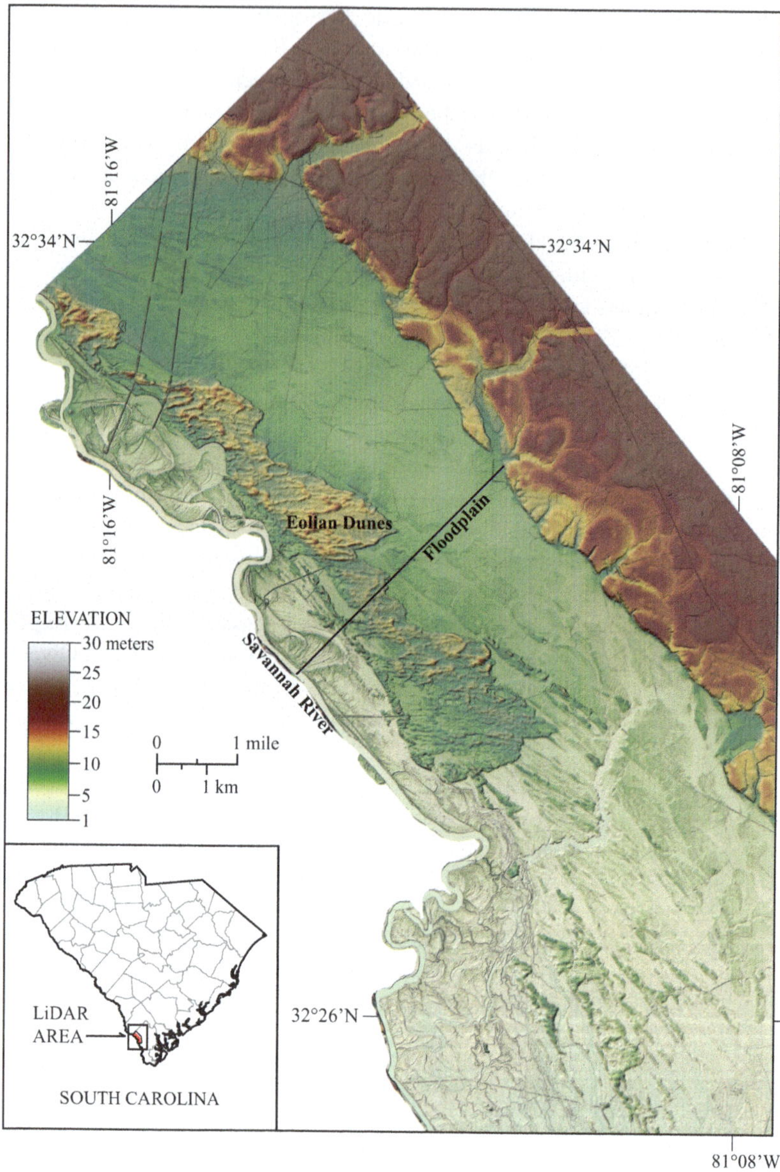


**Fig. 2.3** LiDAR image of eolian dunes in the valley of the Savannah River, Jasper County, South Carolina (modified from Swezey et al. 2013). These dunes are located ~50–70 km inland from the coast, on the east side of the modern river channel. Elevations are given relative to sea level

River and Little Pee Dee River (Thom 1970), and 1–10 m in the valley of the Savannah River (Swezey et al. 2013). In Georgia, eolian dune height is 4–12 m in the valley of the Flint River (Ivester and Leigh 2003), 8 m in the valley of the Satilla River (Markewich and Markewich 1994), and 4–14 m in the valley of the Altamaha River (Ivester and Leigh 2003). Eolian dune height is typically 2–5 m in the valley of the Ohoopee River, and 2–4 m in the valley of the Canoochee River (Ivester et al. 2001; Ivester and Leigh 2003).

The eolian dunes in river valleys display many similarities with respect to grain size, sorting, and composition (Pickering and Jones 1974; Denny et al. 1979; Soller 1988; Markewich and Markewich 1994; Ivester and Leigh 2003; Swezey et al. 2013). Most dunes are composed of 95–100% sand and 0–5% mud, and the sand size is predominantly fine to medium sand. The eolian dune sand is typically moderately sorted to well sorted (following terminology of Folk and Ward 1957). Sand composition is typically 95–100% quartz, 0–5% feldspar (although feldspar content is usually <1%), and < 1% mica and opaque minerals. Most of the quartz sand grains have roundness values that range from subangular to rounded, and sphericity values that range from high sphericity to low sphericity (following terminology of Powers 1953).

Most eolian dunes in river valleys do not display obvious sedimentary structures, except for bioturbation by plant roots. However, large cross-bedding, parallel laminations, reactivation surfaces, and buried paleosols have been reported from some eolian dunes in the river valleys of Georgia (Ivester and Leigh 2003). Likewise, east-dipping cross-bedding has been reported from eolian dunes in the valley of the Congaree River in South Carolina (Johnson 1961), and cross-bedding has also been reported from eolian dunes in the valley of the Great Pee Dee River in South Carolina (Thom 1970).

Forty-six OSL ages have provided an absolute chronology for eolian sand dunes within river valleys of South Carolina and Georgia (Table 2.1, Fig. 2.4). These OSL ages range from circa (ca.) 54–5 thousand years ago (ka), although most of the ages range from ca. 35–15 ka. Only two ages are younger than 11 ka, and these two ages are from river valleys in southern Georgia. No OSL ages from eolian dunes in river valleys are younger than ca. 5 ka.

### 2.4.2 *Eolian Dunes and Sand Sheets of the Carolina Sandhills*

The Carolina Sandhills, which has long been recognized as a distinct geomorphologic province (e.g., Holmes 1893), is a 15–60 km wide physiographic region that extends ~700 km from the western border of Georgia across South Carolina to central North Carolina along the updip (inland) part of the Atlantic Coastal Plain province (Fig. 2.1). As described by Swezey et al. (2016b), the region is characterized by: (1) vegetated (stabilized) eolian sand dunes and sand sheets that are mapped as the Quaternary Pinehurst Formation; and (2) outcrops of Cretaceous strata consisting of sand, slightly indurated sandstone, conglomerate, and mud. The region is also

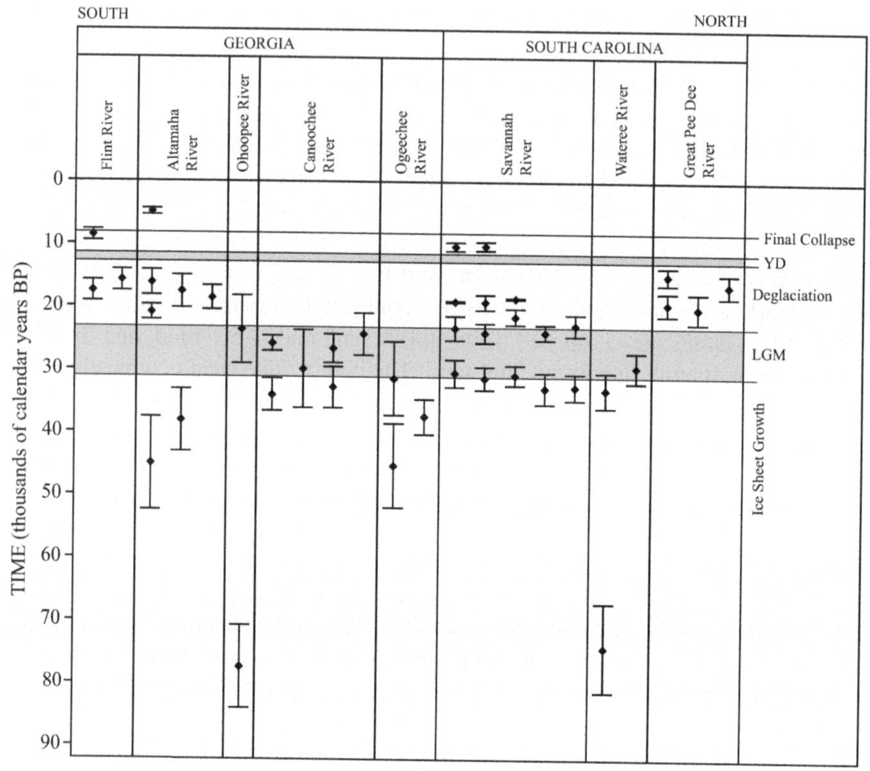


**Fig. 2.4** OSL ages from eolian dunes in river valleys of the U.S. Atlantic Coastal Plain province. *LGM* Last glacial maximum, *YD* Younger Dryas event, *Final Collapse* Final collapse of the Laurentide Ice Sheet. Detailed age data are given in Table 2.1

notable for supporting an endangered ecosystem of longleaf pine (*Pinus palustris*) and wiregrass (*Aristida stricta*) that is maintained by frequent low-intensity fires (Earley 2004; Askins 2010).

The geology of the Carolina Sandhills has been studied in greatest detail in Chesterfield County, South Carolina (Fitzwater 2016; Swezey et al. 2016a, b). In this area, the Carolina Sandhills region is a relatively high plateau that is bounded to the west by Paleozoic schist of the Piedmont province and bounded to the east by the east-facing Orangeburg Scarp (Fig. 2.5), which is interpreted as having formed by marine wave erosion during the middle Pliocene (Dowsett and Cronin 1990). Throughout this area, Quaternary eolian sand (Pinehurst Formation) is present above an unconformity that caps the Cretaceous Middendorf Formation, which is a unit of sand, sandstone, conglomerate, and mud. Relief on this unconformity ranges up to 5 m in places, and the immediately underlying Cretaceous strata display pedogenic mottling and other paleosol characteristics.

In Chesterfield County, the Quaternary eolian sand (Pinehurst Formation) forms both dunes and sand sheets (Fitzwater 2016; Swezey et al. 2016a, 2016b). At most

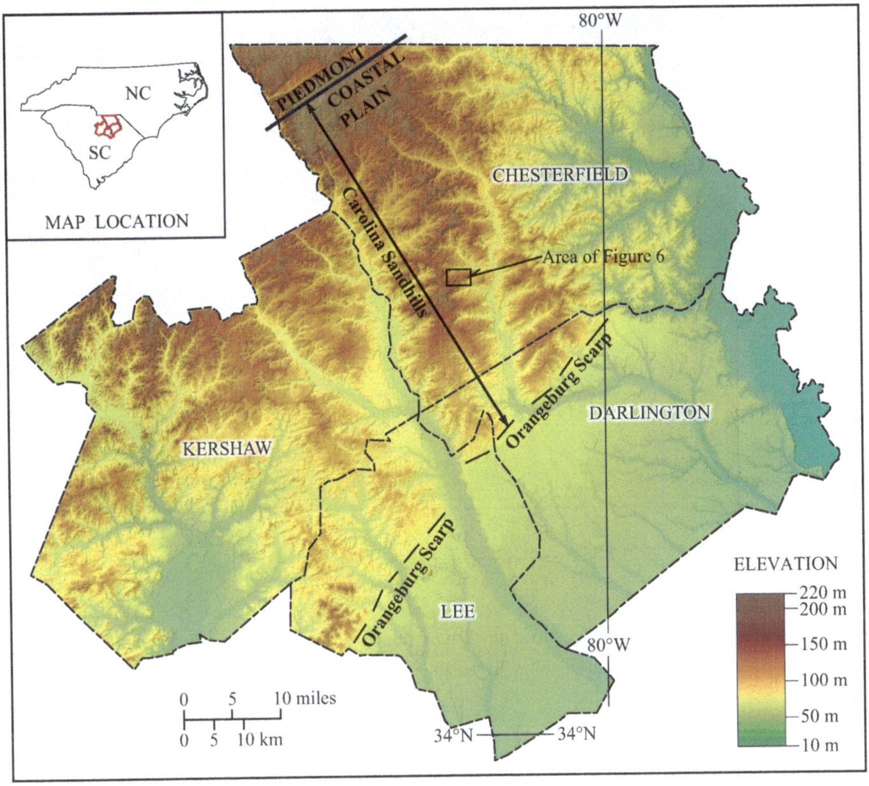


**Fig. 2.5** LiDAR image of four counties in northern South Carolina, showing the Carolina Sandhills region in the inland (updip) part of the Coastal Plain province (modified from Swezey et al. 2016b). In this area, the Carolina Sandhills region is bounded on the west by the Piedmont province and is bounded on the east by the Orangeburg Scarp. Elevations are given relative to sea level

locations the sand is <2 m thick and forms a sand sheet of low relief, but in areas of higher elevation the sand can be up to 10 m thick and can form subdued hills of up to 6 m relief with steeper sides on the east and southeast (Fig. 2.6). The sand is grayish orange (10 YR 7/4; color nomenclature of Goddard et al. 1963). Grain sizes range from fine (lower) sand to coarse (lower) sand, but the most frequently occurring grain size of individual samples ranges from medium (upper) to coarse (lower) sand (0.35–0.59 mm diameter). Sorting values ($\sigma_\phi$) range from moderately sorted to poorly sorted. The sand-size grains consist predominantly of quartz (99%) with 1% mica and opaque minerals. Most of the quartz grains of medium sand size and coarser are subrounded to subangular, ranging from high sphericity to low sphericity. Exposures of the sand display evidence of bioturbation by vegetation (plant roots), and pedogenic features such as soil lamellae and argillic horizons. Exposures do not display primary sedimentary structures, although ground-penetrating radar

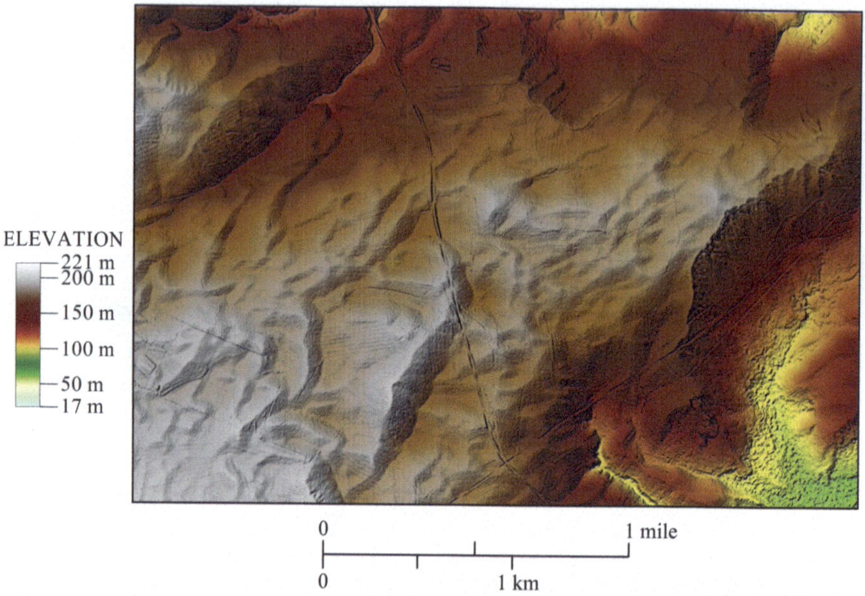


**Fig. 2.6** Detailed LiDAR image showing eolian dune morphology in the Carolina Sandhills region, Chesterfield County, South Carolina (from Swezey et al. 2016b). Elevations are given relative to sea level

(GPR) traverses have revealed 2–5-m thick sets of southeast-dipping cross-bedding at depths below 2 m.

The thickness of the eolian sand (Pinehurst Formation) displays great variability across South Carolina. In contrast with Chesterfield County where the sand reaches a thickness of 10 m, in western Lexington County the sand can be up to 24 m thick (Doar and Howard 2010). In these Lexington County locations, exposures of the sand display cross-bedding, wind ripple laminations, deformed bedding, and paleosols with organic matter.

OSL data from the sand of the Pinehurst Formation have yielded Quaternary ages (Table 2.2, Fig. 2.7). Fourteen OSL ages have been obtained from the Pinehurst Formation in Chesterfield County of South Carolina (Swezey et al. 2016b), one OSL age (published here for the first time) has been obtained from an eolian sand dune on the south side of White Pond in Kershaw County of South Carolina, and two OSL ages have been reported from an eolian sand dune within the Carolina Sandhills at Fort Bragg in Cumberland County, North Carolina (Leigh 2008). Most of these OSL ages range from ca. 42–9 ka, although the full range of reported ages is ca. 98–6 ka. At the older end of the age spectrum, the OSL age of 92.3 ± 5.2 ka from White Pond is the oldest age yet obtained from the Pinehurst Formation. No reported OSL ages from the Pinehurst Formation are younger than ca. 6 ka.

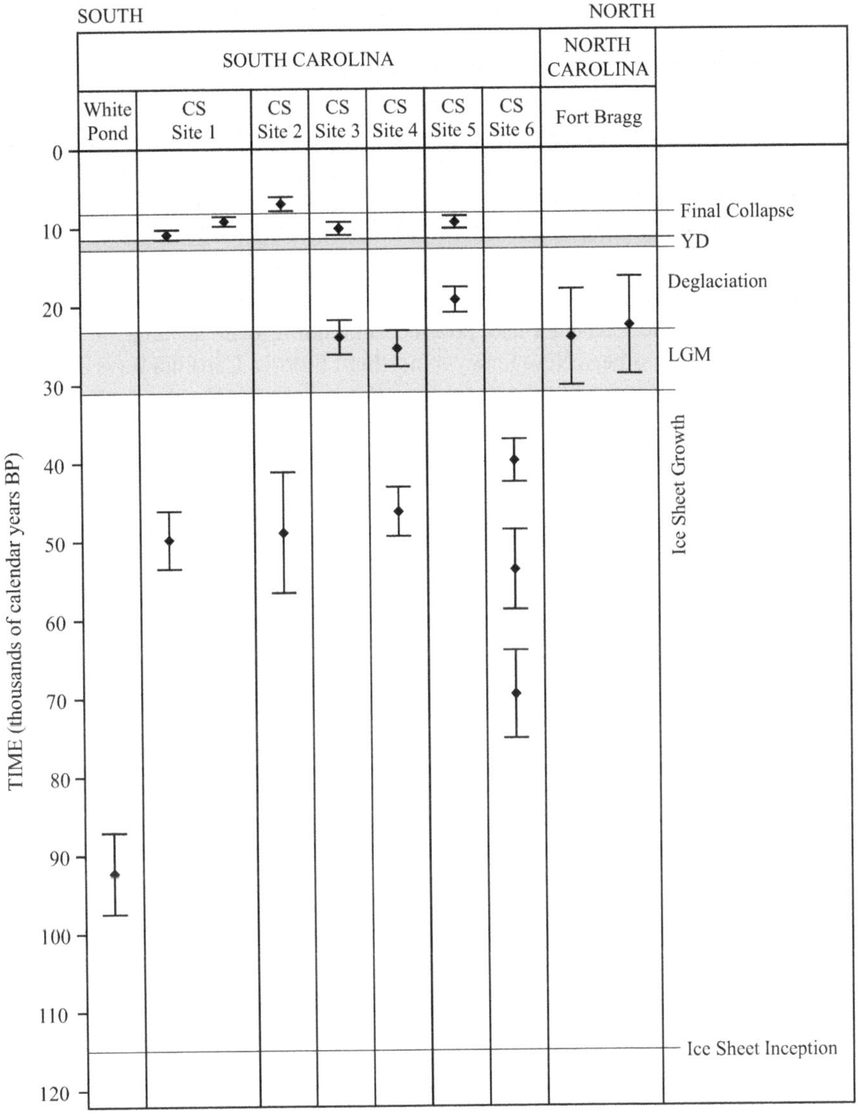


**Fig. 2.7** OSL ages from the Carolina Sandhills region of the U.S. Atlantic Coastal Plain province. *LGM* Last glacial maximum, *YD* Younger Dryas event, *Final Collapse* Final collapse of the Laurentide Ice Sheet. Detailed age data are given in Table 2.2

### 2.4.3 Eolian Dunes Associated with Carolina Bays

Oriented low-relief oval depressions referred to as "Carolina Bays" are present throughout most of the U.S. Atlantic Coastal Plain province, and arcuate ridges (referred to as "sand ridges" or "sand rims") of sand (interpreted as eolian and lacustrine deposits) are present on the south and east margins of many of these depressions (e.g., Glenn 1895; Melton and Schriever 1933; Prouty 1952; Thom 1970; Bliley and Pettry 1979; Stolt and Rabenhorst 1987a, b; Bliley and Burney 1988; Grant et al. 1998; Moore et al. 2016). The Carolina Bays are situated on different geomorphologic surfaces of the coastal plain, although they are not present in recent sediments at the modern coast. Some Carolina Bays are isolated features, whereas others exhibit cross-cutting and nested relations whereby one Carolina Bay cuts across other Carolina Bays. Most Carolina Bays range from a few hundred meters to ~12 km across their long axis, and have relief ranging from 1–3 m (not including the sand ridges). Prouty (1952) estimated that there are ~500,000 Carolina Bays in the Atlantic Coastal Plain province extending over an area of ~83,000 square miles from southern New Jersey to northern Florida. Carolina Bays are especially abundant to the east of the Orangeburg Scarp in South Carolina and adjacent areas of North Carolina and Georgia. In this region, most Carolina Bays are oval with an NW-SE orientation, and many are occupied by ponds. Farther south in Georgia, Carolina Bays are less abundant and many are circular rather than oval.

Carolina Bays are primarily surficial features that do not have much of a subsurface expression. In a publication about several Carolina Bays in North Carolina, Gamble et al. (1977, p. 199) stated, "Power auger drilling indicates that the bedding and sediments underlying a bay are undisturbed." Subsurface studies with cores and augers of numerous Carolina Bays have revealed that they consist of a few meters of sand and (or) muddy sand that overlie an unconformity on an older fine-grained substrate that does not show any sign of disturbance or interruption (Thom 1970; Gamble et al. 1977; Bliley and Burney 1988; Rodríguez et al. 2012). The nature of the underlying substrate varies from location to location, but specific identified substrates include Pleistocene clay (Gamble et al. 1977; Rodríguez et al. 2012), saprolite that formed from felsic gneiss (Bliley and Burney 1988), sandy clay of the Pliocene Duplin Formation (Brooks et al. 2001), marl and shell-bearing limestone of the Pliocene Duplin Formation? (Thom 1970), Eocene sandy silt and clay (Brooks et al. 1996), and mud of the Cretaceous Black Creek Formation (Moore et al. 2016).

In North Carolina, detailed descriptions have been published for the following three Carolina Bays: (1) Lake Mattamuskeet; (2) Wilson's Bay; and (3) Herndon Bay. OSL samples from the sand ridges of Herndon Bay have yielded ages ranging from ca. 37–27 ka (Table 2.3).

Lake Mattamuskeet (Hyde County, North Carolina; Fig. 2.1) is a conglomeration of multiple Carolina Bays that form a lake. According to Rodríguez et al. (2012), the eastern margin of the lake is a 2.9–km-wide plain with several parabolic sand ridges that exhibit relief of ca. 0.5–2.0 m. The taller ridges are located farther to the east. Cores from the bay and from the sand ridges have revealed the presence of an underlying unit of Pleistocene gray clay to sandy clay (with marine shells and burrows) that is capped by an unconformity. Within the Carolina Bay, this gray clay to sandy clay is overlain by a 0.3–1.2 m thick unit of sand and sandy silt (interpreted as lacustrine deposits and paleosols). Cores from the sand ridges have revealed the presence of the same unit of sand and silty sand, overlain by a separate 2.6–2.9 m thick unit of silt, sandy silt, and silty sand (interpreted as paleosols, loess, prograding shoreline deposits, and eolian dune deposits). The unit of silt, sandy silt, and silty sand in a western sand ridge (closer to the bay margin) contained samples of charcoal and wood that yielded ages of ca. 5760 and 1270 $^{14}C$ year BP. The unit of silt, sandy silt, and silty sand in an eastern sand ridge (farther from the bay margin) contained samples of organic sediment and charcoal that yielded ages ranging from ca. 7750–2780 $^{14}C$ year BP.

Wilson's Bay (Johnston County, North Carolina; Fig. 2.1) is a Carolina Bay with a slightly oval NW-SE orientation (~750 m long and 600 m wide), and parabolic sand ridges of ~0.5–2 m relief on the northeast and southeast margins. According to Bliley and Burney (1988), hand augers and borings from the bay and from the sand ridges have revealed the presence of an underlying unit of saprolite that formed from felsic gneiss. This saprolite is capped by an unconformity with a thin bed (lag) of quartz gravel in some places. Within the bay, the saprolite (and quartz gravel lag) is overlain by a 1.5–3.2 m thick unit of sand, sandy silt, and silty sand (interpreted as lacustrine deposits). A sample of organic material within this unit yielded a radiocarbon age of ca. 21,920 $^{14}C$ year BP. Cores from the sand ridges revealed that the saprolite (and quartz gravel lag) at these locations is overlain by a 1.5–4.0 m thick unit of muddy sand, sand, and gravel.

Herndon Bay (Robeson County, North Carolina; Fig. 2.1) is an oval Carolina Bay that is oriented NW-SE (~1 km long and 0.65 km wide), with several parabolic sand ridges of 1.5–4 m relief on the southeast margin (Fig. 2.8). According to Moore et al. (2016), four cores each drilled into different sand ridges at successively greater distances from the bay margin revealed that the sand ridges are 2.5–4.5 m thick accumulations of predominantly fine to coarse sand that rest on an unconformity above mud of the Cretaceous Black Creek Formation. Coarse sand and slightly gravelly sand are present in the two more proximal sand ridges, whereas muddy sand and sandy mud are present in the two more distal sand ridges. Three OSL ages have been reported from these sand ridges (Table 2.3, Fig. 2.9). These four cores through the sand ridges are described as follows: (1) The core in the sand ridge closest to the bay (core 1) reached a total depth of 3.6 m in mud of the Black Creek Formation (drilling depths: 3.6–2.5 m), above which was a 0.7 m thick unit of laminated fine sand (drilling depths: 2.5–1.8 m) that yielded an OSL age of ca. 27 ka (sample UW2786; Table 3). Above the fine sand, the core recovered a 1.0 m thick unit of predominantly coarse sand (drilling depths: 1.4–0.4 m); (2) The core in the next sand ridge away from the bay margin (core 2) reached a total depth of 3.6 m and recovered a 0.8 m thick unit of predominantly coarse sand (drilling depths: 3.6–2.8 m) that yielded an OSL age of ca. 30 ka (sample UW2787; Table 2.3). Above the coarse sand, the core recovered another 0.8 m thick unit of predominantly coarse sand (drilling depths: 2.6–1.8 m), above which the core recovered a 0.7 m thick unit of medium sand (drilling depths: 1.1–0.4 m); (3) The core in the third sand ridge away from the bay margin (core 3) reached a total depth of 4.8 m in mud of the Black Creek Formation (drilling depths: 4.8–4.5 m), above which was a 1.6 m thick unit of predominantly medium sand (drilling depths: 4.5–2.9 m). This unit of medium sand was overlain by a 2.7 m thick unit of predominantly fine sand (drilling depths: 2.9–0.2 m); and (4) The core in the fourth (most distal) sand ridge

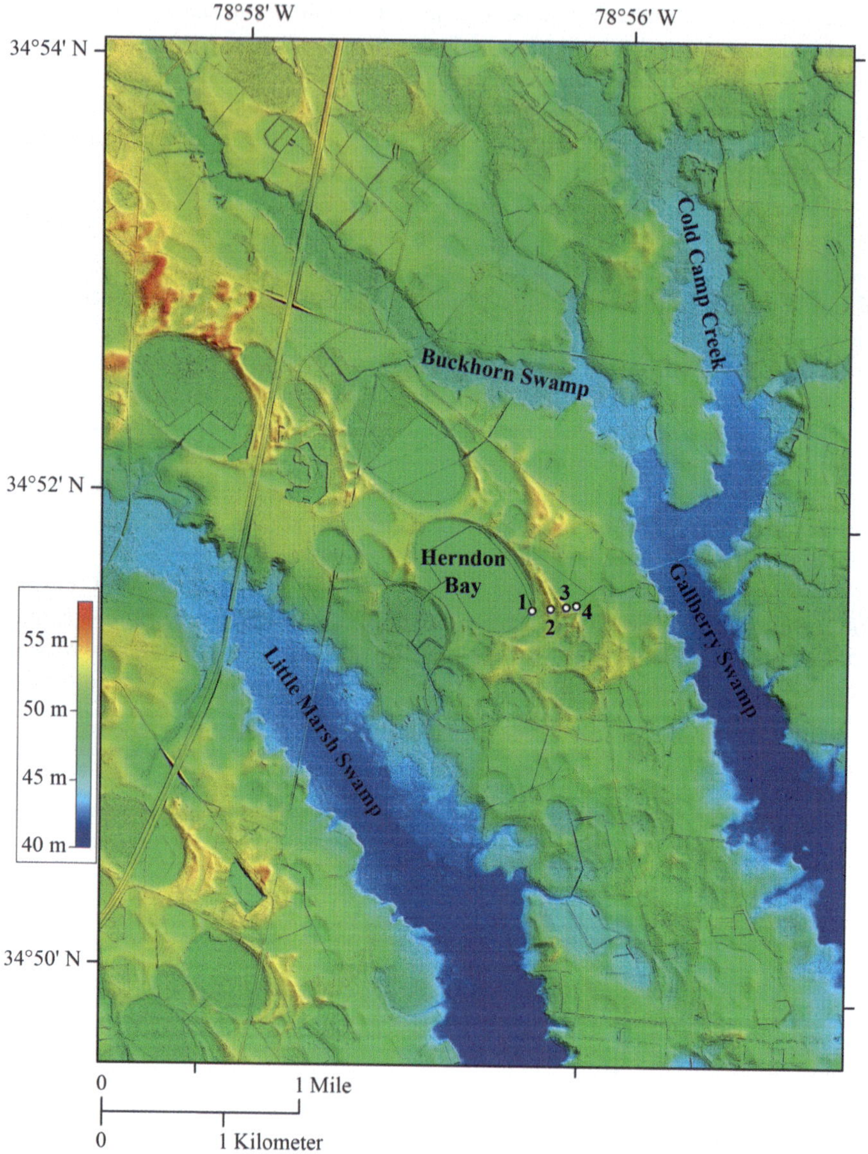


**Fig. 2.8** LiDAR image of Herndon Bay, Robeson County, North Carolina. The white circles and adjacent numbers denote to sand ridge cores obtained by Moore et al. (2016). Elevations are given relative to sea level

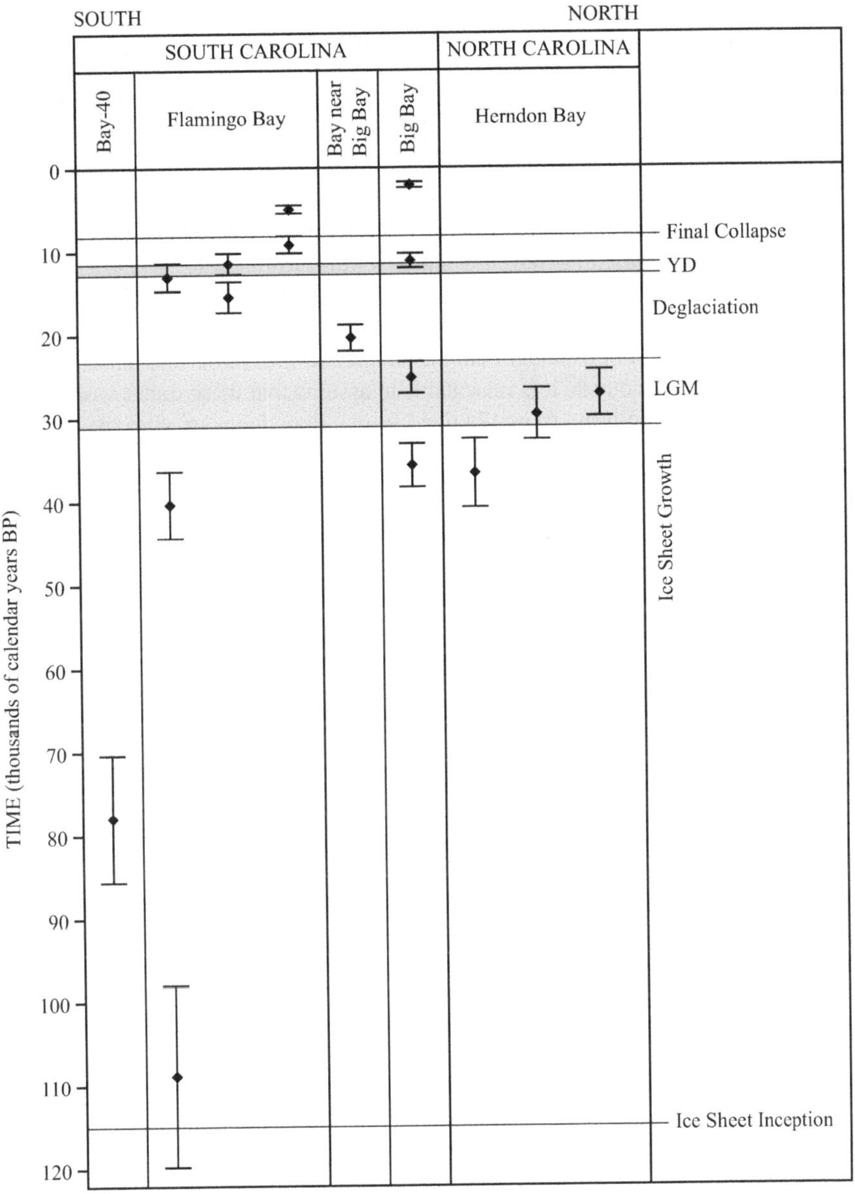


**Fig. 2.9** OSL ages from sand ridges of Carolina Bays of the U.S. Atlantic Coastal Plain province. *LGM* Last glacial maximum, *YD* Younger Dryas event, *Final Collapse* Final collapse of the Laurentide Ice Sheet. Detailed age data are given in Table 2.3

from the bay margin (core 4) reached a total depth of 3.6 m in mud of the Black Creek Formation (drilling depths: 3.6–2.5 m), above which was a 1.1 m thick unit of predominantly fine to medium sand (drilling depths: 2.3–1.2 m) that yielded an OSL age of ca. 37 ka (sample UW2788; Table 3). This unit of fine to medium sand was overlain by 0.7 m thick unit of predominantly medium sand (drilling depths: 1.1–0.4 m).

In South Carolina and Georgia, several Carolina Bays are present within river valleys, and they exhibit cross-cutting relations with eolian dunes in these valleys. For example, Bear Swamp (Marion County, South Carolina; Fig. 2.1) is a Carolina Bay that is oriented NW-SE (ca. 1.2 km long and 0.8 km wide). This Carolina Bay is inset into a field of eolian dunes in the valley of the Great Pee Dee River (Fig. 2.10), and thus this bay must be younger than the dunes. Although OSL ages have not been published from these dunes, it is reasonable to assume that these dunes would yield OSL ages that are similar to those obtained from eolian dunes in other river valleys of the coastal plain. Ivester et al. (2001) described a similar setting in the valley of the Ohoopee River (Tattnall County, Georgia), where a Carolina Bay named Dukes Pond (Fig. 2.1) is inset within eolian dunes that have yielded an OSL age of ca. 23.6 ka (Table 2.1).

In South Carolina, detailed descriptions have been published for the following three Carolina Bays: (1) Big Bay; (2) an unnamed Carolina Bay immediately southwest of Big Bay; and (3) Flamingo Bay. OSL ages obtained from samples from the Carolina Bay sand ridges range from ca. 108.7–2.2 ka (Table 2.3). Where multiple sand ridges are present, older ages have been obtained from the sand ridges that are farther from the modern bay.

Big Bay (Sumter County, South Carolina; Fig. 2.1) is an oval Carolina Bay in the valley of the Wateree River on the north side of Big Bay Road in Sumter County, South Carolina (Fig. 2.11). The bay is oriented NW-SE (~5 km long and 3 km wide), and has several parabolic sand ridges of 1.5–3.5 m relief on the south and east margins. The western margin of Big Bay is covered by eolian sand in the form of a sand sheet with a few parabolic dunes. According to Ivester et al. (2002) and Brooks et al. (1996, 2010), this eolian sand has yielded three OSL ages. Within the confines of Big Bay, a sample from the sand sheet yielded an OSL age of ca. 74.3 ka and a sample from a parabolic dune yielded an OSL age of ca. 29.6 ka. At a distance ~1 km west of the inferred margin of Big Bay, a sample from a parabolic dune that is part of the same eolian sand deposit yielded an OSL age of ca. 33.2 ka. Brooks et al. (2001) described a core (drill hole D1/2; Fig. 2.11) drilled through the sand sheet within the confines of Big Bay that reached a total depth of 10.6 m in sandy clay of the Pliocene Duplin Formation (drilling depths: 10.6–9.0 m), above which was a 4.5 m thick unit of silty sand and sandy mud with abundant organic material (drilling depths: 9.0–4.5 m), above which was a 4.5 m thick unit of quartz sand (drilling depths 4.5–0 m).

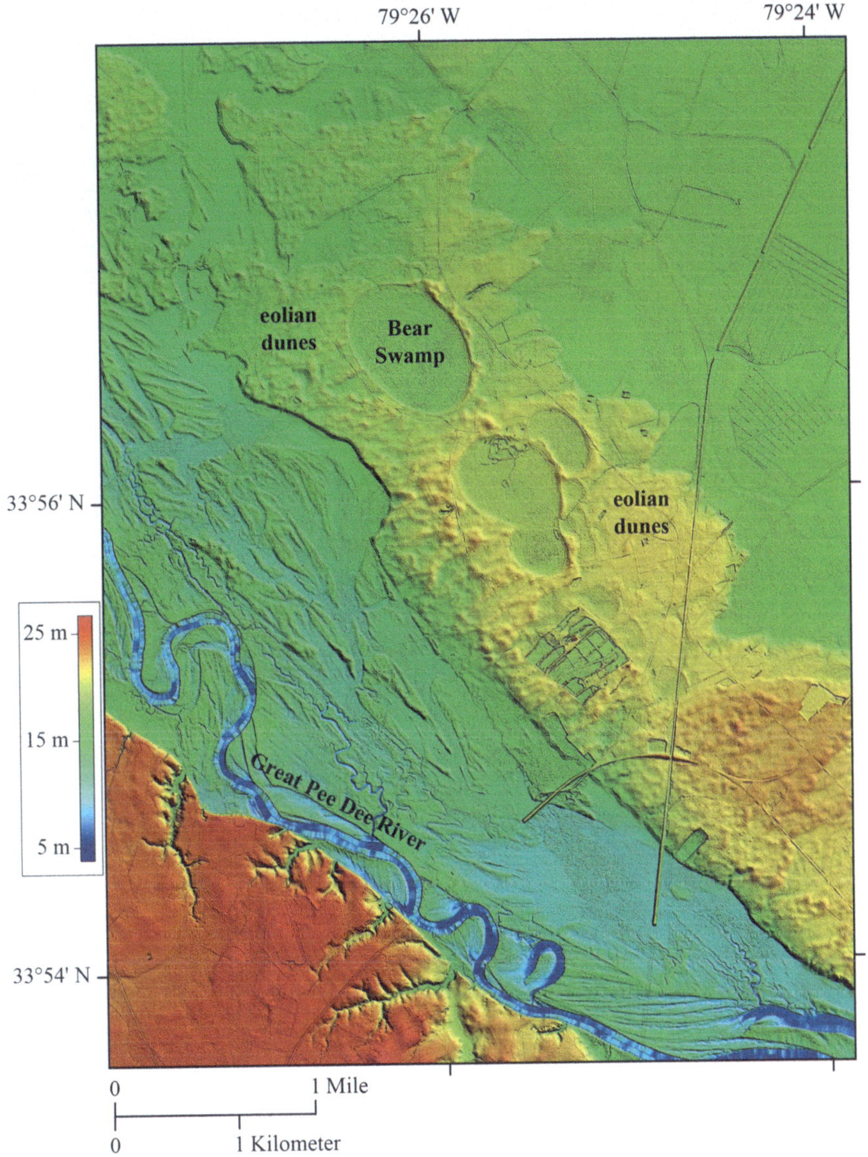


**Fig. 2.10** LiDAR image of Bear Swamp, Marion County, South Carolina. Bear Swamp is a Carolina Bay inset into eolian dunes in the valley of the Great Pee Dee River. Elevations are given relative to sea level

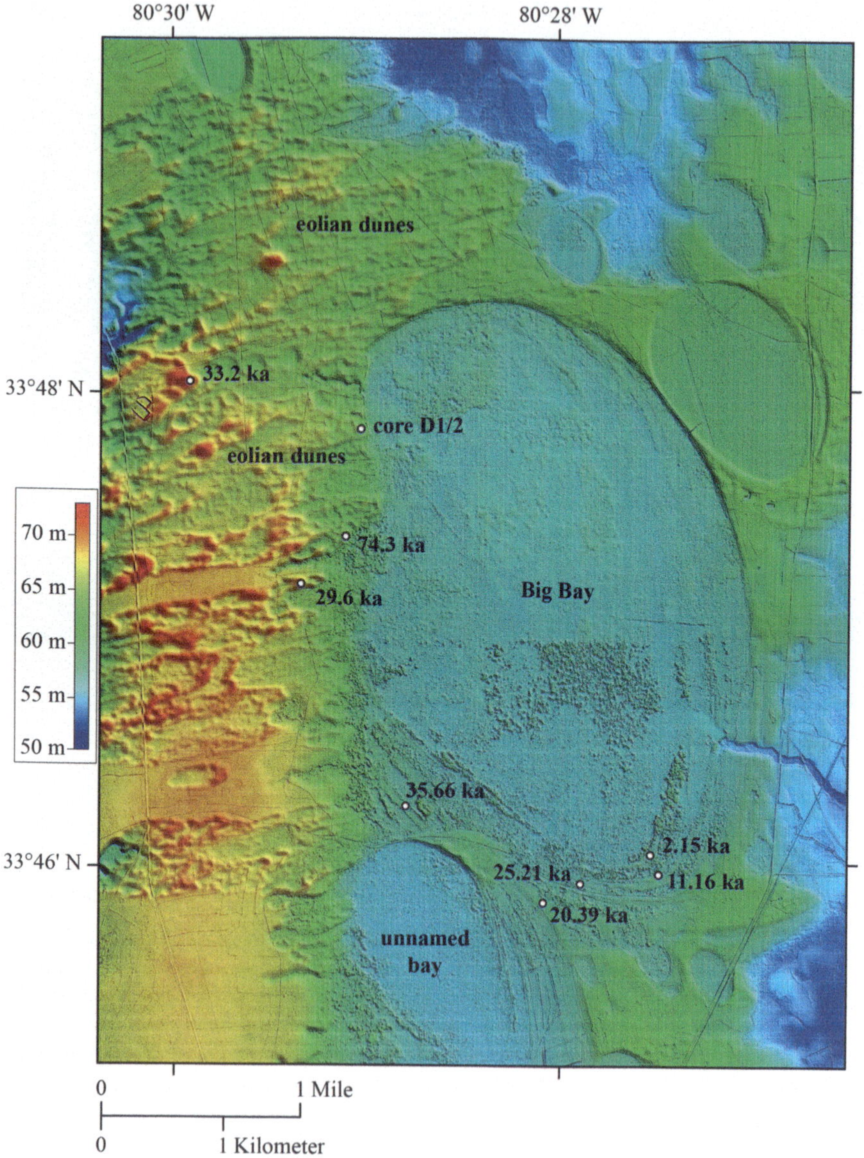


**Fig. 2.11** LiDAR image of Big Bay, Sumter County, South Carolina. The white circles and adjacent numbers denote cores and OSL ages reported by Ivester et al. (2002) and Brooks et al. (1996, 2010). Elevations are given relative to sea level

Several OSL ages have been reported from sand ridges along the south and east margins of Big Bay (Brooks et al. 1996, 2010; Ivester et al. 2003). These OSL data are described as follows: (1) From the sand ridge closest to the bay, a sediment sample at 60–75 cm depth yielded an OSL age of 2.15 ± 0.30 ka (this age was reported as 2.2 ka by Brooks et al. 2010); (2) From the next sand ridge away from the bay margin, a sediment sample at 60–75 cm depth yielded an OSL age of 11.16 ± 0.90 ka (this age was reported as 11.3 ka by Brooks et al. 2010); (3) From the third sand ridge away from the bay margin, a sediment sample at 60–75 cm depth yielded an OSL age of 25.21 ± 1.9 ka (this age was reported as 25.4 ka by Brooks et al. 2010); and (4) From the fourth (most distal) sand ridge from the bay margin, a sediment sample at 60–75 cm depth yielded an OSL age of 35.66 ± 2.60 ka (this age was reported as 35.9 ka by Brooks et al. 2010).

Additional data are available from an unnamed Carolina Bay located immediately to the southwest of Big Bay, on the south side of Big Bay Road in Sumter County (Fig. 2.11). This unnamed bay is oriented NW-SE (~2 km long and 1 km wide), and has several parabolic sand ridges of 0.8–1 m relief on the south and east margins of the bay. Ivester et al. (2003) stated that sediment from a sand ridge associated with this bay yielded an OSL age of 20.39 ± 1.60 ka (this age was reported as 20.5 ka by Brooks et al. 2010). It is possible, however, that this sand ridge might be associated more appropriately with Big Bay than with the unnamed bay.

Flamingo Bay (Aiken County, South Carolina; Fig. 2.1) is an oval Carolina Bay that is oriented NW-SE (~0.6 km long and 0.5 km wide), with several parabolic sand ridges of 4–5 m relief on the east and south margins (Fig. 2.12). According to Brooks et al. (1996), a core (C1) taken within Flamingo Bay penetrated a 94-cm thick unit of quartz sand above a unit of sandy silt and clay that they referred to as a "BC soil horizon," which they interpreted as Eocene sediments that were subsequently altered by lateritic weathering. Charcoal samples from the unit of quartz sand within the Flamingo Bay yielded ages ranging from ca. 4505–2550 $^{14}$C year BP. Brooks et al. (1996) also reported that a core (P25) through the highest part of a sand ridge on the eastern side of Flamingo Bay revealed a 1.85 m-thick unit of Quaternary quartz-rich medium sand, overlying an unconformity with evidence of lateritic weathering ("BC Soil Horizon") on a unit of Eocene sandy silt and clay. Additional cores on the western flank of this sand ridge revealed a similar stratigraphy, but with the addition of quartz pebbles on the unconformity. Moore et al. (2012) reported the following five OSL ages from a Flamingo Bay sand ridge composed of medium sand: (1) 5.0 ± 0.5 ka at ~35 cm depth; (2) 9.2 ± 1.0 ka at ~50 cm depth); (3) 11.5 ± 1.3 ka at ~65 cm depth; (4) 15.5 ± 1.8 ka at ~78 cm depth; and (5) 13.1 ± 1.7 ka at ~95 cm depth. In addition, Ivester et al. (2002) and Brooks et al. (2010) reported the following two OSL ages from sand ridges of Flamingo Bay: (1) 108.7 ± 10.9 ka; and (2) 40.3 ± 4.0 ka. Ivester et al. (2002) and Brooks et al. (2010) also reported an OSL age of 77.9 ± 7.6 ka from a sand ridge of a nearby Carolina Bay named Bay-40.

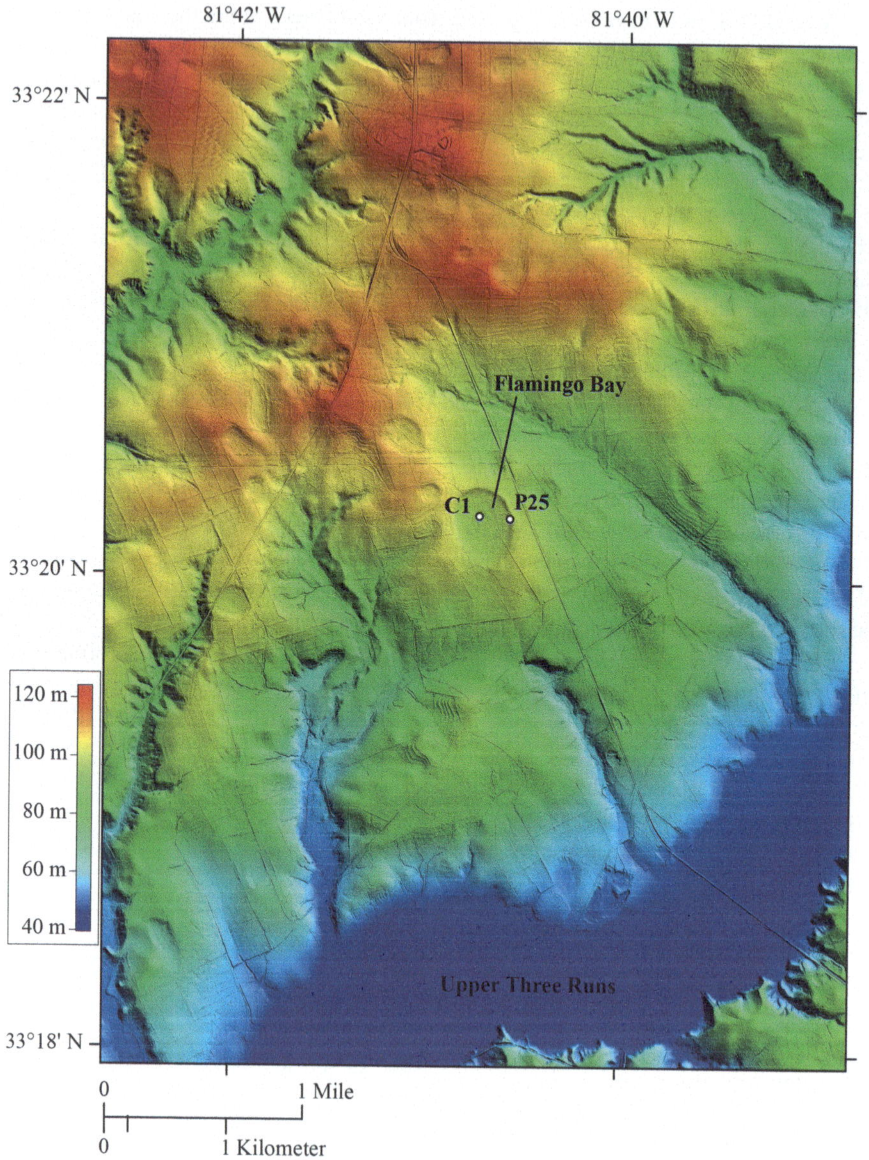


**Fig. 2.12** LiDAR image of Flamingo Bay, Aiken County, South Carolina. C1 and P25 denote cores obtained by Brooks et al. (1996). Elevations are given relative to sea level

### 2.4.4 *Eolian Dunes and Sand Sheets on Upland Areas of the Northern Atlantic Coastal Plain*

Vegetated (stabilized) eolian sand dunes and sand sheets are present on many upland areas of the northern Atlantic Coastal Plain (Figs. 2.1 and 2.13). These eolian deposits are extensive and particularly well developed on the Delmarva Peninsula (Delaware, Maryland, Virginia) east of the Chesapeake Bay (Rasmussen and Slaughter 1955; Denny and Owens 1979; Denny et al. 1979; Mixon 1985; Lowery et al. 2010; Newell and DeJong 2011). Similar eolian sand and silt have also been identified on the Atlantic Coastal Plain west of the Chesapeake Bay (west of the Delmarva Peninsula) in Maryland (Hack 1955; Markewich et al. 2009). Farther north in Sussex County (Delaware), an approximately correlative unit of fine sand, silt, and clayey silt is mapped as the Cypress Swamp Formation and is interpreted as deposits of fresh-water bogs, swamps, marshes, ponds, floodplains, and eolian

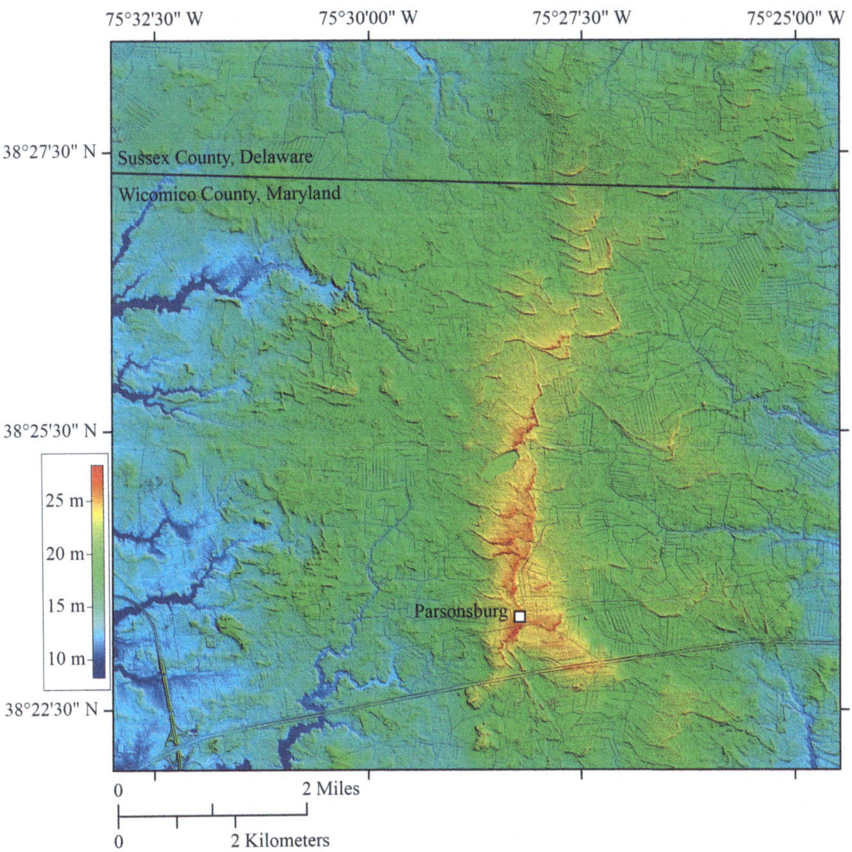


**Fig. 2.13** LiDAR image showing parabolic dunes on the Delmarva Peninsula (Wicomico County, Maryland; Sussex County, Delaware). Elevations are given relative to sea level

dunes (Andres and Howard 2000). Quaternary eolian sand and (or) silt are also present in the Pine Barrens region of southern New Jersey (Newell and DeJong 2011; French and Demitroff 2012), around the western margin of the Atlantic Coastal Plain province in Mercer County of New Jersey (Stanford 1993), and in nearby areas west of the Atlantic Coastal Plain province in southeastern Pennsylvania (Carey et al. 1976).

According to Denny and Owens (1979) and Denny et al. (1979), eolian sand on the upland areas of the Delmarva Peninsula is typically 1.2–6 m thick, and overlies a variety of older units of sand and mud that range in age from Miocene to Quaternary (e.g., Pensauken Formation, Beaverdam Sand, Walston Silt, Omar Formation, Kent Island Formation). At most locations the eolian sand is composed of light brown to dark gray fine to medium sand, although clayey silt is present at some locations. The sand consists predominantly of quartz, with a small fraction of feldspar and opaque minerals. Most exposures of the sand display evidence of bioturbation and pedogenic features such as soil lamellae and argillic horizons. Evidence of primary sedimentary structures is absent in most exposures, although low-angle bedding is visible at the base of some exposures. Thin units of peaty sand within the Parsonsburg Sand have yielded several radiocarbon ages ranging from ca. 30,560–13,420 $^{14}$C year BP. The upper surface of the eolian sand in the upland area of the central Delmarva Peninsula varies from a relatively flat surface to irregular mounds and sinuous ridges (Denny and Owens 1979; Denny et al. 1979). Some of the ridges are in the shape of parabolic dunes with tails that point to the northwest (Fig. 2.13).

On the west side of the Delmarva Peninsula (east side of the Chesapeake Bay), there are several exposures of eolian sand and sandy silt that may be correlative to the Parsonsburg Sand. In Northampton County (Virginia), for example, Mixon (1985) described irregular fields of 0.6–3 m high eolian sand dunes. Farther north in Kent County (Maryland), Foss et al. (1978) described a 0.5–2.1 m thick unit of Quaternary sandy silt (interpreted as loess) that overlies a "buried paleosol." In Talbot County (Maryland), Lowery et al. (2010, p. 1475) described two units of sandy silt (interpreted as loess) that overlie an unconformity (paleosol) on a "fine-sandy sequum formed in either fluvial/estuarine sediments or possibly coarse eolian deposits." The lower unit of sandy silt is ~1 m thick and is capped by an unconformity (paleosol), and the upper unit of sandy silt is ~0.8 m thick. Lowery et al. (2010) named the lower unit as the Miles Point Loess, upper unit as the Paw Paw Cove Loess, and the intervening paleosol as the Tilghman Soil. At a locality named Miles Point, the "fine-sandy sequum" immediately below the Miles Point Loess yielded two OSL ages of ca. 41 ka, the Miles Point Loess yielded two OSL ages of ca. 41 and 35 ka, and the Tilghman Soil yielded two OSL ages of ca. 29 and 28 ka (Table 2.4, Fig. 2.14). The Tilghman Soil also yielded four radiocarbon ages ranging from ca. 27,249–21,490 $^{14}$C year BP (Lowery et al. 2010).

On the west side of the Chesapeake Bay (west of the Delmarva Peninsula), vegetated (stabilized) eolian sand and silt are present across several upland areas of the Maryland coastal plain. In upland areas of Prince Georges and Charles Counties of Maryland, Hack (1955) identified a < 2 ft. (0.6 m) thick unit of eolian fine sand that rests on an unconformity on Miocene sand, gravel, and loam. With the availability of LiDAR data, Markewich et al. (2009) were able to discern the presence of parabolic dunes among this eolian sand in Charles County. Relief of these parabolic dunes is 8–12 m, the axis length of the dunes is ~3 km, and the dune tails point to the northwest. At a site named Fenwick Shores/Piscataway, one sample of eolian sand yielded an OSL age of ca. 27 ka (Table 2.4, Fig. 2.14). At a site named Goose Bay, a ~4 m thick quartz-rich medium sand of eolian origin yielded two OSL ages of ca. 30 and 24 ka (Table 2.4, Fig. 2.14). This eolian unit rests on an unconformity above a ~0.9 m-thick estuarine deposit of quartz-rich medium to fine sand that fines up to clayey silt with organic-rich sediment and wood that yielded radiocarbon ages of ca. 26,270 and 20,500 $^{14}$C year BP. Markewich et al. (2009) also reported OSL ages from eolian sand at two other sites in Charles County (Chapman Landing, Brandywine2). Eolian sand from the Chapman Landing site yielded an OSL age of 27 ka, and eolian sand from the Brandywine2 site yielded an OSL age of ca. 19 ka (Markewich et al. 2009).

In the Piedmont province immediately west of the Atlantic Coastal Plain province, sand and silty sand of possible eolian origin are present at several locations in central and northern Virginia. For example, on many of the interfluvial upland areas of Louisa County, the author has noted a < 0.5 m thick unit of possible eolian origin composed of quartz-rich very fine sand and silt that overlies deeply weathered Paleozoic metamorphic rocks. Similarly, on an interfluvial upland area in Prince William County, Feldman et al. (2000) described a ~0.5–1 m thick unit of silt (interpreted as loess) that overlies granitic saprolite. This silt yielded a thermoluminescence (TL) age of ca. 14 ka.

## 2.5 Interpretations of Eolian Dunes and Sand Sheets

When viewed as an ensemble, the eolian dunes and sand sheets in inland settings of the U.S. Atlantic Coastal Plain province show some interesting differences and similarities. For example, the eolian sediments are thought to have been derived from distinctly different sources in each of the four identified settings (river valleys; Carolina Sandhills; Carolina Bays; upland areas of the northern Atlantic Coastal Plain). Yet the eolian sediments show many similarities with respect to inferred timing of eolian mobilization, thus providing a common framework for interpretations of paleoclimate variables such as wind direction, wind velocity, and air temperature.

### 2.5.1 *Eolian Dunes in River Valleys*

Several previous studies have interpreted the eolian dunes in the river valleys as being derived from fluvial sand from the nearby river channels (e.g., Pickering and Jones 1974; Carver and Brook 1989; Markewich and Markewich 1994; Ivester et al.

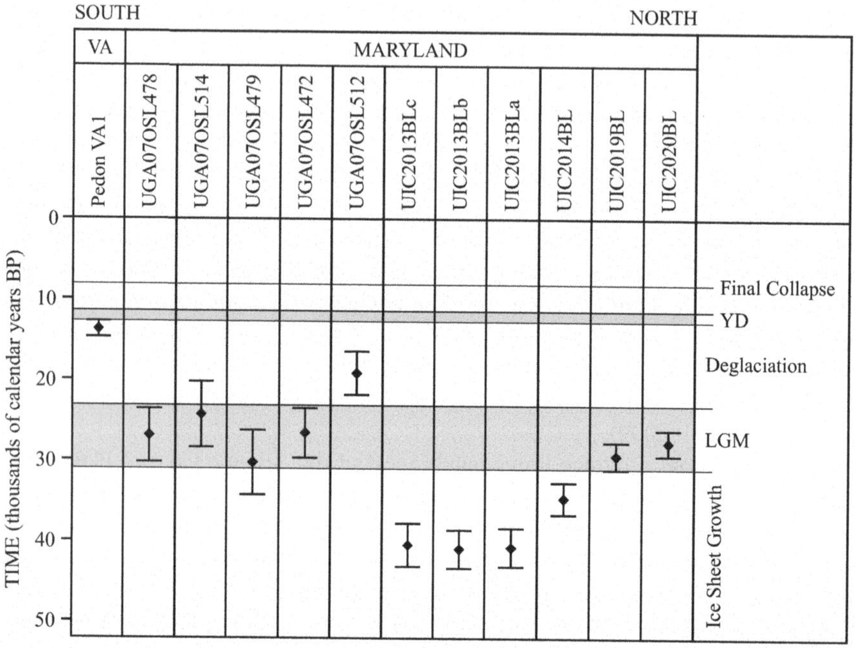


**Fig. 2.14** OSL ages from eolian dunes on upland areas of the northern Atlantic Coastal Plain province. *LGM* Last glacial maximum, *YD* Younger Dryas event, *Final Collapse* Final collapse of the Laurentide Ice Sheet. Detailed age data are given in Table 2.4

2001; Ivester and Leigh 2003; Swezey et al. 2013). There must have been enough variability in climate and river discharge such that the fluvial sand could be mobilized into eolian dunes, which migrated away from the river channels until they became stabilized. The eolian dunes, however, are located only 1–5 km from the modern river channels, indicating that the dunes did not travel very far from their sediment sources before becoming stabilized.

OSL ages from eolian dunes in the river valleys range from ca. 84–5 ka and the majority of these ages are approximately coincident with the last glaciation in the northern hemisphere (Fig. 2.4), assuming the date of inception of the Laurentide Ice Sheet to be ca. 115 ka (Mix 1992; Kleman et al. 2010) and the date of final collapse of the Laurentide Ice Sheet to be ca. 8.2 ka (Barber et al. 1999; Shuman et al. 2002). The great range of OSL ages suggests that eolian sediment mobilization was episodic during this time. Most of the OSL ages range from ca. 35–14 ka and are approximately coincident with the last glacial maximum (LGM) and the time of deglaciation before the Younger Dryas (YD) event, assuming the LGM to be ca. 31,100 to 23,200 calibrated years before present [reported by Clark et al. 2009 as 26.5 to 19–20 ka in radiocarbon years before present, and converted to calibrated years using the program CALIB 6.1.1 (available at http://calib.qub.ac.uk/calib) in conjunction with Stuiver and Reimer 1993 and Reimer et al. 2009], and assuming the YD event to be 12,800 to 11,500 cal year BP (Alley et al. 1993).

At the older end of the age spectrum, OSL ages of 77.4 ± 6.6 ka from the valley of the Ohoopee River and 74.3 ± 7.1 ka from the valley of the Wateree River provide evidence of eolian sand mobilization well before the LGM (Fig. 2.4). For some reason, the eolian sand at these locations was not reworked during the LGM or later. There are apparent gaps in the river valley OSL ages from ca. 67–14 ka, and from ca. 14–11 ka. Eolian sand may have been stabilized during these times, but it is also possible that additional OSL ages may fill in these gaps. It is also important to realize that the OSL ages indicate not the total time of eolian sand mobilization but only the time that eolian sand was last exposed to sunlight. At the young end of the age spectrum, an OSL age of ca. 5 ka from the Altamaha River may attest to some isolated eolian sand remobilization in this area, or perhaps may be attributed to the effects of post-depositional bioturbation exposing sand grains to sunlight.

No OSL ages from eolian sand in the river valleys are younger than ca. 5 ka (Fig. 2.4), and thus it appears that since this date the sand has been stabilized by vegetation and subjected to pedogenic processes. This final stabilization of the dunes is thought to have been caused by a change to a less arid climate along with a general increase in air temperature, decrease in wind velocity, and increase in vegetation density (Swezey et al. 2013). Most exposures of the dunes do not display sedimentary structures except for traces of bioturbation by plant roots, suggesting that the dunes have been stabilized for a duration long enough for vegetation to obliterate primary sedimentary structures.

Most eolian dunes in the river valleys are parabolic dunes, which is a morphology that is usually associated with an adequate sand supply, moderate vegetation cover, and a unidirectional wind regime for sand-mobilizing winds (McKee and Bigarella 1979; Lancaster 1995; Hugenholtz et al. 2008; Hugenholtz 2010). Pollen data also support the interpretation of some vegetation being present during the last glaciation and deglaciation (e.g., Watts 1980a, b; Delcourt and Delcourt 1984, 1985; LaMoreaux et al. 2009; Spencer et al. 2017). With regards to wind directions, the tails of parabolic dunes point to the northwest in the northern coastal plain (Delaware, Maryland), suggesting that the winds that mobilized the sand in this area blew from the northwest. In contrast, the tails of parabolic dunes point to the west in the central and southern coastal plain (North Carolina, South Carolina, Georgia), suggesting that the winds that mobilized the sand in this area blew from the west. Some climate models (e.g., Kutzbach et al. 1998) have proposed that surface winds in the southeastern United States blew from the west during the LGM winter and from the southeast during the LGM summer (e.g., Kutzbach et al. 1998), prompting speculation that eolian dune mobilization may have occurred preferentially during the winter (Swezey et al. 2016b).

Very few eolian dunes have been reported from river valleys in central and southern Virginia, which is the area where winds from the northwest (Delaware, Maryland) are presumed to have converged with winds from the west (North Carolina, South Carolina, Georgia). Quaternary eolian sand has been reported from the Cactus Hill and Rubis Pearsall archeological sites, both of which are located in the upper coastal plain part of the Nottoway River valley in Sussex County, Virginia (Wagner and McAvoy 2004; Feathers et al. 2006; Macphail and McAvoy 2008;

Markewich et al. 2009). These sands have yielded OSL ages ranging from ca. 23.7–7.4 ka (Feathers et al. 2006), which are generally coincident with the ages of numerous eolian sands described in this paper. However, the criteria cited for determining an eolian origin of these sands are now considered to be non-diagnostic criteria (Swezey 1998; Swezey et al. 2016b), and LiDAR data in this area show primarily fluvial sand bars rather than fields of large and obvious eolian dunes such as are present in North Carolina, South Carolina, and Georgia. If eolian sand is present at the Cactus Hill and Rubis Pearsall sites, then this sand is likely to be a relatively thin deposit reworked from underlying fluvial sediments. Some fields of eolian dunes, however, may be present in the Nottoway River valley even farther downstream near the Virginia–North Carolina border (Powars et al. 2016).

### 2.5.2 *Eolian Dunes and Sand Sheets of the Carolina Sandhills*

The unconsolidated sand (Quaternary Pinehurst Formation) of the Carolina Sandhills is interpreted as eolian dunes and sand sheets derived from sand of the underlying Cretaceous strata (Fitzwater 2016; Swezey et al. 2016a, 2016b). This interpretation of sand source is supported by the spatial association of the Quaternary and Cretaceous units, and the fact that the two units have similar grain sizes and similar abundance and composition of opaque minerals. Furthermore, the poor sorting suggests that sand of the Pinehurst Formation has not traveled far from its source.

OSL ages from the eolian sand (Pinehurst Formation) are approximately coincident with the last glaciation in the northern hemisphere (Fig. 2.7). The great range of OSL ages (ca. 98–6 ka) suggests that eolian sediment mobilization was episodic during this time. Most of the OSL ages, however, range from ca. 42–9 ka and are approximately coincident with the interval from the latter stages of ice sheet growth (leading to the LGM) to the Final Collapse of the Laurentide Ice Sheet. At the older end of the age spectrum, an OSL age of 92.3 ± 5.2 ka from White Pond (South Carolina) provides evidence of eolian sand mobilization well before the LGM. The southern location (greater distance from the influences of the Laurentide Ice Sheet) may have helped to prevent eolian sand at this location from being reworked during the LGM or later. There are apparent gaps in the Carolina Sandhills OSL ages from ca. 87–75 ka, 64–59 ka, 37–30 ka, and 17–12 ka. Eolian sediment may have been stabilized during these times, but it is also possible that additional OSL ages may fill in these gaps. As stated above, the OSL ages indicate not the total time of eolian sand mobilization but only the time that the sand was last exposed to sunlight.

No OSL ages from the Pinehurst Formation are younger than ca. 6 ka, and thus it appears that since this date the sand has been stabilized by vegetation and subjected to pedogenic processes. As with eolian dunes in the river valleys, this final stabilization of eolian sand in the Carolina Sandhills region is thought to have been caused by a change to a less arid climate along with a general increase in air temperature, decrease in wind velocity, and increase in vegetation density (Swezey et al. 2016b). Most exposures of the dunes do not display sedimentary structures except for traces of bioturbation by plant roots, suggesting that the dunes have been stabilized for a duration long enough for vegetation to obliterate primary sedimentary structures.

Several paleoclimate variables may be inferred from characteristics of the Pinehurst Formation (Swezey et al. 2016b). For example, the predominance of sand sheets over dunes is attributed to the coarse grain size and to the likely presence of some vegetation when the sand was mobilized. Eolian mobilization of coarse sand would have been facilitated by colder air temperatures, such as are inferred for the LGM. The dune morphologies and cross-bedding suggest that winds that mobilized the sand blew from the west and (or) northwest. These inferred wind directions are most consistent with both modern January wind directions (Baldwin 1975) and inferred LGM January wind directions (Kutzbach et al. 1998), suggesting that eolian sand mobilization may have occurred preferentially during the winter (when the air temperatures would have been lower, thus facilitating the eolian mobilization of coarse sand). Furthermore, the relatively coarse grain size suggests that eolian sand mobilization during the LGM winter would have required wind velocities of at least 4–6 m/s, after taking into account the effects of colder air temperatures on eolian sand transport (Swezey et al. 2016b).

### *2.5.3 Eolian Dunes Associated with Carolina Bays*

Most geologists interpret the Carolina Bays as relict geomorphologic features that formed via a combination of eolian and lacustrine processes (e.g., Livingstone 1954; Thom 1970; Stolt and Rabenhorst 1987a, b; Bliley and Burney 1988; Grant et al. 1998; Moore et al. 2016). Some of the sediments from Wilson's Bay are interpreted specifically as deposits that formed when ice in a lacustrine basin expanded and was thrust against the bay shoreline (Bliley and Burney 1988). The evidence that Carolina Bays are relict features comes from observations that: (1) modern drainages cut across some Carolina Bays; (2) some Carolina Bays are nested within other bays and show various cross-cutting relations with other Carolina Bays; and (3) Carolina Bays are not present in recent sediments at the modern coast. Furthermore, the wide range of OSL ages and the observation that some Carolina Bays show cross-cutting relations with other Carolina Bays and with eolian dunes in river valleys demonstrates that the Carolina Bays did not form during a single event. Theories of origin related to meteorite impacts (e.g., Melton and Schriever 1933; MacCarthy 1937; Prouty 1952) may be ruled out because the OSL ages show a wide range of values and because there is no evidence of disturbance of the underlying strata. Likewise, theories of origin related to traditional karst phenomena (e.g., Smith 1931) may be ruled out because of the absence of limestone beneath most Carolina Bays.

OSL ages from eolian sand ridges of the Carolina Bays range from ca. 109–2 ka, but most of the ages are approximately coincident with the last glaciation in the northern hemisphere (Fig. 2.9). The great range of OSL ages suggests that eolian sediment mobilization was episodic during this time. Most of the OSL ages range from ca. 40–11 ka and are approximately coincident with the LGM through the Younger Dryas (YD) event. At the older end of the age spectrum, OSL ages of 108.7 ± 10.9 ka from a sand ridge of Flamingo Bay and 77.9 ± 7.6 ka from a sand ridge of Bay-40 (both located in South Carolina) provide evidence of eolian sand mobilization well before the LGM. The southern location (greater distance from the influences of the Laurentide Ice Sheet) may have helped to prevent the eolian sand at these two locations from being reworked during the LGM or later. There are apparent gaps in the sand ridge OSL ages from ca. 98–86 ka, and from ca. 70–44 ka. Eolian sediment may have been stabilized during these times, but it is also possible that additional OSL ages may fill in these gaps. At the young end of the age spectrum, an OSL age of 2.1 ± 0.3 ka from a sand ridge of Big Bay may attest to some isolated eolian sediment remobilization in this area, or perhaps may be attributed to the effects of post-depositional bioturbation exposing sand grains to sunlight. Bioturbation seems likely because this ca. 2 ka age is much younger than any other OSL ages reported from eolian sand of the inland coastal plain.

In addition to the range of OSL ages, stratigraphic relations provide exceptionally clear evidence that the Carolina Bays did not form during one specific event of limited duration. For example, there are certain locations where eolian sand dunes in a river valley overlie parts of a Carolina Bay, and thus that Carolina Bay must be older than the overlying eolian sand (e.g., Big Bay; Fig. 2.11). There are other locations where a Carolina Bay is inset into eolian sand dunes in a river valley, and thus that Carolina Bay must be younger than the underlying eolian sand (e.g., Bear Swamp; Fig. 2.10). Additional evidence of the episodic nature of Carolina Bay genesis comes from the observation by Ivester et al. (2007) that the sand ridges of any specific Carolina Bay yield systematically younger OSL ages towards the center of the bay, confirming that the bays are not the product of a single event of limited duration.

The OSL ages indicate that the Carolina Bays formed during approximately the same time interval as the eolian dunes and sand sheets in the river valleys and in the Carolina Sandhills region. In other words, the Carolina Bays are relict features that formed when air temperatures were cooler, wind velocities were greater, and vegetation density was reduced. Very similar oriented lakes are present today in many high-latitude regions such as Alaska (Fig. 2.15), Canada, Russia, and the Falkland Islands (e.g., Rex 1961; Carson and Hussey 1962; Coté and Burn 2002; Wilson et al. 2002; Hinkel et al. 2005, 2012; Arp et al. 2011; Karlsson et al. 2012; Morgenstern et al. 2013; Zhan et al. 2014). These high-latitude oriented lakes are thermokarst lakes that developed as a result of thaw and collapse of frozen ground, with subsequent modification by lacustrine and (or) eolian processes. Most studies of thermokarst lakes have been conducted in the continuous permafrost zone of northern Alaska, but thermokarst lakes are also present in zones of discontinuous permafrost such as the boreal forest regions of central Alaska and northern Quebec (Jorgenson et al. 2012; Coulombe et al. 2016). In the Barrow region of northern Alaska, the dominant wind direction is from the east to northeast, and this wind generates gyres that promote erosion of the north and south shores of the

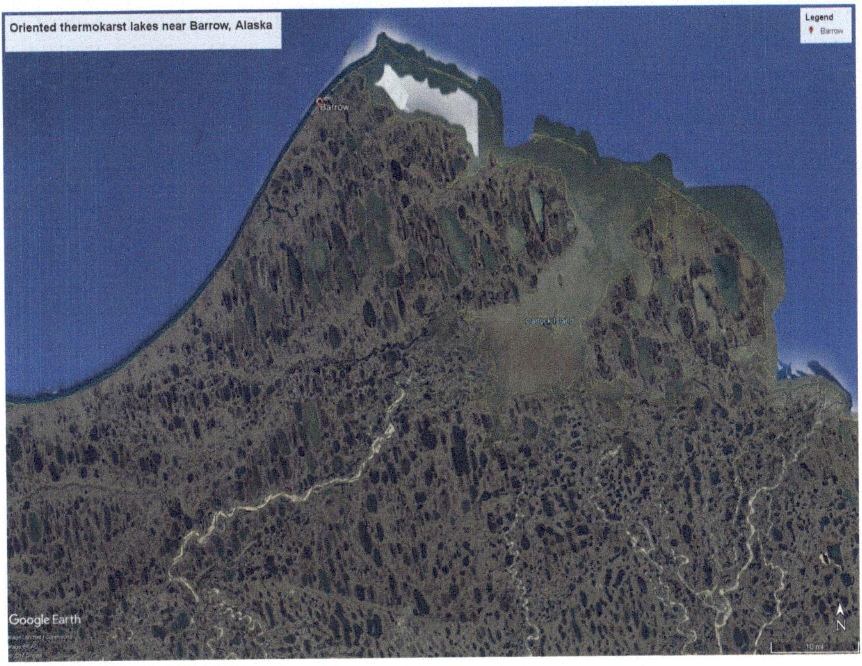


**Fig. 2.15** Google Earth image of oriented thermokarst lakes near Barrow, Alaska

thermokarst lakes, leading to the elongate orientation of the lakes (Zhan et al. 2014). In other words, the orientation of thermokarst lakes is governed by patterns of wind-driven circulation and longshore drift, whereby sediment accumulates on the downwind (lee) side of the lakes and the lakes expand by erosion in zones that are oriented 50° to the wave approach or approximately perpendicular to the dominant wind direction (Rex 1961; Carson and Hussey 1962; Zhan et al. 2014). Many of these thermokarst lakes in northern Alaska have modified an older, stabilized substrate of eolian sand (Carter 1981).

### *2.5.4 Eolian Dunes and Sand Sheets on Upland Areas of the Northern Atlantic Coastal Plain*

Eolian sand dunes and sand sheets on upland areas of the northern Atlantic Coastal Plain were deposited relatively close to the southern margin of the Laurentide Ice Sheet during the LGM. These sands were probably remobilized from any loose sediment that was available in the area, and their location near the southern margin of the ice sheet is similar to extensive deposits of Quaternary eolian sand and loess in the midwest United States, as well as in Europe and in China (e.g., Sun et al. 1998; Zeeberg 1998; Kasse 2002; Zhou et al. 2009; Yang and Ding 2013; Kalińska-Nartiša et al. 2015; Bertran et al. 2016).

OSL ages from eolian sand on upland areas of the northern Atlantic Coastal Plain range from ca. 44–3 ka, but most of the ages are approximately coincident with the last glaciation in the northern hemisphere (Fig. 2.14). The great range of OSL ages suggests that eolian sediment mobilization was episodic during this time. Most of the OSL ages, however, range from ca. 44–20 ka and are approximately coincident with the interval from the latter stages of ice sheet growth (leading to the LGM) to the LGM. No OSL ages older than ca. 44 ka have been reported from eolian sands in this area, and it may be that the northern location (closer distance to the influences of the Laurentide Ice Sheet) may have facilitated the remobilization any older (pre-44 ka) eolian sand during the LGM or later. At the young end of the age spectrum, there is an individual OSL age of 13.8 ± 1.0 ka from Virginia (Feldman et al. 2000), but this age is not anomalously younger than other OSL ages from eolian sands in the area.

Although most of the eolian sediment of the northern Atlantic Coastal Plain has the morphology of sand sheets, some parabolic dunes are present on the Delmarva Peninsula. As mentioned above, the parabolic dune morphology is usually associated with an adequate sand supply, moderate vegetation cover, and a unidirectional wind regime for sand-mobilizing winds (McKee and Bigarella 1979; Lancaster 1995; Hugenholtz et al. 2008; Hugenholtz 2010). The tails of these parabolic dunes point to the northwest, suggesting that the winds that mobilized the sand in this area blew from the northwest. This orientation of the parabolic dunes is consistent with that of other vegetated parabolic dunes even farther north and west in central Pennsylvania (Chase 1977).

## 2.6 Discussion

The various Quaternary inland eolian dunes and sand sheets of the U.S. Atlantic Coastal Plain province have yielded a broadly similar range of OSL ages (ca. 92–5 ka), suggesting that there are regional controls on eolian sediment behavior in the area (Fig. 2.16). These controls appear to be certain climate-related parameters because most of the OSL ages are approximately coincident with growth of the Laurentide Ice Sheet and the LGM in the northern hemisphere. Eolian sediment mobilization appears to have been more common during times of cooler air temperatures (coincident with drier air and greater velocities of surface winds). Despite the broadly synchronous chronology, however, it is likely that eolian sediment mobilization was episodic at any given site. This episodic mobilization of eolian sand may have occurred on a variety of millennial and sub-millennial scales. For example, the orientations of parabolic dunes south of Virginia are consistent with climate models of surface wind directions during the LGM winter, and are opposite the inferred wind directions during the LGM summer (Fig. 2.17), thus suggesting

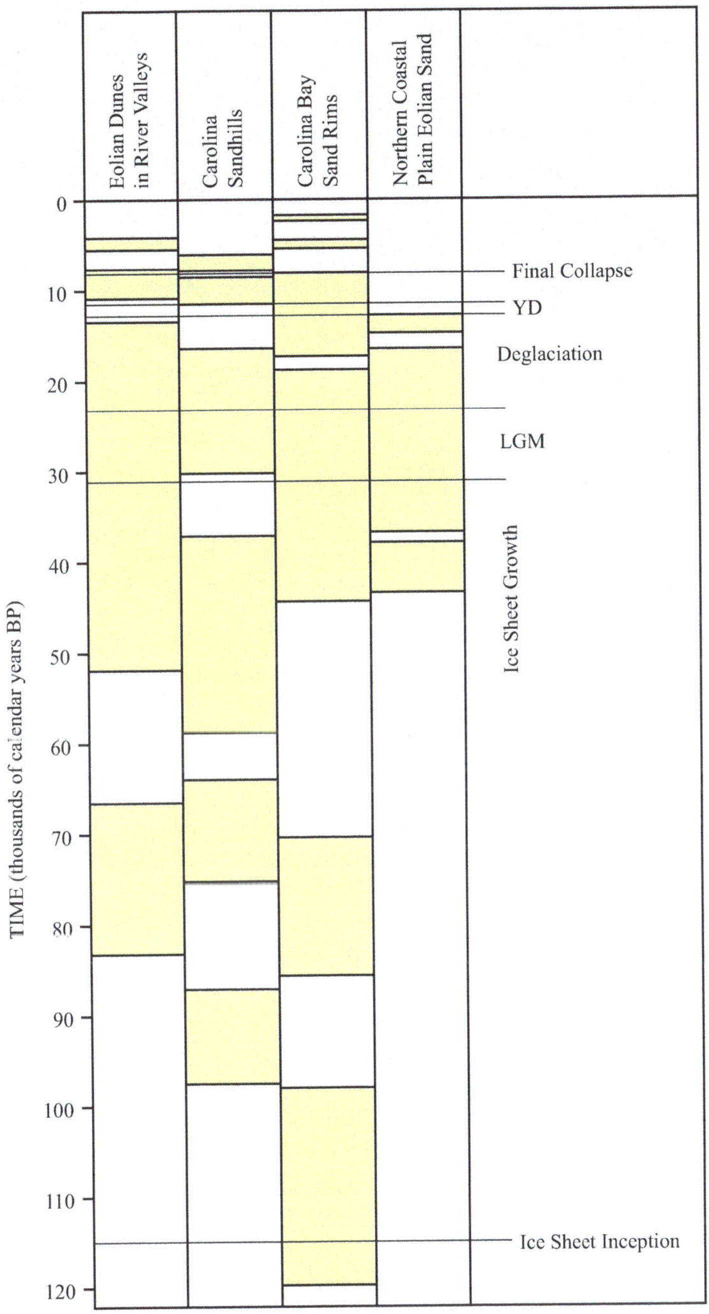


**Fig. 2.16** Correlation chart showing the time span of OSL ages (including ranges of uncertainty) from eolian sand associated with Atlantic Coastal Plain river valleys, the Carolina Sandhills region, Carolina Bays, and upland areas of the northern coastal plain. These data are compiled from Tables 2.1, 2.2, 2.3, and 2.4. *LGM* Last glacial maximum, *YD* Younger Dryas event, *Final Collapse* Final collapse of the Laurentide Ice Sheet

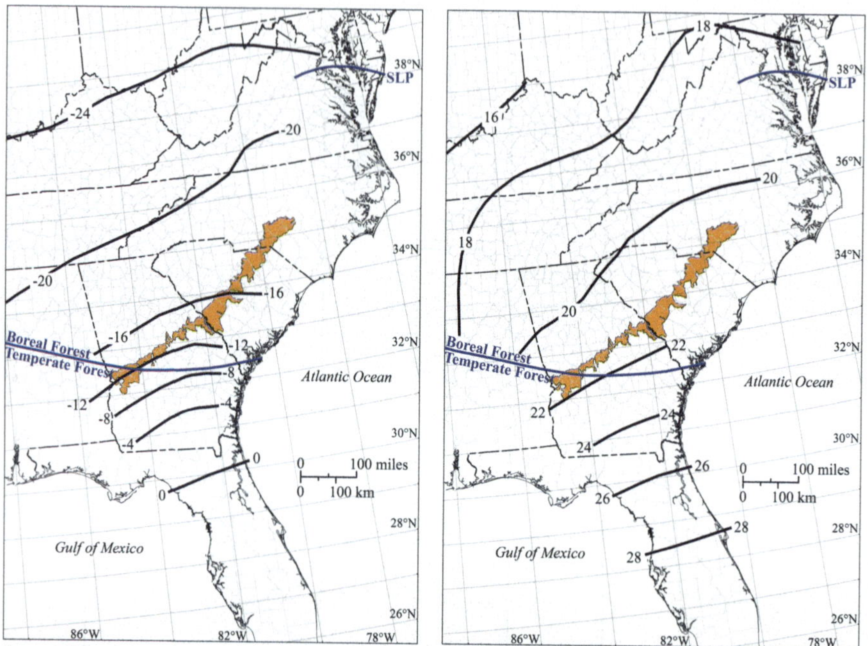


January LGM
(Surface winds generally from the west; Kutzbach et al., 1998)

July LGM
(Surface winds generally from the southeast; Kutzbach et al., 1998)

**Fig. 2.17** Last glacial maximum (LGM) climate data of the southeastern United States. Mean temperature data in degrees Celsius are from Jackson et al. (2000), and surface wind data are from Kutzbach et al. (1998). *SLP* Southern limit of LGM permafrost in the Atlantic Coastal Plain province (from French et al. 2009). Although the southern limit of permafrost during the LGM is typically thought to have been located in central or northern Virginia (French et al. 2009; French and Miller 2014), discontinuous and sporadic permafrost probably extended much farther south. Boreal forest-temperate forest boundary is from Woodcock and Wells (1990)

that eolian dune mobilization may have occurred preferentially during the winter (Swezey et al. 2016b).

Although the OSL ages from eolian sand of the inland Atlantic Coastal Plain range from ca. 92–5 ka, these ages alone do not allow one to distinguish many separate and distinct episodes of eolian sediment mobilization (Fig. 2.16). There is an apparent gap in the OSL ages from ca. 64–59 ka, but it is also possible that additional OSL ages may fill in this gap. There is also an anomalously young OSL age of ca. 2 ka from Big Bay (South Carolina) that is possibly the result of post-depositional bioturbation exposing sand grains to sunlight. At the older end of the age spectrum, it is somewhat unusual to have abundant Quaternary eolian sand of pre-Younger Dryas age preserved across a broad geographical area. In contrast, most OSL ages reported from Quaternary eolian sand in the northern and western United States are younger than the Younger Dryas event (e.g., Arbogast et al. 2015; Halfen et al. 2016). A similar pattern is seen in North Africa, where many Quaternary eolian sands have yielded relatively young ages (Swezey 2001; Bristow and

Armitage 2016). Although there is evidence of widespread eolian sand mobilization in North Africa during the LGM (e.g., Sarnthein 1978), much of this eolian sand was subsequently remobilized during the YD event. This pattern of widespread remobilization, however, is not the case in the U.S. Atlantic Coastal Plain province, where many eolian sands have yielded OSL ages that are substantially older than the YD event. Likewise, numerous OSL ages that are substantially older than the YD event have been reported from Quaternary eolian sand of the Gulf of Mexico Coastal Plain (Otvos 2004). In other words, the OSL ages reveal a general pattern of greater preservation of older eolian sediment with increasing distance from the LGM glacial front, demonstrating that the landscape of the southeastern United States is a generally older landscape that has not been reworked as readily as some other landscapes. During the YD event, the nature of the vegetation that stabilized eolian sand in the southeastern United States may have changed, but the landscape did not lose all of the stabilizing vegetation such that there was complete reworking of older eolian sand.

Despite the broadly synchronous OSL ages, the different settings of Quaternary inland eolian sand display some differences in geomorphology and sediment grain size. Eolian dune morphologies are primarily parabolic in river valleys and on upland areas of the northern Atlantic Coastal Plain, whereas eolian dune morphologies are linear but of relatively short length in the Carolina Sandhills region. Eolian sand sheets are common in the northern Atlantic Coastal Plain (e.g., Delmarva Peninsula) and in the Carolina Sandhills region, and arcuate ridges of eolian sand are common adjacent to many of the Carolina Bays. Most of these eolian sediments are fine to medium sand, with the exception of: (1) the Carolina Sandhills, where the eolian sediments are primarily medium (upper) to coarse (lower) sand; and (2) the northern Atlantic Coastal Plain, where the eolian sand sheets contain a substantial silt component. The relatively coarse size of eolian sand of the Carolina Sandhills is attributed to the relatively coarse size of the underlying Cretaceous sediments from which the eolian sand was derived, and the substantial silt component of the eolian sediments of the northern Atlantic Coastal Plain is attributed to greater proximity to the Laurentide Ice Sheet, which would have generated abundant silt.

The parabolic dune morphologies and the relative abundance of eolian sand sheets suggest that some vegetation was present when the eolian sand was mobilized. Both parabolic dunes and eolian sand sheets are often associated with moderate vegetation cover (McKee and Bigarella 1979; Kocurek and Nielson 1986; Lancaster 1995; Hugenholtz et al. 2008; Hugenholtz 2010). Furthermore, in the case of the Carolina Sandhills, the presence of paleosols with organic matter within the eolian sand implies the presence of vegetation that was capable of producing organic litter (Swezey et al. 2016b). Pollen data also support the interpretation of some vegetation being present during the last glaciation and deglaciation (e.g., Watts 1980a, b; Delcourt and Delcourt 1984, 1985; LaMoreaux et al. 2009; Spencer et al. 2017). This vegetation is thought to have been primarily a boreal forest of spruce and pine, with a tree cover that was much less dense than in modern boreal forests (Watts 1980a, b; Taylor et al. 2011). The southern limit of this boreal forest during the LGM is estimated to have been located near 33°N latitude (Woodcock and Wells 1990), which is near the southern limit of both the Carolina Sandhills and the Carolina Bays (Fig. 2.17).

Additional paleoclimate data are available from the orientations of the parabolic dunes. The tails of parabolic dunes point to the northwest in the northern Atlantic Coastal Plain (Delaware, Maryland), suggesting that the winds that mobilized the sand in this area blew from the northwest. In contrast, the tails of parabolic dunes point to the west in the central and southern Atlantic Coastal Plain (North Carolina, South Carolina, Georgia), suggesting that the winds that mobilized the sand in this area blew from the west. These wind directions inferred from dune shapes are consistent with climate models by Kutzbach et al. (1998) suggesting that surface winds in the southeastern United States blew generally from the west during the LGM winter and from the southeast during the LGM summer (Fig. 2.17). Thus, the results of these climate models have prompted speculation that eolian dune mobilization may have occurred on a seasonal basis, preferentially during winter (Swezey et al. 2016b).

Information about the velocity of paleo-winds may be obtained from the grain sizes of the eolian sediment. In the Carolina Sandhills region, the most frequently occurring grain size of individual eolian sand samples ranges from 0.35–0.59 mm diameter. In relatively warm low-latitude climates, 0.25–0.50 mm diameter quartz sand typically requires threshold wind velocities of 4–6 m/s for sustained eolian mobilization (e.g., Hsu 1974). For 0.25–0.33 mm diameter quartz sand, a threshold wind velocity of 6 m/s has been used for calculations of eolian sediment drift potential (e.g., Fryberger and Dean 1979). In the southeastern United States, however, the mean velocity of modern surface winds is 1.3–2.2 m/s. Therefore, although there is some variability around these mean values, the mean velocity of modern surface winds is generally not sufficient for sustained eolian mobilization of sand in inland locations of the southeastern United States. Cold winds, however, are more effective at eolian transport than warm winds (Selby et al. 1974; McKenna Neuman 1989, 1993, 2003, 2004), and wind tunnel experiments have shown that -12 °C air can entrain particles 40–50% larger in diameter than +32 °C air (McKenna Neuman 2003). Thus, eolian mobilization of sand in inland locations of the U.S. Atlantic Coastal Plain province would have required much greater wind velocities and (or) colder air temperatures than are present today. Swezey et al. (2016b) estimated that sustained eolian mobilization of sand in the Carolina Sandhills region would have required wind velocities of at least 4–6 m/sec during the LGM winter, and would have required even greater wind velocities during the LGM summer. Furthermore, they postulated that eolian dune mobilization may have occurred preferentially during the winter, because the dune morphologies are not consistent with the inferred wind directions proposed by climate models of the LGM summer (Swezey et al. 2016b).

Finally, the distribution of the Carolina Bays (and their associated eolian sand ridges) may provide information about both paleo-wind directions, the former distribution of frozen ground, and an upper boundary on temperature in the southeastern United States during the last glaciation. The OSL ages indicate that Carolina

Bays are relict features that formed during approximately the same time as the eolian dunes and sand sheets in the river valleys and in the Carolina Sandhills region. Both the formation of thermokarst lakes and the mobilization of eolian sand in the Atlantic Coastal Plain province would have required cooler air temperatures, greater wind velocities, and reduced vegetation density. The southern limit of continuous permafrost during the LGM is generally thought to have been located in central or northern Virginia approximately 230–320 km south of the LGM ice margin (French et al. 2009; French and Miller 2014), but the distribution of Carolina Bays suggests that permafrost (possibly discontinuous and/or sporadic) may have extended much farther south. For comparison, during the last glaciation, permafrost in Europe is thought to have extended 800–1200 km south of the LGM ice margin, and permafrost in Asia is thought to have extended 2000–4500 km south of the LGM ice margin (Vandenberghe et al. 2014; Ballantyne 2018). Furthermore, pollen data from the southeastern United States suggest that the southern limit of boreal forest during the LGM was located in central to southern Georgia (Fig. 2.17). These interpretations are consistent with evidence of LGM iceberg scour on the upper continental slope offshore the coasts of South Carolina, Georgia, and eastern Florida (Hill et al. 2008; Hill and Condron 2014). In other words, all of these features suggest that the southeastern United States became very cold, dry, and windy during the LGM. Much of the landscape appears to have been covered by a sparse boreal forest, discontinuous and sporadic permafrost probably extended south into Georgia (where Carolina Bays are present), and eolian sand mobilization appears to have occurred episodically (preferentially during winter) wherever loose sediment was available throughout most of the U.S. Atlantic Coastal Plain province.

## 2.7 Conclusions

The modern Atlantic Coastal Plain province of the eastern United States is not very conducive to widespread eolian sediment mobilization because of relatively low surface wind velocities, relatively dense vegetation, and a humid and mesothermal climate with average air temperatures ranging from ~0 °C to ~30 °C (depending upon location and season). Quaternary eolian dunes and sand sheets that are stabilized by vegetation, however, are present at many inland locations throughout the coastal plain. These locations include river valleys, the Carolina Sandhills region, adjacent to Carolina Bays, and upland areas of the northern coastal plain. In river valleys, eolian dunes are primarily parabolic and are located to the east of the modern river channels. In Maryland and Delaware, these parabolic dunes are composed of fine to medium sand and silty sand, and dune tails point northwest. In North Carolina, South Carolina, and Georgia, these parabolic dunes are composed of fine to medium sand, and dune tails point west. In the Carolina Sandhills region, eolian dunes are linear, and are composed of medium (upper) to coarse (lower) sand. Arcuate ridges of fine to medium eolian sand are located on the east and south sides of many shallow depressions known as Carolina Bays, which are most abundant east of the Orangeburg Scarp in South Carolina and in adjacent areas of North Carolina and Georgia. In upland areas of the northern coastal plain (e.g., Delmarva Peninsula), eolian dunes and sand sheets are composed of fine to medium sand and silty sand, and the dunes are parabolic with tails pointing to the northwest.

OSL ages reveal that these eolian dunes and sand sheets are relict features that were active episodically from ca. 92–5 ka. These dunes and sand sheets have been degraded by vegetation and pedogenic processes, and are stabilized under modern environmental conditions. Most of the OSL ages from these sands are approximately coincident with the LGM, when conditions would have been generally colder, drier, and windier. During this time, discontinuous and sporadic permafrost may have extended south into Georgia (where Carolina Bays are present). The presence of parabolic dunes and the abundance of eolian sand sheets suggest that some vegetation was present when the eolian sand was mobilized (and pollen studies suggest that this vegetation was a sparse boreal forest). Furthermore, the orientations of the parabolic dunes and the locations of the Carolina Bay eolian sand ridges suggest that the winds that mobilized the sand blew from the northwest in the coastal plain region of Maryland and Delaware, and that these winds blew from the west in the coastal plain region of North Carolina, South Carolina, and Georgia. These inferred wind directions are consistent with climate models for the LGM winter, suggesting that eolian sand mobilization may have occurred preferentially during winter. Eolian mobilization of the relatively coarse sand in the Carolina Sandhills region would have been facilitated by colder air temperatures during winter.

The OSL ages reveal a general pattern of greater preservation of older eolian sand towards the south (farther from the glacial front). In other words, the southern location (greater distance from the influences of the Laurentide Ice Sheet) may have helped to prevent some eolian sand from being reworked during the LGM or later. This observation suggests that the landscape of the southeastern United States is a generally older landscape that has not been reworked as readily as some other landscapes. During the YD event, for example, the nature of the vegetation that stabilized eolian sand in the southeastern United States may have changed, but the landscape did not lose all of the stabilizing vegetation such that there was complete reworking of all older eolian sand (and resetting of OSL ages). At any given site, however, eolian sand mobilization appears to have occurred episodically from ca. 92–5 ka, facilitated by conditions of sand availability, stronger wind velocity, lower air temperature, lower air humidity, and (or) reduced vegetation cover.

**Acknowledgements** The author extends sincere thanks to Nicholas Lancaster for the invitation to contribute this chapter. He also thanks Steven Cahan, Kelsey Ciarrocca, and Christopher Garrity for their invaluable work with the LiDAR images, Bradley Fitzwater and G. Richard Whittecar (Old Dominion University) for collaborative field work in the Carolina Sandhills region of Chesterfield County (South Carolina), William R. Doar III (South Carolina Geological Survey) for guidance among the sandhills of Lexington County (South Carolina), and Shannon Mahan (USGS) for answering many questions about OSL analyses. Many thanks also go to Andrew Ivester and Christopher Moore for general discussion and exchange of ideas during the past few years. This manuscript benefitted from reviews by Selene Deike, by USGS geologists Miriam Jones and Kevin Kincare, and by journal editors Nicholas Lancaster and Patrick Hesp. Any use of trade, firm, or product names is for descriptive purposes only and does not imply endorsement by the U.S. Government.

## References

Alley R, Meese D, Shuman C, Gow A, Taylor K, Grootes P, White J, Ram M, Waddington E, Mayewski P, Zielinski G (1993) Abrupt increase in Greenland snow accumulation at the end of the Younger Dryas event. Nature 362:527–529

Andres AS, Howard CS (2000) The Cypress Swamp Formation, Report of Investigations, vol 62. Delaware Geological Survey, Newark, 13p

Arbogast AF, Luehmann MD, Miller BA, Wernette PA, Adams KM, Waha JD, O'Neil GA, Tang Y, Boothroyd JJ, Babcock CR, Hanson PR, Young AR (2015) Late-Pleistocene paleowinds and aeolian sand mobilization in north-central lower Michigan. Aeolian Res 16:109–116

Arp CD, Jones BM, Urban FE, Grosse G (2011) Hydrogeomorphic processes of thermokarst lakes with grounded-ice and floating-ice regimes on the Arctic coastal plain, Alaska. Hydrol Process 25:2422–2438

Askins AH (2010) Carolina Sandhills National Wildlife Refuge comprehensive conservation plan. U.S. Fish and Wildlife Service–Southeast Region, Atlanta, Georgia, 234p

Baldwin JL (1975) Weather Atlas of the United States. Gale Research Company, Detroit, 262p

Ballantyne CK (2018) Periglacial Geomorphology. Wiley, New York, 465p

Barber DC, Dyke A, Hillaire-Marcel C, Jennings AE, Andrews JT, Kerwin MW, Bilodeau G, McNeely R, Southon J, Morehead MD, Gagnon J-M (1999) Forcing of the cold event of 8,200 years ago by catastrophic drainage of Laurentide lakes. Nature 400:344–348

Bertran P, Liard M, Sitzia L, Tissoux H (2016) A map of Pleistocene aeolian deposits in western Europe, with special emphasis on France. J Quat Res 31:844–856

Bliley DJ, Burney DA (1988) Late Pleistocene climatic factors in the genesis of a Carolina Bay. Southeast Geol 29:83–101

Bliley DJ, Pettry DE (1979) Carolina bays on the eastern shore of Virginia. Soil Sci Soc Am J 43:558–564

Bristow CS, Armitage SJ (2016) Dune ages in the sand deserts of the southern Sahara and Sahel. Quat Int 410:46–57

Brooks MJ, Taylor BE, Grant JA (1996) Carolina Bay geoarchaeology and Holocene landscape evolution on the Upper Coastal Plain of South Carolina. Geoarchaeology 11:481–504

Brooks MJ, Taylor BE, Stone PA, Gardner LR (2001) Pleistocene encroachment of the Wateree River sand sheet into Big Bay on the Middle Coastal Plain of South Carolina. Southeast Geol 40:241–257

Brooks MJ, Taylor BE, Ivester AH (2010) Carolina Bays: time capsules of culture and climate change. Southeast Archaeol 29:146–163

Carey JB, Cunningham RL, Williams EG (1976) Loess identification in soils of southeastern Pennsylvania. Soil Sci Soc Am J 40:745–750

Carson CE, Hussey KM (1962) The oriented lakes of Arctic Alaska. J Geol 70:417–439

Carter LD (1981) A Pleistocene sand sea on the Alaskan Arctic Coastal Plain. Science 211:381–383

Carver RE, Brook GA (1989) Late Pleistocene paleowind directions, Atlantic Coastal Plain, U.S.A. Palaeogeogr Palaeoclimatol Palaeoecol 74:205–216

Chase CM (1977) Central Pennsylvania sand dunes. Pa Geol 8(3):9–12

Clark PU, Dyke AS, Shakun JD, Carlson AE, Clark J, Wohlfarth B, Mitrovica JX, Hostetler SW, McCabe AM (2009) The last glacial maximum. Science 325:710–714

Coté MM, Burn CR (2002) The oriented lakes of Tuktoyaktuk Peninsula, western Arctic coast, Canada: a GIS-based approach. Permafr Periglac Process 13:61–70

Coulombe O, Bouchard F, Pienitz R (2016) Coupling of sedimentological and limnological dynamics in subarctic thermokarst ponds in northern Québec (Canada) on an interannual basis. Sediment Geol 340:15–24

Court A (1974) The climate of the conterminous United States. In: Bryson RA, Kare FK (eds) Climates of North America. Elsevier, Amsterdam, pp 193–343

Daniels RB, Gamble EE, Boul SW (1969) Eolian sands associated with coastal plain river valleys—some problems in their age and source. Southeast Geol 11:97–110

Davis RE, Hayden BP, Gay DA, Phillips WL, Jones GV (1997) The North Atlantic subtropical anticyclone. J Clim 10:728–744

Delcourt PA, Delcourt HR (1984) Late Quaternary paleoclimates and biotic responses in eastern North America and the western North Atlantic Ocean. Palaeogeogr Palaeoclimatol Palaeoecol 48:263–284

Delcourt HR, Delcourt PA (1985) Quaternary palynology and vegetational history of the southeastern United States. In: Bryant VM Jr, Holloway RG (eds) Pollen Records of Late-Quaternary North American Sediments. American Association of Stratigraphic Palynologists Foundation, Dallas, pp 1–37

Denny CS, Owens JP (1979) Sand dunes of the central Delmarva Peninsula, Maryland and Delaware. U.S. Geological Survey Professional Paper 1067-C, 15p

Denny CS, Owens JP, Sirkin LA, Rubin M (1979) The Parsonsburg Sand in the central Delmarva Peninsula, Maryland and Delaware. U.S. Geological Survey Professional Paper 1067-B, 16p

Doar WR III, Howard CS (2010) Geologic map of the Steedman quadrangle, Aiken and Lexington Counties, South Carolina. South Carolina Geological Survey Geologic Quadrangle Map GQM-49, 1 sheet

Dowsett HJ, Cronin TM (1990) High eustatic sea level during the middle Pliocene: evidence from the southeastern U.S. Atlantic Coastal Plain. Geology 18:435–438

Earley LS (2004) Looking for Longleaf: The Fall and Rise of an American Forest. The University of North Carolina Press, Chapel Hill, 336p

Farnsworth RK, Thompson ES, Peck EL (1982) Evaporation Atlas for the Contiguous 48 United States. National Oceanic and Atmospheric Administration (NOAA) Technical Report NWS 33, 26p. and 4 plates

Feathers JK, Rhodes EJ, Huot S, Mcavoy JM (2006) Luminescence dating of sand deposits related to late Pleistocene human occupation at the Cactus Hill Site, Virginia, USA. Quat Geochronol 1:167–187

Feldman SB, Zelazny LW, Pavich MJ, Millard HT Jr (2000) Late Pleistocene eolian activity and post-depositional alteration on the Piedmont of northern Virginia. Southeast Geol 39:183–198

Fitzwater BA (2016) Reevaluating the geologic formations of the Upper Coastal Plain in Chesterfield County, South Carolina: M.S. thesis, Old Dominion University, Norfolk, Virginia, 189p

Folk RL, Ward WC (1957) Brazos River point bar: a study in the significance of grain size parameters. J Sediment Petrol 27:3–26

Foss JE, Fanning DS, Miller FP, Wagner DP (1978) Loess deposits of the eastern shore of Maryland. Soil Sci Soc Am J 42(2):329–334

French H, Demitroff M (2012) Late-Pleistocene paleohydrography, eolian activity and frozen ground, New Jersey Pine Barrens, eastern USA. Neth J Geosci 91:25–35

French HM, Miller SWS (2014) Permafrost at the time of the Last Glacial Maximum (LGM) in North America. Boreas 43:667–677

French HM, Demitroff M, Newell WL (2009) Past permafrost on the mid-Atlantic Coastal Plain, eastern United States. Permafr Periglac Process 20:285–294

Fryberger SG, Dean, G (1979) Dune forms and wind regime. In: McKee E (ed) A Study of Global Sand Seas. U.S. Geological Survey Professional Paper 1052, pp 137–169

Galbraith RF, Laslett GM (1993) Statistical models for mixed fission track ages. Nucl Tracks Radiat Meas 21:459–470

Galbraith RF, Roberts RG (2012) Statistical aspects of equivalent dose and error calculation and display in OSL dating: an overview and some recommendations. Quat Geochronol 11:1–27

Galbraith RF, Roberts RG, Laslett GM, Yoshida H, Olley JM (1999) Optical dating of single and multiple grains of quartz from Jinmium rock shelter, northern Australia: part I experimental design and statistical models. Archaeometry 41:339–364
Gamble EE, Daniels RB, Wheeler WH (1977) Primary and secondary rims of Carolina Bays. Southeast Geol 18:199–212
Glenn LC (1895) Some notes on Darlington (S. C.), 'bays.'. Science 2(41):472–475
Goddard EN, Trask PD, De Ford RK, Rove ON, Singewald JT Jr, Overbeck RM (1963) Rock-Color Chart. Geological Society of America, New York, New York
Grant JA, Brooks MJ, Taylor BE (1998) New constraints on the evolution of Carolina Bays from ground-penetrating radar. Geomorphology 22:325–345
Griffith GE, Omernik JM, Comstock JA, Lawrence S, Martin G, Goddard A, Hulcher VJ, Foster T (2001) Ecoregions of Alabama and Georgia. U.S. Geological Survey, Reston, Virginia, 1:1,700,000 scale map, 1 sheet
Griffith GE, Omernik JM, Comstock JA, Schafale MP, McNab WH, Lenat DR, MacPherson TF, Glover JB, Shelburne VB (2002) Ecoregions of North Carolina and South Carolina. U.S. Geological Survey, Reston, Virginia, 1:1,500,000 scale map, 1 sheet
Hack JT (1955) Geology of the Brandywine area and origin of the upland of southern Maryland. U.S. Geological Survey Professional Paper 267-A, 41p
Halfen AF, Lancaster N, Wolfe S (2016) Interpretations and common challenges of aeolian records from North American dune fields. Quat Int 410:76–95
Harman JR (1991) Synoptic climatology of the Westerlies: process and patterns. Association of American Geographers, Washington, DC, 80p
Hill JC, Condron A (2014) Subtropical iceberg scours and meltwater routing in the deglacial western North Atlantic. Nat Geosci 7:806–810
Hill JC, Gayles PT, Driscoll NW, Johnstone EA, Sedberry GR (2008) Iceberg scours along the southern U.S. Atlantic margin. Geology 36:447–450
Hinkel KM, Frohn RC, Nelson FE, Eisner WR, Beck RA (2005) Morphometric and spatial analysis of thaw lakes and drained thaw lakes in the western Arctic Coastal Plain, Alaska. Permafr Periglac Process 16:327–341
Hinkel KM, Sheng Y, Lenters JD, Lyons EA, Beck RA, Eisner WR, Wang J (2012) Thermokarst lakes on the Arctic Coastal Plain of Alaska: geomorphic controls on bathymetry. Permafr Periglac Process 23:218–230
Holmes JA (1893) Geology of the sand-hill country of the Carolinas. Geol Soc Am Bull 5:33–35
Hsu SA (1974) Computing eolian sand transport from routine weather data. In: Proceedings of the 14th Coastal Engineering Conference, 24–28 June 1974, vol 2. Copenhagen, Denmark, pp 1619–1626
Hugenholtz CH (2010) Topographic changes of a supply-limited inland parabolic sand dune during the incipient phase of stabilization. Earth Surf Process Landf 35:1674–1681
Hugenholtz CH, Wolfe SA, Moorman BJ (2008) Effects of sand supply on the morphodynamics and stratigraphy of active parabolic dunes, Bigstick Sand Hills, southwestern Saskatchewan. Can J Earth Sci 45:321–335
Ivester AH, Leigh DS (2003) Riverine dunes on the Coastal Plain of Georgia, USA. Geomorphology 51:289–311
Ivester AH, Leigh DS, Godfrey-Smith DI (2001) Chronology of inland eolian dunes on the Coastal Plain of Georgia. Quat Res 55:293–302
Ivester AH, Godfrey-Smith DI, Brooks MJ, Taylor BE (2002) Carolina Bays and inland dunes of the southern Atlantic Coastal Plain yield new evidence for regional paleoclimate. Geol Soc Am Abstr Programs 34(6):273–274
Ivester AH, Godfrey-Smith DI, Brooks MJ, Taylor BE (2003) Concentric sand rims document the evolution of a Carolina Bay in the middle coastal plain of South Carolina. Geol Soc Am Abstr Programs 35(6):169
Ivester A.H., Brooks, M.J., Taylor, B.E., 2007. Sedimentology and ages of Carolina Bay sand rims: Geol Soc Am Abstr Programs 39(2): 5

Jackson ST, Webb RS, Anderson KH, Overpeck JT, Webb T III, Williams JW, Hansen BCS (2000) Vegetation and environment in eastern North America during the last glacial maximum. Quat Sci Rev 19:489–508
Johnson HS Jr (1961) Fall line stratigraphy Northeast of Columbia, S.C. Geol Notes 5:81–87
Jorgenson MT, Kanevskiy M, Shur Y, Osterkamp T, Fortier D (2012) Thermokarst lake and shore fen development in boreal Alaska. In: Proceedings of the 10th International Conference on Permafrost, vol 1, pp 179–184
Kalińska-Nartiša E, Thiel C, Nartišs M, Buylaert J-P, Murray AS (2015) Age and sedimentary record of inland eolian sediments in Lithuania, NE European Sand Belt. Quat Res 84:82–95
Karlsson JM, Lyon SW, Destouni G (2012) Thermokarst lake, hydrological flow and water balance indicators of permafrost change in Western Siberia. J Hydrol 464-465:459–466
Kasse C (2002) Sandy aeolian deposits and environments and their relation to climate during the Last Glacial Maximum and Lateglacial in northwest and central Europe. Prog Phys Geogr 26:507–532
Katz RW, Parlange MB, Tebaldi C (2003) Stochastic modeling of the effects of large-scale circulation on daily weather in the southeastern U.S. Clim Chang 60:189–216
Kleman J, Jansson K, De Angelis H, Stroeven AP, Hättestrand C, Alm G, Glasser N (2010) North American ice sheet build-up during the last glacial cycle, 115–21 kyr. Quat Sci Rev 29:2036–2051
Kocurek G, Nielson J (1986) Conditions favorable for the formation of warm climate aeolian sand sheets. Sedimentology 33:795–816
Kutzbach J, Gallimore R, Harrison S, Behling P, Selin R, Laarif F (1998) Climate and biome simulations for the past 21,000 years. Quat Sci Rev 17:473–506
LaMoreaux HK, Brook GA, Knox JA (2009) Late Pleistocene and Holocene environments of the southeastern United States from the stratigraphy and pollen content of a peat deposit on the Georgia Coastal Plain. Palaeogeogr Palaeoclimatol Palaeoecol 280:300–312
Lancaster N (1995) Geomorphology of Desert Dunes. Routledge, London, 290p
Leigh DS (2006) Terminal Pleistocene braided to meandering transition in rivers of the southeastern USA. Catena 66:155–160
Leigh DS (2008) Late Quaternary climates and river channels of the Atlantic Coastal Plain, southeastern USA. Geomorphology 101:90–108
Livingstone DA (1954) On the orientation of lake basins. Am J Sci 252:547–554
Lowery DL, O'Neal MA, Wah JS, Wagner DP, Stanford DJ (2010) Late Pleistocene upland stratigraphy of the western Delmarva Peninsula, USA. Quat Sci Rev 29:1472–1480
MacCarthy GR (1937) The Carolina Bays. Bull Geol Soc Am 48:1211–1226
Macphail RI, McAvoy JM (2008) A micromorphological analysis of stratigraphic integrity and site formation at Cactus Hill, an early Paleoindian and hypothesized pre-Clovis occupation in south-central Virginia, USA. Geoarchaeology 23:675–694
Mallinson D, Burdette K, Mahan S, Brook G (2008) Optically stimulated luminescence age controls on late Pleistocene and Holocene coastal lithosomes, North Carolina, USA. Quat Res 69:97–109
Markewich HW, Markewich W (1994) An overview of Pleistocene and Holocene inland dunes in Georgia and the Carolinas – morphology, distribution, age, and paleoclimate, U.S. Geological Survey Bulletin 2069. 32p
Markewich HW, Litwin RJ, Pavich MJ, Brook GA (2009) Late Pleistocene eolian features in southeastern Maryland and Chesapeake Bay region indicate strong WNW-NW winds accompanied growth of the Laurentide ice sheet. Quat Res 71:409–425
Markewich HW, Litwin RJ, Wysocki DA, Pavich MJ (2015) Synthesis on Quaternary aeolian research in the unglaciated eastern United States. Aeolian Res 17:139–191
McKee ED, Bigarella JJ (1979) Sedimentary structures in dunes. In: McKee E (ed) A Study of Global Sand Seas. U.S. Geological Survey Professional Paper 1052, pp 83–134
McKenna Neuman C (1989) Kinetic energy transfer through impact and its role in entrainment by wind of particles from frozen surfaces. Sedimentology 36:1007–1015

McKenna Neuman C (1993) A review of aeolian transport processes in cold environments. Prog Phys Geogr 17:137–155
McKenna Neuman C (2003) Effects of temperature and humidity upon the entrainment of sediment particles by wind. Bound-Layer Meteorol 108:61–89
McKenna Neuman C (2004) Effects of temperature and humidity upon the transport of sedimentary particles by wind. Sedimentology 51:1–17
Melton FA, Schriever W (1933) The Carolina "Bays"–are they meteorite scars? J Geol 41:52–66
Miller W III (1979) Stratigraphic framework of the Wharton Station dune field, easternmost Beaufort County, North Carolina. Southeast Geol 20:261–273
Mix AC (1992) The marine oxygen isotope record: constraints on timing and extent of ice-growth events (120–65 ka). In: Clark PU, Lea PD (eds) The Last Integlacial-Glacial Transition in North America. Geological Society of America Special Paper 270, pp 19–30
Mixon RB (1985) Stratigraphic and geomorphic framework of uppermost Cenozoic deposits in the southern Delmarva Peninsula, Virginia and Maryland. U.S. Geological Survey Professional Paper 1067-G, 53p
Moore CR, Brooks MJ, Ivester AH, Ferguson T, Feathers JK (2012) Radiocarbon and luminescence dating at Flamingo Bay (38AK469): implications for site formation processes and artifact burial at a Carolina Bay. Legacy 16(1):16–21
Moore CR, Brooks MJ, Mallinson DJ, Parham PR, Ivester AH, Feathers JK (2016) The Quaternary evolution of Herndon Bay, a Carolina Bay on the Coastal Plain of North Carolina (USA): implications for paleoclimate and oriented lake genesis. Southeast Geol 51:145–171
Morgenstern A, Ulrich M, Günther F, Roessler S, Fedorova IV, Rudaya NA, Wetterich S (2013) Evolution of thermokarst in East Siberian ice-rich permafrost: a case study. Geomorphology 201:363–379
Newell WL, DeJong BD (2011) Cold-climate slope deposits and landscape modifications of the mid-Atlantic Coastal Plain, eastern USA. In: Martini IP, French HM, Alberti AP (eds) Ice-Marginal and Periglacial Processes and Sediments, Geological Society, London, Special Publication 354. pp 259–276
Otvos EG (2004) Prospects for interregional correlations using Wisconsin and Holocene aridity episodes, Northern Gulf of Mexico coastal plain. Quat Res 61:105–118
Parham PR, Riggs SR, Culver SJ, Mallinson DJ, Rink WJ, Burdette K (2013) Quaternary coastal lithofacies, sequence development and stratigraphy in a passive margin setting, North Carolina and Virginia, USA. Sedimentology 60:503–547
Peek KM, Mallinson DJ, Culver SJ, Mahan SA (2014) Holocene geologic development of the Cape Hatteras region, Outer Banks, North Carolina, USA. J Coast Res 30:41–58
Pickering SM, Jones RC (1974) Morphology of aeolian parabolic sand features along streams in southeast Georgia. Geol Soc Am Abstr Programs 6(4):387–388
Powars DS, Edwards LE, Johnson GH, Berquist CR (2016) Geology of the Virginia Coastal Plain: new insights from continuous cores and geophysical surveys. In: Bailey CM, Sherwood WC, Eaton LS, Powars DS (eds) The Geology of Virginia, Virginia Museum of Natural History, Special Publication 18. Martinsville, pp 193–240
Powers MC (1953) A new roundness scale for sedimentary particles. J Sediment Petrol 23:117–119
Prouty WF (1952) Carolina Bays and their origin. Geol Soc Am Bull 63:167–224
Rasmussen WC, Slaughter TH (1955) The ground-water resources. In: The Water Resources of Somerset, Wicomico, and Worcester Counties, Maryland Department of Geology, Mines and Water Resources Bulletin 16, pp 1–170
Reimer PJ, Baillie MGL, Bard E, Bayliss A, Beck JW, Blackwell PG, Bronk Ramsey C, Buck CE, Burr GS, Edwards RL, Friedrich M, Grootes PM, Guilderson TP, Hajdas I, Heaton TJ, Hogg AG, Hughen KA, Kaiser KF, Kromer B, McCormac FG, Manning SW, Reimer RW, Richards DA, Southon JR, Talamo S, Turney CSM, van der Plicht J, Weyhenmeyer CE (2009) IntCal09 and Marine09 radiocarbon age calibration curves, 0–50,000 years cal BP. Radiocarbon 51:1111–1150

Rex RW (1961) Hydrodynamic analysis of circulation and orientation of lakes in northern Alaska. In: Raasch GO (ed) Geology of the Arctic – Proceedings of the First International Symposium on Arctic Geology, vol 2. University of Toronto Press, Toronto, pp 1021–1043
Rodríguez AB, Waters MN, Piehler MF (2012) Burning peat and reworking loess contribute to the formation and evolution of a large Carolina-bay basin. Quat Res 77:171–181
Sahsamanoglou HS (1990) A contribution to the study of action centres in the North Atlantic. Int J Climatol 10:247–261
Sarnthein M (1978) Sand deserts during glacial maximum and climatic optimum. Nature 272:43–46
Scott TW, Swift DJP, Whittecar GR, Brook GA (2010) Glacioisostatic influences on Virginia's late Pleistocene coastal plain deposits. Geomorphology 116:175–188
Selby MJ, Rains RB, Palmer RW (1974) Eolian deposits of the ice-free Victoria Valley, southern Victoria Land, Antarctica. N Z J Geol Geophys 17:543–562
Seminack CT, Buynevich IV (2013) Sedimentological and geophysical signatures of a relict tidal inlet complex along a wave-dominated barrier: Assateague Island, Maryland, U.S.A. J Sediment Res 83:132–144
Shuman B, Bartlein P, Logar N, Newby P, Webb T III (2002) Parallel climate and vegetation responses to the early Holocene collapse of the Laurentide ice sheet. Quat Sci Rev 21:1793–1805
Smith LL (1931) Solution depressions in sandy sediments of the coastal plain in South Carolina. J Geol 39:641–652
Soller DR (1988) Geology and tectonic history of the lower Cape Fear River valley, southeastern North Carolina. U.S. Geological Survey Professional Paper 1466-A, 60p
Soulé PT (1998) Some spatial aspects of southeastern United States climatology. J Geogr 97:142–150
Spencer J, Jones KB, Gamble DW, Benedetti MM, Taylor AK, Lane CS (2017) Late-Quaternary records of vegetation and fire in southeastern North Carolina from Jones Lake and Singletary Lake. Quat Sci Rev 174:33–53
Stanford SD (1993) Late Cenozoic surficial deposits and valley evolution of unglaciated northern New Jersey. Geomorphology 7:267–288
Stolt MH, Rabenhorst MC (1987a) Carolina Bays on the eastern shore of Maryland: I. Soil characterization and classification. Soil Sci Soc Am J 51:394–398
Stolt MH, Rabenhorst MC (1987b) Carolina Bays on the eastern shore of Maryland: II. Distribution and origin. Soil Sci Soc Am J 51:399–405
Stuiver M, Reimer P (1993) Extended $^{14}$C data base and revised CALIB 3.0 $^{14}$C age calibration program. Radiocarbon 35:215–230
Sun J, Ding Z, Liu T (1998) Desert distributions during the glacial maximum and climatic optimum: example of China. Episodes 21(1):28–31
Suther BE, Leigh DS, Brook GA (2011) Fluvial terraces of the Little River Valley, Atlantic Coastal Plain, North Carolina. Southeast Geol 48:73–93
Swezey C (1998) The identification of eolian sands and sandstones. Comptes Rendus de l Academie des Sciences 327(8):513–518
Swezey C (2001) Eolian sediment responses to late Quaternary climate changes: temporal and spatial patterns in the Sahara. Palaeogeogr Palaeoclimatol Palaeoecol 167:119–155
Swezey CS, Schultz AP, Alemán González W, Bernhardt CE, Doar WR III, Garrity CP, Mahan SA, McGeehin JP (2013) Quaternary eolian dunes in the valley of the Savannah River, Jasper County, South Carolina. Quat Res 80:250–264
Swezey C.S, Fitzwater BA, Whittecar GR (2016a) Geology and geomorphology of the Carolina Sandhills, Chesterfield County, South Carolina. In: Doar III WR (ed) Gold, Structures, and Landforms in Central South Carolina – Field Guides for the GSA Southeastern Section Meeting, Columbia, South Carolina, 2016. Geological Society of America Field Guide 42, pp 9–36
Swezey CS, Fitzwater BA, Whittecar GR, Mahan SA, Garrity CP, Alemán González WB, Dobbs KM (2016b) The Carolina Sandhills: Quaternary eolian sand sheets and dunes along the updip margin of the Atlantic Coastal Plain province, southeastern United States. Quat Res 86:271–286

Taylor BE, Rich FJ, Brooks MJ, Ivester AH, Clement CO (2011) Late Pleistocene and Holocene vegetation changes in the Sandhills, Fort Jackson, South Carolina. Southeast Geol 48:147–163

Thom BG (1970) Carolina Bays in Horry and Marion Counties, South Carolina. Geol Soc Am Bull 81:783–814

Thornthwaite CW (1931) The climate of North America according to a new classification. Geogr Rev 21:633–655

Thornthwaite CW (1948) An approach toward a rational classification of climate. Geogr Rev 38:55–94

Timmons EA, Rodriguez AB, Mattheus CR, DeWitt R (2010) Transition of a regressive to a transgressive barrier island due to back-barrier erosion, increased storminess, and low sediment supply: Bogue Banks, North Carolina, USA. Mar Geol 278:100–114

UNESCO (1979) Map of World Distribution of Arid Regions. United Nations Educational, Scientific and Cultural Organization (UNECSO) Programme on Man and the Biosphere (MAB) Technical Notes 7, Explanatory Note (54p.) and Map (1 sheet at scale of 1:25,000,000)

Vandenberghe J, French HM, Gorbunov A, Marchenko S, Velichko AA, Jin H, Cui Z, Zhang T, Wan X (2014) The last permafrost maximum (LPM) map of the Northern Hemisphere: permafrost extent and mean annual air temperatures, 25–17 ka BP. Boreas 43:652–666

Wagner DP, McAvoy JM (2004) Pedoarchaeology of Cactus Hill, a sandy Paleoindian site in southeastern Virginia, U.S.A. Geoarchaeology 19:297–322

Watts WA (1980a) Late-Quaternary vegetation history at White Pond on the inner Coastal Plain of South Carolina. Quat Res 13:187–199

Watts WA (1980b) The late Quaternary vegetation history of the southeastern United States. Annu Rev Ecol Syst 11:387–409

Webb T III, Bartlein PJ, Harrison SP, Anderson KH (1993) Vegetation, lake levels, and climate in eastern North America for the past 18,000 years. In: Wright JE Jr, Kutzbach JE, Webb T III, Ruddiman WF, Street-Perrott FA, Bartlein PJ (eds) Global Climates since the Last Glacial Maximum. University of Minnesota Press, Minneapolis, pp 415–467

Wilson P, Clark R, Birnie J, Moore DM (2002) Late Pleistocene and Holocene landscape evolution and environmental change in the Lake Sulivan area, Falkland Islands, South Atlantic. Quat Sci Rev 21:1821–1840

Woodcock DW, Wells PV (1990) Full-glacial summer temperatures in eastern North America as inferred from Wisconsinan vegetational zonation. Palaeogeogr Palaeoclimatol Palaeoecol 79:305–312

Yang L-R, Ding Z-L (2013) Expansion and contraction of Hulun Buir Dunefield in north-eastern China in the last late glacial and Holocene as revealed by OSL dating. Environ Earth Sci 68:1305–1312

Zeeberg J (1998) The European sand belt in eastern Europe–and comparison of late glacial dune orientation with GCM simulation results. Boreas 27:127–139

Zhan S, Beck RA, Hinkel KM, Liu H, Jones BM (2014) Spatio-temporal analysis of gyres in oriented lakes on the Arctic Coastal Plain of northern Alaska based on remotely sensed images. Remote Sens 6:9170–9193

Zhou Y, Lu H, Zhang J, Mason JA, Zhou L (2009) Luminescence dating of sand-loess sequences and response of Mu Us and Otindag sand fields (North China) to climatic changes. J Quat Res 24:336–344

# Chapter 3
# Dunes of the Laurentian Great Lakes

**Edward Hansen, Suzanne DeVries-Zimmerman, Robin Davidson-Arnott, Deanna van Dijk, Brian Bodenbender, Zoran Kilibarda, Todd Thompson, and Brian Yurk**

**Abstract** Most dunes along the 16, 000-km long shorelines of the Laurentian Great Lakes formed after lake levels fell from highs at 5 to 3.5 ka. Foredunes develop during decadal scale variations in lake levels, forming during low and eroding during high lake levels. More permanent foredune plain complexes, with parallel dune ridges separated by swales, form on constructional shores. Transgressive and transitional complexes containing inland-migrating blowouts, parabolic dunes, and dune ridges occur on shores with high onshore sand drift potential and abundant sand supply. Vegetation cover is highest and wind energy and aeolian transport lowest in summer. Vegetation die back and high wind energy accompanying the passage of extratropical cyclones cause an autumnal increase in aeolian activity. Activity decreases as shoreline ice and surface snow develop during the winter. In the spring, niveo-aeolian structures collapse, vegetation increases, and storminess decreases. Great Lakes dune complexes host many ecological communities, including open dune communities, mesic forests and interdunal wetlands. The 34 million people living near Great Lakes shorelines generate heavy anthropogenic pressure. Two national governments, a Canadian province and seven US states have jurisdiction over Great Lakes shorelines, complicating the regulation of coastal activities.

E. Hansen (✉) · S. DeVries-Zimmerman · B. Bodenbender
Geological and Environmental Sciences Department, Hope College, Holland, MI, USA
e-mail: hansen@hope.edu

R. Davidson-Arnott
Department of Geography, University of Guelph, Guelph, ON, Canada

D. van Dijk
Geology, Geography and Environmental Studies Department, Calvin College, Grand Rapids, MI, USA

Z. Kilibarda
Department of Geosciences, Indiana University Northwest, Gary, IN, USA

T. Thompson
Indiana Geological Survey, Indiana University, Bloomington, IN, USA

B. Yurk
Mathematics Department, Hope College, Holland, MI, USA



N. Lancaster, P. Hesp (eds.), *Inland Dunes of North America*,
Dunes of the World, https://doi.org/10.1007/978-3-030-40498-7_3

**Keywords** Coastal dunes · Foredunes · Foredune plains · Transgressive dunes · Great Lakes shorelines · Open dunes

## 3.1 Introduction

The Laurentian Great Lakes (Fig. 3.1) consist of five large freshwater lakes, Lakes Superior, Michigan, Huron, Erie, and Ontario (Table 3.1) that, together with Lake St Claire, form an interconnected network draining through the Saint Lawrence Seaway into the Atlantic Ocean. The total shoreline length of these lakes is 16,000 km, leading some to call this region North America's Third Coast. Geologically, the Great Lakes terrane is diverse with Precambrian crystalline rocks dominating the bedrock along northern Lake Huron and most of Lake Superior and Paleozoic sedimentary rocks dominating the bedrock further south. Glacial outwash, till, and glaciolacustrine sediments occur throughout the Great Lakes Basin, but are particularly thick and well developed towards the south. This geological

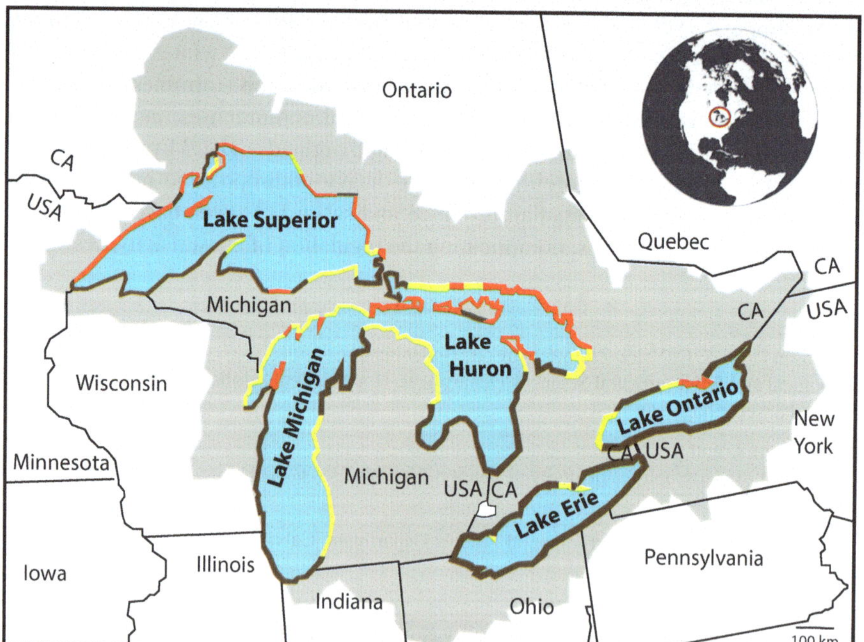


**Fig. 3.1** Overview of the Great Lakes Region. The lakes are shown in blue, the watershed of the Great Lakes is shown in gray. Yellow designates shores made up predominantly of sandy sediments. Brown designates shores composed predominantly of till or fine-grained (silt to clay) sediments. Red designates shores made up predominantly of bedrock. Data for this diagram were taken from Leverett (1911), Cadwell and Pair (1991), and Neff et al. (2005)

**Table 3.1** Characteristics of the Great Lakes

| Lake Name | Surface Area[a] ($km^2$) | Shoreline Length[a] (km) | Length[a] | Breadth[a] | Drainage Basin Area[b] ($km^2$) | Depth[a] | Population within watershed[b] |
|---|---|---|---|---|---|---|---|
| Lake Superior | 82,000 | 4393 | 563 | 257 | 127,700 | 149 m average 406 m maximum | 444,000 U.S. 229,000 Canada |
| Lake Michigan | 58,000 | 2639 | 494 | 190 | 118,095 | 825 m average 281 m maximum | 12.1 million U.S.[c] |
| Lake Huron | 60,000 | 6164 | 332 | 295 | 131,303 | 59 m average 229 m maximum | 1.5 million U.S. 1.5 million Canada |
| Lake Erie | 26,000 | 1402 | 388 | 92 | 58,788 | 19 m average 64 m maximum | 10.6 million U.S. 1.9 million Canada |
| Lake Ontario | 19,000 | 1146 | 311 | 85 | 60,601 | 86 m average 244 m maximum | 2.8 million U.S. 2.8 million Canada |

[a]GLERL (2017a)
[b]Michigan Sea Grant (2018)
[c]Includes metropolitan Chicago, which is not part of the drainage basin, but uses Lake Michigan for drinking water

diversity has created a wealth of coastal landforms with rocky sea cliffs, bluffs in unconsolidated glacial sediments, and sandy beaches all occurring along the Great Lakes shores. Among the coastal landforms are dunes which occur to a greater or lesser extent along all of the Great Lakes (GLERL 2017a). Low ephemeral foredunes are common throughout the area. Sets of low parallel dune ridges occur in coastal embayments, and along spits and cuspate forelands. In places, large dune fields extend a kilometer or more inland and contain parabolic dunes more than 50 m high.

All of the dune processes and landforms along the Great Lakes have analogies in processes and landforms along ocean coasts. Consequently, studies of the Great Lakes dunes have contributed to our understanding of coastal dunes in general. However, despite these similarities, there are also differences between the processes shaping Great Lakes dunes and most oceanic dunes. Great Lakes coasts are relatively young. The oldest recognized shorelines are found in southern Lake Michigan and Lake Erie and formed at roughly 16 ka during the retreat of the Pleistocene age glaciers (Larson and Schaetzl 2001). Most of the currently active coastal dunes occur along shorelines developed after the fall from the Nipissing high lake levels from 5 to 3.5 ka. Tides in the Great Lakes are very small. The lake basins respond to changes in lake level which are not tied to eustatic sea level changes, but to regional climate fluctuations and changes in the elevations of the drainage outlets driven by erosion or glacioisostatic rebound (Mainville and Craymer 2005). Great Lakes dunes are also not affected by salt crusts on the dune surface, and local water table dynamics are not affected by interactions between saline and fresh groundwater. The continental climate means there are large seasonal variations in the climate conditions. Relatively weak winds and extensive vegetation cover during the summer months contrast with stronger winds, persistent snow cover, coastal ice, and decreased vegetation cover during the winter. Storminess in the Great Lakes is tied to the passage of extratropical cyclones enhanced by "lake-effect" temperature contrasts between the lakes and the surrounding land. Finally, because of the restricted size of the lakes compared to ocean basins, each of the Great Lakes has shores in every possible orientation with respect to strong sand-transporting winds.

Roughly 34 million people live on or near the shores of the Great Lakes (Michigan Sea Grant 2018), including Canada's most populous city, Toronto, the United States' third most populous city, Chicago, and many smaller metropolitan centers (Table 3.1). The relatively dense population, extensive coastal development, and heavy industrialization in some areas have put significant anthropogenic pressure on most Great Lakes shores. The coastal dunes are well known and valued by the region's residents, and there have been significant efforts to protect them. In publicly owned dune complexes a critical question has been how to best manage the dunes to preserve their natural character and ecological diversity while still providing recreational access. Development of privately owned dune properties has also been regulated in many jurisdictions, an approach that has inevitably led to conflict between landowners, conservation groups, and government agencies.

Hundreds of scientific papers have been written on the Great Lakes coastal dunes, a large sample of which can be found in the reference section of this paper. Almost all of these have focused on a local area, either a single dune complex or the dune complexes along a portion of the shoreline of a single lake. A few review papers have addressed dunes along the entire Canadian shore (Martini 1981) or the entire shore of the State of Michigan (Arbogast 2009), which includes significant portions of the Lake Superior, Michigan and Huron shores. However, to the best of our knowledge, no paper has yet examined dunes throughout the entire Great Lakes region. After reviewing the climate and hydrology of the region, this paper presents the dune field morphology and classification of Great Lakes dunes. It then looks at aeolian sand drift potential along Great Lakes shorelines and how it affects the distribution of dunes along these same shores. An analysis of the environment and processes influencing the dunes, including the impact of storms and seasonal processes, is provided together with an examination of the sedimentology and ecological communities of the dunes. Holocene lake level changes in the Laurentian Great Lakes and the contribution of these changes to the geomorphic history of the dunes are reviewed. Lastly, dune management in the United States and Canada is discussed.

## 3.2 Climate and Hydrology

The climate of the Great Lakes is temperate and continental characterized by hot summers and cold winters. The region spans 7.5 degrees in latitude from the northernmost point on Lake Superior (~49°N) to the southernmost point on Lake Erie (~41.5°N) and 16 degrees of longitude from the easternmost point on Lake Ontario (~76° W) to the westernmost point on Lake Superior (~92°W). This wide range leads to a large variation in climate especially evident in a latitudinal temperature gradient (Fig. 3.2). Even in the southern portions of the Great Lakes, average air temperatures in the winter are low enough to create winter ice which typically reaches its maximum extent in February. There is considerable interannual variation in ice cover from a maximum recorded coverage of 95% in 1979 to a minimum

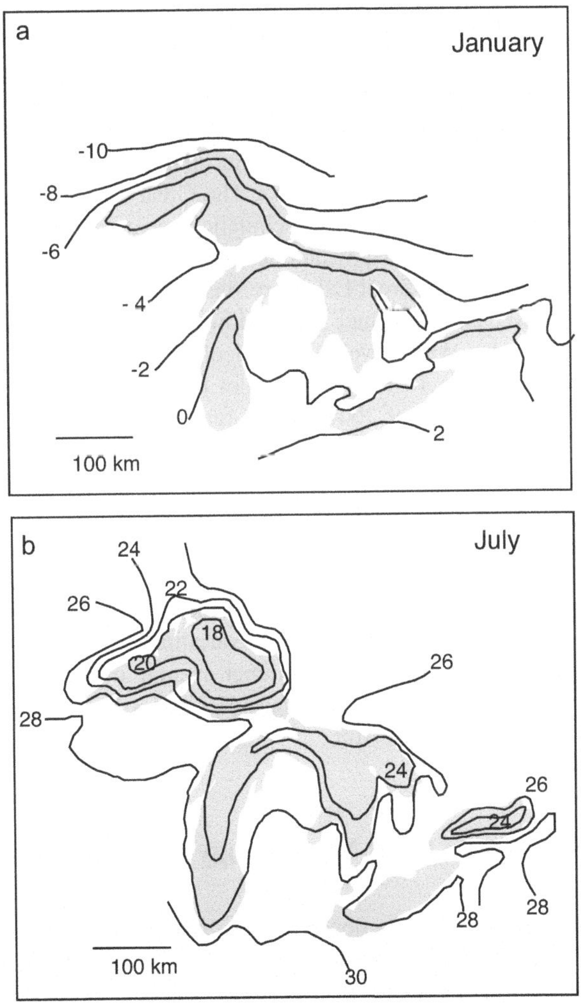


**Fig. 3.2** Isotherms in degrees Celsius for average maximum daily temperatures in (**a**) January and (**b**) July in the Great Lakes region. (Modified from Eichenlaub 1979)

recorded coverage of 11% in 2002 (Wang et al. 2012). The accumulation of ice along the shore fixes the width of the beach and the beach protects foredunes from wave erosion (Sect. 3.6.2).

The Great Lakes lie roughly equidistant between Hudson Bay and the Gulf of Mexico with no major topographic barriers to the movement of air between the two. As a consequence climate and weather in the Great Lakes region is influenced by the migration of both relatively cold and dry air masses from northern Canada and relatively warm and moist air masses from the Gulf of Mexico (Eichenlaub 1979). The movement of these air masses is often linked to the movement of extratropical (mid-latitude) cyclones which are responsible for the majority of the strong wind events in the Great Lakes region (Angel 1996; Angel and Isard 1998). The pattern of cyclone activity is seasonal with the majority of the strong cyclones occurring between November and April (Angel and Isard 1998).

The Great Lakes have a significant effect on the climate of the region (Scott and Huff 1996; Notaro et al. 2013). The thermal effects of the large water mass in moderating summer and winter temperatures are shown by the deflection of the isotherms in Fig. 3.2. During the summer the lakes are cooler than the surrounding land, causing an increase in atmospheric stability. During the winter the lakes are warmer, causing a decrease in atmospheric stability. These effects enhance the seasonal variability of storminess. The Great Lakes also act as moisture sources and increased cloudiness and precipitation occur downwind from the lakes, especially in the autumn and early winter, leading to enhanced "lake-effect" snowfall.

Tides in the Great Lakes are small. For example, in Lake Michigan, they rarely exceed 6 cm (Mortimer 2004). However, water levels vary on time periods of hours to decades. Water levels at a particular location on a Great Lakes coast may change by more than 1 m over periods of hours to days in response to storm surges and seiches (Mortimer 2004). The Great Lakes have relatively small watersheds and large lake surfaces (Table 3.1). As a consequence, lake levels are very sensitive to seasonal, annual and decadal changes in precipitation and evaporation. Relatively good records of direct lake levels extend back to 1860 (Fig. 3.3) (National Ocean Service 1986). Over the last 150 years, all lakes have shown seasonal variations in average monthly lake levels typically ranging from 0.15 to 0.6 m with the highest levels occurring in the summer and the lowest in the winter. These seasonal changes are superimposed on longer term variations in lake levels (Fig. 3.3) (Sect. 3.7). Over a period of a decade, the summer average monthly lake levels have varied by as much as 1.4 m for Lake Ontario, 1.2 m for Lakes Erie, Huron, and Michigan and 0.5 m for Lake Superior.

## 3.3 Dune Field Morphology and Classification

Numerous classification schemes have been proposed for Great Lakes coastal dunes, beginning with Cowles (1899) who divided dunes along Lake Michigan into active dunes and established dunes. Many of the attempts to describe and classify

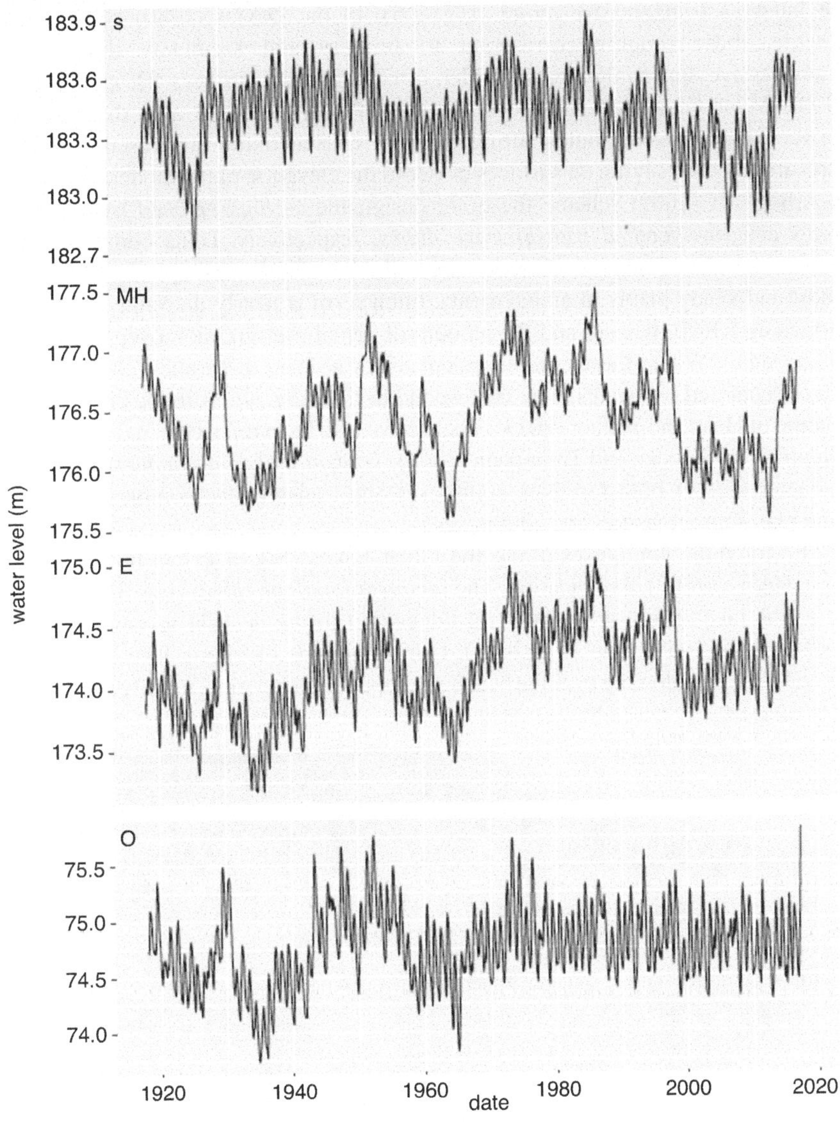


**Fig. 3.3** Monthly average lake levels in meters (International Great Lakes Datum (IGLD) 1985) for the Laurentian Great Lakes. *S* Lake Superior, *MH* Lakes Michigan and Huron, *E* Lake Erie, *O* Lake Ontario. (Data from GLERL 2017b)

Great Lakes dunes, such as that proposed by Buckler (1979), used a regional terminology often different from that of the general coastal dune community. In order to avoid confusion, we follow Arbogast (2009) in adapting more standard terminology for Great Lakes coastal dunes. Almost all attempts to classify these dunes make a distinction between the dunes along the lakeshore and the dunes somewhat further inland. Arbogast (2009) designated the former as *foredune complexes* (Fig. 3.4) and the latter as *transgressive dunes* (Fig. 3.5). In the Davies (1980) classification scheme, these roughly correspond to primary and secondary dune fields. Transgressive dunes migrate over older, often vegetated, surfaces (Hesp and Thom 1990). Most Great Lakes dune classification schemes have also distinguished between transgressive dunes sitting on high coastal bluffs composed of glacial deposits and dunes lying on surfaces closer to the elevation of the modern shoreline. In Arbogast's (2009) scheme these are designated as *high-perched transgressive dunes* and *low-perched transgressive dunes*, respectively. Dune complexes that form on constructional shoreline features, like baymouth bars, spits, cuspate forelands, and sandy plains in embayments, often form at nearly the same time as the surface on which they sit, and the amount of actual transgression over this surface can be small. Where these complexes consist of a series of parallel relict foredune ridges separated by swales, they correspond to *foredune plains* in the classification scheme of Hesp and Walker (2013). These have also been referred to as *strand-plain complexes* by Baedke and Thompson (2000). *Transitional dune complexes* share the characteristics of both foredune complexes and transgressive complexes or foredune plains and transgressive complexes.

The foredune complexes along the Great Lakes (Fig. 3.4) can be subdivided using Hesp's (2002) classification. The *incipient foredune ridges* (Fig. 3.4) are closest to the lake. These are ephemeral dunes tied to the decadal variations in lake levels (Olson 1958c; van Dijk 2014). During periods of falling lake levels, sand moved by winds blowing over the beach is trapped by vegetation, usually *Ammophila*

**Fig. 3.4** Examples of a Great Lakes foredune complex at P.J Hoffmaster State Park, Michigan showing **a** incipient foredune ridges and **b** established foredune ridge with **c** blowouts

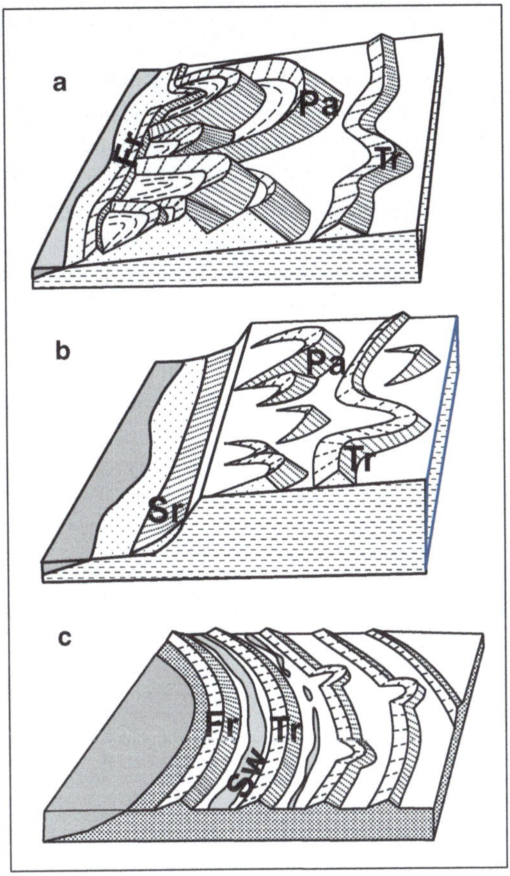


**Fig. 3.5** Schematics of coastal dune field categories except foredune complexes in the Great Lakes. These dune fields contain the following individual dune forms: *Fr* established foredune ridges, *Pa* parabolic dunes, *Tr* transverse ridges which in many cases may be relict foredune ridges, *Sr* sand ramp, *Sw* interdunal swale. (Modified from Hansen et al. 2010). (**a**) Low-perched transgressive dune complexes. (**b**) High-perched transgressive dune complexes during a period of low lake levels. (**c**) Foredune plain

*breviligulata*, growing just beyond the reach of the typical storm waves. The resulting incipient foredunes often coalesce into incipient foredune ridges (Davidson-Arnott and Law 1996) that continue to grow lakeward as long as lake levels drop during shoreline progradation. During periods of lake level rise, storm waves erode the lakeward edge of the incipient foredune ridges, causing them to become narrower, and, in periods of high lake levels, to disappear altogether. In some Great Lakes dune complexes, an *established foredune ridge* (Fig. 3.4) occurs landward of the incipient foredune ridge. Compared to the incipient foredune ridges, the established ridges are taller, persist through periods of high lake levels, and have a more complex ecological community, including shrubs and trees. Olson (1958c) suggested that these ridges grow during periods of high lake levels when some of the sand exposed by the erosion of the incipient foredune ridge is blown inland onto the established foredune ridges. Sand may also be added to the established foredune ridges during the winter when vegetation on the incipient foredune ridges is covered by ice, snow, or sand deposited earlier in the year (Yurk et al. 2013). This decrease in surface roughness allows windblown sand to move across the incipient foredune ridges onto the established foredune ridge.

Low-perched transgressive dune complexes (Fig. 3.5a) can occur inland of foredune complexes. Blowouts, and parabolic dunes, with relief that can exceed 50 m, are the characteristic landforms of these complexes (Martini 1981; Arbogast 2009; Hansen et al. 2010; Rawling and Hanson 2014). Where an established foredune ridge is present, the arms of the parabolic dunes closest to the lake often merge with it, suggesting that the parabolic dunes originated as blowouts on this ridge. Sets of overlapping and nested parabolic dunes occur in many of these complexes (Fig. 3.6) with mobile dunes closer to the lakeshore migrating upon and partially burying forested dunes further inland. Sinuous dune ridges with relief up to about 15 m occur in many low-perched transgressive complexes. The cuspate form of many of these ridges (Fig. 3.6) gives the appearance of merged parabolic dunes. In some larger complexes there are several parallel dune ridges. A particularly good example of this occurs at P.J. Hoffmaster State Park shown in Fig. 3.6. Heavily forested ridges often form the inland edge of low-perched transgressive complexes. OSL ages from these ridges are typically older that those from the forested parabolic

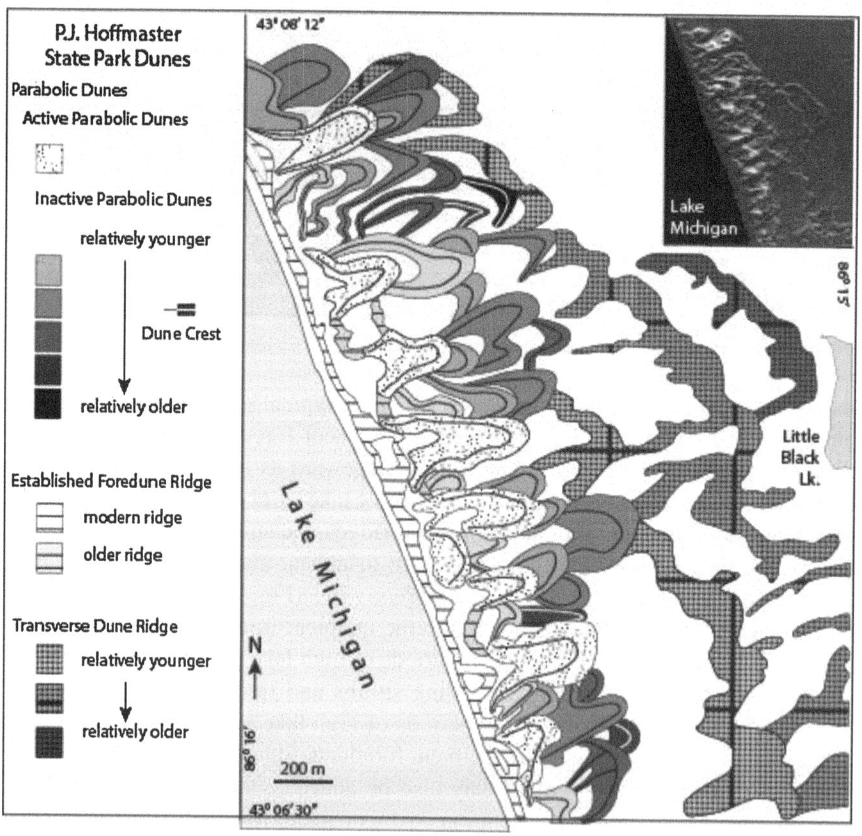


**Fig. 3.6** An example of multiple generations of overlapping dunes from a low-perched transgressive complex at P.J. Hoffmaster State Park, Michigan. (Modified from Hansen et al. 2010)

dunes closer to the lakes (Hansen et al. 2002, 2010), suggesting that these ridges were the first part of the complex to form and stabilize.

Transgressive dunefields, as defined by Hesp and Thom (1990), are "broad, active (free-moving) sand surfaces migrating landwards or alongshore." In addition to parabolic dunes and blowouts, transgressive dunefields can contain a number of aeolian landforms that are rare or absent in most Great Lakes dune complexes. These include actively migrating transverse ridges, barchanoid dunes, gegenwalle ridges, aklé and precipitation ridges (Hesp and Thom 1990; Hesp et al. 1989, 2011; Martinho et al. 2010). Low-perched transgressive dune complexes along the Great Lakes tend to be dominated by a combination of active and stable forested, parabolic dunes. The range of OSL ages on sand from below vegetated dune surfaces and radiocarbon ages from dune paleosols (e.g., Hansen et al. 2010) suggests that different dunes within these complexes were active at different times. In many complexes, actively migrating parabolic dunes closer to the lake overlap and partially bury stabilized parabolic dunes further from the lake. Some complexes contain up to five generations of overlapping parabolic dunes (Fig. 3.6). It appears as if many Great Lakes transgressive dune complexes did not evolve from a single large migrating sand surface, developing instead by the asynchronous inland migration of individual parabolic dunes. Although they can overlap, forming a nearly continuous aeolian landform, the individual dunes in these complexes usually still preserve their original parabolic form. Hesp (2013) outlines different scenarios by which transgressive dune fields can form and develop. Morphologically most Great Lakes low-perched complexes most closely resemble the morphology expected from the dunefields in the earlier stages of Hesp's third scenario: the formation of transgressive dunefields by the breakdown or merging of parabolic dunes. However, some Great Lakes transgressive dune complexes do consist of larger migrating sand surfaces. A good example of this is the Silver Lake complex along the eastern shore of Lake Michigan (Fisher and Loope 2005; Fisher et al. 2007; Hansen et al. 2010) which contains an actively migrating sand sheet approximately 0.5 km in width and 4 km in length. This sand sheet contains dune forms similar to those in transgressive dunefields in other parts of the world, including actively migrating transverse ridges within the sand sheet and a precipitation ridge along its outer margin.

High-perched transgressive dunes (Fig. 3.5b), also known as clifftop dunes, were first described in the Lake Michigan area by Cowles (1899). They were later described in more detail by Dow (1937) who examined the dunes atop the Manistee Moraine on the lake's eastern coast. High-perched transgressive dunes along the southern shore of Lake Superior were described by Bergquist (1936) and later Marsh and Marsh (1987). High-perched dunes also occur along the Canadian shores of the Great Lakes (Martini 1981). The typical high-perched dune sits on top of a coastal bluff composed of glacial sediments. The most characteristic landforms in these complexes are shoreline parallel transverse ridges with blowouts (Blumer et al. 2012) and parabolic dunes (Arbogast 2009). As in the case of low-perched transgressive complexes, the parabolic dunes can form sets of overlapping dunes. During low lake levels, incipient foredunes and foredune ridges may form on the beach at the base of the bluff and, in some cases, sand ramps or climbing dunes occur directly against the bluff (Buckler 1979; Arbogast 2009). During periods of high lake levels, these protective barriers are destroyed and waves are able to reach the base of the bluff. Dow (1937) suggested that high-perched transgressive dunes form during high lake level periods when erosion exposes sediment along the bluff, a suggestion that was elaborated upon by Marsh and Marsh (1987), Anderton and Loope (1995) and Loope et al. (2004). In this model, erosion at the base of the bluff triggers slumps that expose the sediment towards the top of the bluff. This sediment is picked up by onshore winds and transported to the top of the bluff where it initiates a period of dune growth and migration. A somewhat similar model appears to account for the formation of perched dunes on bluffs made of glaciolacustrine and glacial fluvial deposits at Rubjerg Knude, Denmark (Saye et al. 2006). Geochemical, textural, and chronological data from this area suggest that sand was transported by wind from the eroding faces of the bluffs to the dunes during a period of relatively rapid sea bluff retreat. In contrast, perched dunes in some parts of the world, e.g., Brittany, France (Haslett et al. 2000), sit on top of consolidated bedrock sea cliffs and models for the formation of these dunes (Jennings 1967) do not involve aeolian transport of sand from an eroding cliff face.

Descriptions of foredune plains in embayments that trap sediment (Figs. 3.5c, 3.7a) extend back at least to Calver (1946), and are generally referred to as strand-plain complexes in the Great Lakes literature. During periods of high lake levels, waves cut a scarp in beach deposits. As lake levels fall, wind blows across the beach, picks up sand and deposits it at the top of these scarps, forming a foredune ridge (Thompson and Baedke 1995). As the shore progrades, the scarp and dune ridge marking the next cycle of rising and falling lake levels forms further lakeward. The result is a series of parallel ridges capped by aeolian sediment separated by swales (Hesp and Walker 2013). Foredune plains also form on constructional platforms like spits and cuspate forelands (Fig. 3.7b) that project into the lakes. Each ridge represents the position of the shore at different stages in the growth of the structure. Although these form to some extent on all of the Great Lakes, particularly prominent examples occur on the northern shore of Lake Erie (Coakley 1976, 1989; Martini 1981). Relatively high rates of isostatic uplift in the northern portions of the Great Lakes can lead to a lakeward progression of the shoreline which can, in turn, lead to one or more relict foredune ridges stranded inland away from the current shoreline (Davidson-Arnott and Pyskir 1988; Lovis et al. 2012).

Low-perched transgressive complexes can evolve from foredune complexes by the inland migration of blowouts on the established foredune ridge (Hesp 2013). Transitional dune fields that combine the characteristics of both foredune and transgressive dune complexes develop during the early stages of this process. Other transitional dune fields along the Great Lakes combine characteristics of transgressive complexes and foredune plains. These form when blowouts within dune ridges in foredune plains develop into parabolic dunes that migrate inland. Thus, in some places, foredune plains give way inland to low-perched transgressive dune fields (Martini 1981). Narrow baymouth bars or spits may contain one or two relict foredune ridges with blowouts and parabolic dunes beyond their inland edges, referred to as baymouth bar complexes by Hansen et al. (2010).

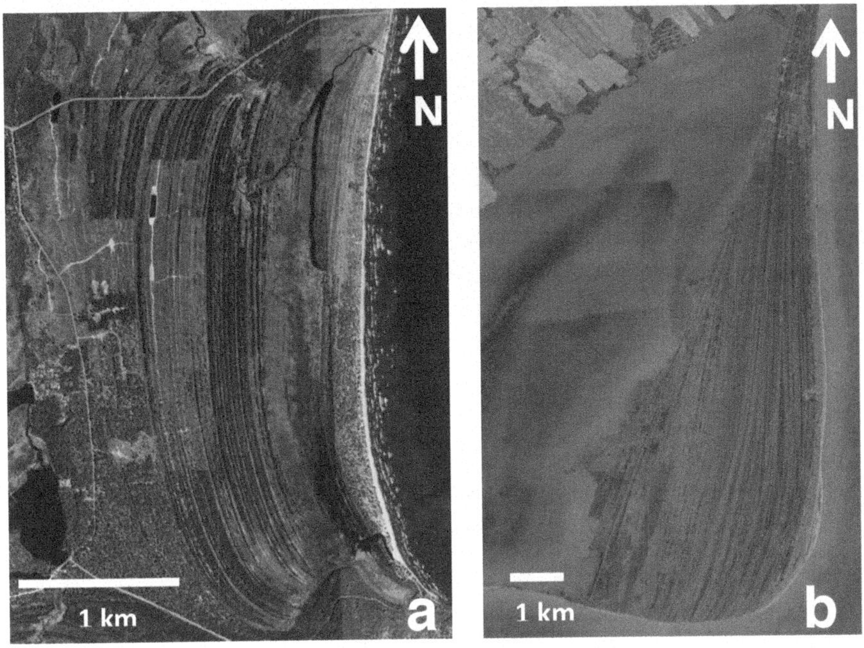


**Fig. 3.7** Aerial views of foredune plains showing the characteristic pattern of multiple parallel alternating dune ridges and swales. (**a**) Foredune plain in a coastal embayment (strand-plain), Big Traverse Bay, Michigan, on the eastern side of the Keweenaw Peninsula, southern shore of Lake Superior. USGS 1998 aerial photograph. (**b**) Foredune plain on a cuspate foreland, Pointe-au-Pins, Ontario, Canada, on the northern shore of Lake Erie. Google Earth satellite image

## 3.4 Sand Drift Potential Along Great Lakes Shores

Variations in the ability of winds to transport sand can be displayed on sand drift potential diagrams. Archived wind speed data suitable for calculating drift potential in the Great Lakes region are available from airport weather stations accessible from the Weather Underground website (2017) and from the United States National Oceanic and Atmospheric Administration's (NOAA) coastal weather stations, accessible through the NOAA website (NOAA 2017). We used these data to calculate sand drift potentials at 66 locations on or near Great Lakes coasts. For our calculations we used a variation of the Fryberger and Dean (1979) equation:

$$\mathrm{DP} = V^2 \left(V - V_t\right) P_V$$

in which DP is the drift potential associated with a wind of a given velocity in a given direction, $V$ is wind velocity, $V_t$ is the threshold velocity, which we set to 7.01 m/s, and $P_v$ is the proportion of time that wind with velocity $V$ blew in that specific direction. All wind velocities were either recorded at 10 m height, or were corrected to 10 m height using the formula of Simiu and Scanlan (1996) as described in Yurk et al. (2014). Drift potential was calculated for the 10-year period from 2006 to 2016. Data gaps of up to several weeks were present in the records for many of these stations, making the actual period over which the drift potential was calculated somewhat less than 10 years.

Sand drift potentials from 21 weather stations, chosen to highlight the general patterns, are shown in Fig. 3.8. The drift potential for each of the 16 directions of the 16-point-compass rose is shown as a solid black line while the resultant sand potential direction is indicated by a dashed line. The Great Lakes are in the belt of prevailing southwesterly regional winds (Eichenlaub 1979). At most sites there is relatively

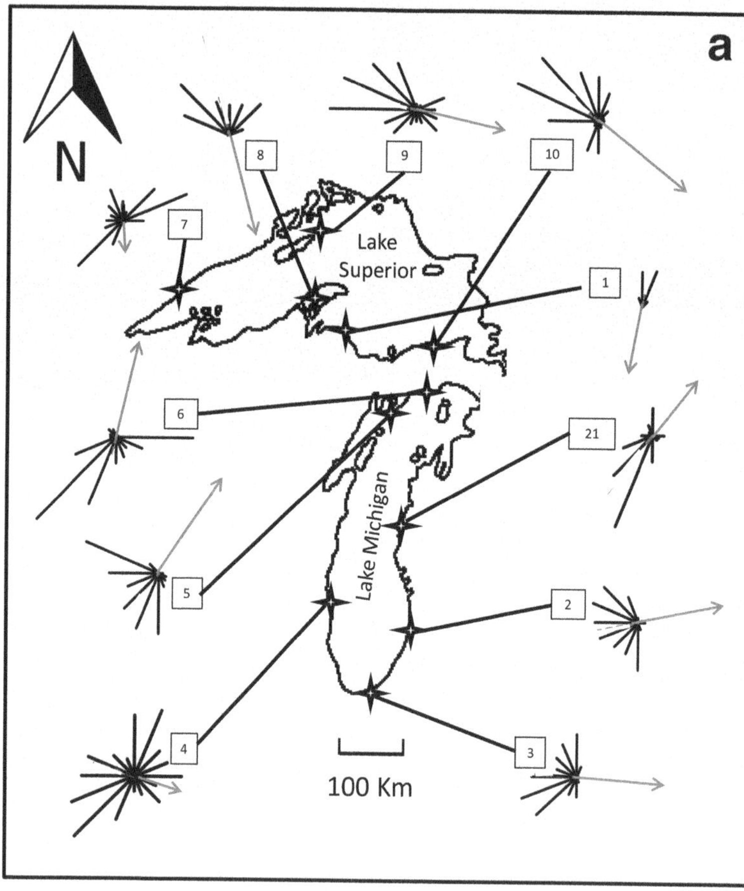


**Fig. 3.8** Sand drift potentials for selected areas along the Great Lakes shorelines. Sand drift potentials were calculated using a modified version of an equation from Fryberger and Dean (1979) as described in the text. Sand drift potentials for each of the directions of the 16 point compass rose are shown as solid black lines extended in the direction from which the sand originates. Resultant sand drift potentials are shown as gray arrows, plotted at half the scale, and extended in the direction towards which the sand moves. (**a**) Sand drift potentials for Lakes Superior and Michigan. (**b**) Sand drift potentials for Lakes Huron, Erie, and Ontario

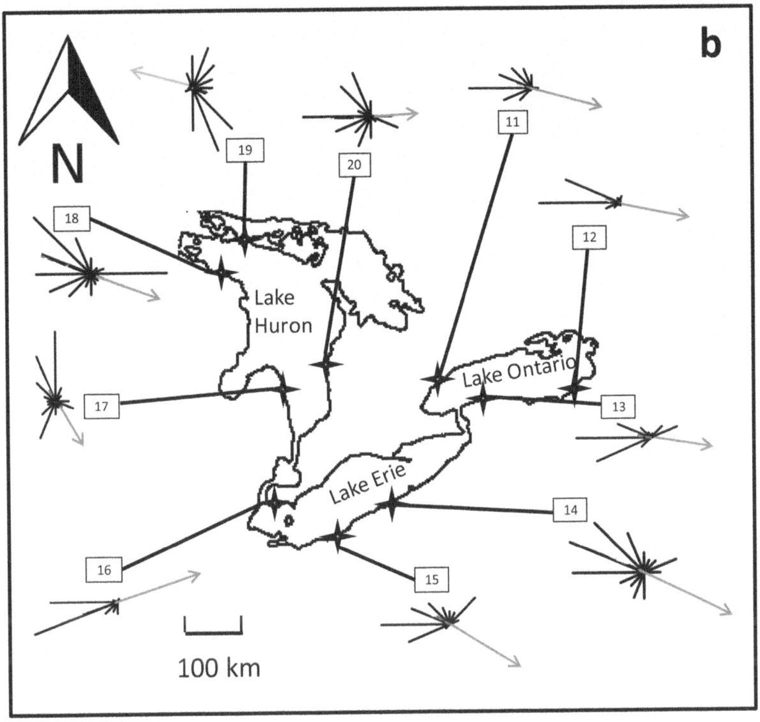


**Fig. 3.8** (continued)

strong drift potential from the west, reflecting the influence of these prevailing westerlies. However, several factors contribute to the development of strong onshore winds deviating from the prevailing wind directions. The directions of strong storm winds associated with extratropical cyclones can vary significantly from the average prevailing direction and will often switch directions as the cyclone moves through the area (Sect. 3.6.1). Moreover the relatively low surface roughness of the lakes imposes less drag on boundary layer airflow, resulting in higher speed winds blowing over the lake. Thus, the fetch leading up to a shore can lead to the development of strong onshore winds, even on shores that do not face the prevailing southwesterly winds. Drift potentials have relatively large north-south components where shores have relatively large north-south fetches. With a few exceptions, easterly drift potentials tend to be low.

## 3.5 Distribution of Dunes Along the Great Lakes Shores

The distribution of dune fields is determined by a combination of the nature of the material that makes up the coast (Fig. 3.1), including the sediment supply from bluff erosion, longshore sediment transport within littoral cells, and the sand drift

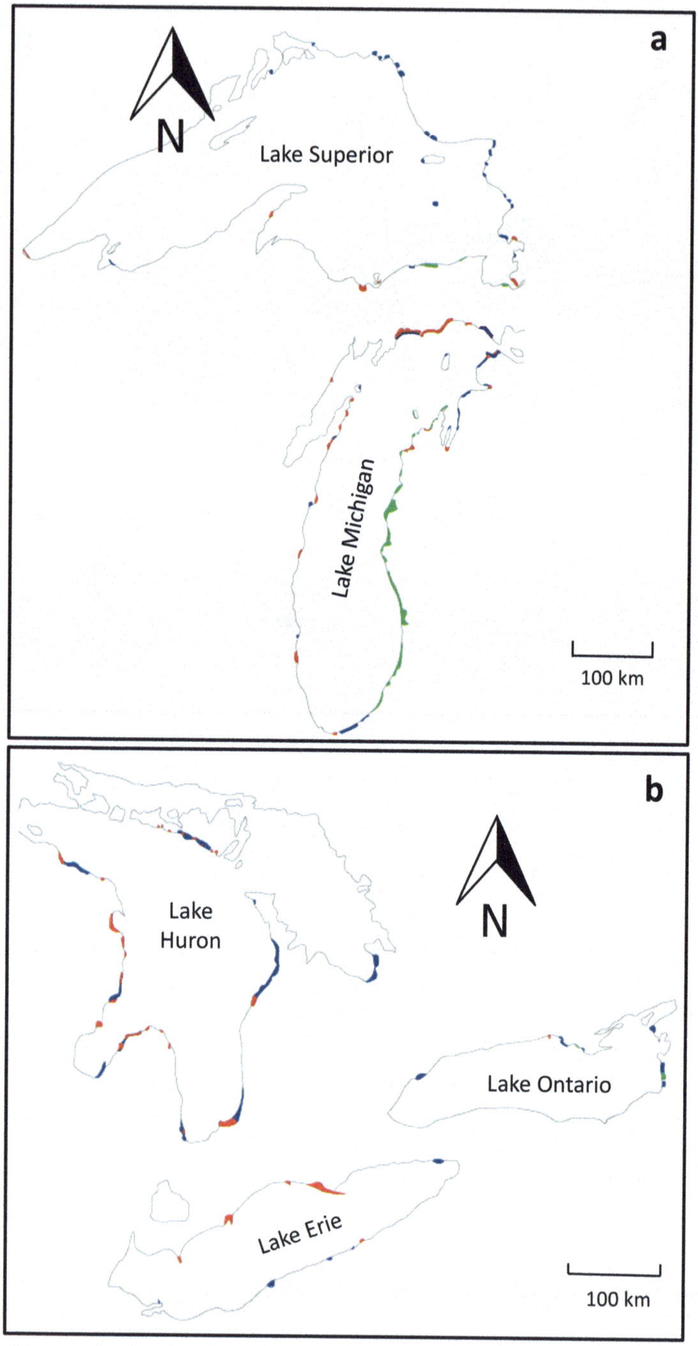


**Fig. 3.9** Distribution of known "secondary" (transgressive, transitional and foredune plain) dune fields along the Great Lakes. Foredune plain complexes are shown in red, transgressive complexes

potential along the directions that connect dunes to the areas that can supply sand (Fig. 3.8). The locations of known "permanent" (transgressive, foredune plain, and transitional complexes) dune fields along the Great Lakes are shown in Fig. 3.9. Ephemeral incipient foredunes and incipient foredune ridges occur to some extent at the inland edges of most sandy beaches along the Great Lakes and we have not attempted to map their locations. Foredune plain complexes are scattered throughout the Great Lakes shorelines and occur even on coasts with relatively moderate onshore sand drift potential (Fig. 3.8). The key ingredients for the development of these complexes are a prograding sedimentary platform (indicative of a positive littoral sediment budget) combined with enough onshore wind to develop foredune ridges. Extensive transitional and transgressional complexes (Fig. 3.9), on the other hand, are limited to shores with large onshore-oriented sand drift potential (Fig. 3.8) and an abundance of sandy sediments (Fig. 3.1). Relatively strong onshore sand drift potential appears to be necessary to cause dunes to migrate inland onto older geomorphic surfaces. The distribution of dune fields along each of the Great Lakes is discussed in the following sections.

**Lake Superior** Relatively high westerly and northwesterly drift potential occurs through most of Lake Superior which is also reflected in a tendency towards northwesterly resultant drift potential (Fig. 3.8a). Along the western and northwestern parts of the lake, the high drift potentials tend to be offshore or parallel to shore. The shores along these parts of the lake also tend to be rocky with limited sand supply. Consequently, dunes are relatively rare (Fig. 3.9a). Parallel ridges on small forelands appear on LiDAR images in a few localities along the northwestern coast (Boerboom 2017) and these may be relict foredune plains. However, they are not included in Fig. 3.9 because of their uncertain status. Along Lake Superior's northeastern shore, scattered dune complexes occur in embayments wherein sediment has accumulated in the generally rocky coast (Bakowsky and Henson 2014). In places, these complexes contain well developed parabolic dunes that are migrating inland over lake terraces formed during earlier stages in the history of Lake Superior (Martini 1981). Along the eastern portion of the southern coast, high-perched transgressive dunes occur where high onshore drift potential (Fig. 3.8a) is combined with sandy glacial sediments (Fig. 3.9a) in high bluffs (e.g., Loope et al. 2004). Prominent foredune plain complexes occur scattered along the eastern and southern shores

**Fig. 3.9** (continued) in green and transitional and unclassified complexes in blue. While an attempt has been made to accurately show the extent of each complex parallel to the shoreline, the width of complexes perpendicular to the shoreline has been exaggerated in places to increase their visibility. Dune complex positions were located using a combination of geological maps, published papers, personal communications, and web pages supplemented by satellite images and historical aerial photographs accessed through Google Earth. Sources of data include: WDNR 2017a (LS), Arbogast et al. 2009 (LM, LH), Arbogast et al. 2010 (LH), Bakowsky and Henson 2014 (LS, LH, LE, LO), Cadwell and Pair 1991 (LO), MDEQ 2017 (LM, LS), Delano 1991 (LE), Earnest et al. 2002 (LO), ODNR 2017 (LE), Leverett 1911 (LM, LH), Jones 2017 (LE), Martini 1981 (LS, LH, LE, LO), Rawling and Hanson 2014 (LM), Rawling 2017 (LM, LS), ODNR 2007 (LE), Thompson et al. 2004 (LM, LS), WDNR 2017b (LM). (LS: Lake Superior; LM: Lake Michigan; LH: Lake Huron; LE: Lake Erie; LO: Lake Ontario). (**a**) Distribution of secondary dune fields along Lake Superior and Lake Michigan. (**b**) Distribution of secondary dune fields along Lake Huron, Lake Erie and Lake Ontario where sediment has accumulated in embayments (e.g., Johnston et al. 2012). Dune fields also occur on spits and cuspate forelands in the Apostle Islands in the western portion of the southern shore (WDNR 2017a).

**Lake Michigan** The drainage basins of streams flowing into eastern Lake Michigan contain extensive areas of sandy sediments in former outwash plains and glacial lake plains which, in many places, extend to the shore (Leverett 1911). The large amounts of sandy sediments (Fig. 3.1) and strong westerly drift potential (Fig. 3.8a) have led to the widespread development of dune complexes lining over 75% of Lake Michigan's eastern shore (Fig. 3.9a). Arbogast et al. (2009) estimates sand volume totaling ~1.7 km$^3$ in these dune complexes. The southern part of the eastern shore is dominated by low-perched transgressive dune complexes (Hansen et al. 2010), while the northern portion has a mixture of high-perched transgressive dunes, low-perched transgressive dunes, and foredune plain complexes (Lovis et al. 2012; Thompson et al. 2004). The resultant drift potential along the southern shore is westerly at a low angle to the shoreline (Fig. 3.8a). Parabolic dunes which stabilized before 3 ka along the southern shore have roughly east-west axes (Argyilan et al. 2014; Kilibarda et al. 2014), and, thus, appear to reflect the orientation of the resultant drift potential. However, blowouts and parabolic dunes active since 1 ka tend to have northwest-southeast axes (Kilibarda et al. 2014). Hence, their orientations may reflect the relatively high drift potential associated with northerly onshore winds (Fig. 3.8a). Foredune plain and transitional dune complexes line most of the eastern portion of Lake Michigan's northern shore. Strong southwesterly drift potential occurs along northern Lake Michigan (Fig. 3.8a) and the axes of blowouts in these dune complexes tend to have southwest-northeast orientations. Foredune plain complexes occur sporadically on Lake Michigan's western margin (Fig. 3.9a). They are especially common in embayments and baymouths on Wisconsin's Door Peninsula (Rawling and Hanson 2014). There is a significant southerly and southwesterly drift potential along the Door Peninsula (Fig. 3.8a). Blowouts and parabolic dunes have formed along dune ridges in south-facing embayments on the peninsula (Rawling and Hanson 2014), leading to the development of transitional complexes.

**Lake Huron** Sandy sediments (Fig. 3.1) and dunes (Fig. 3.9b) are common along the northern portion of Lake Huron's western shore (Fig. 3.1). The resultant sand drift potential is easterly and offshore. Towards the southern portion of Lake Huron's western shore, the highest drift potentials are from the north and northwest. Dunes are particularly common in this area where the onshore winds are also from the north, such as along the southern shore of Saginaw Bay (Figs. 3.8b, 3.9b). The dune fields along the western shore of Lake Huron are relatively small, containing only about 1/20th of the sand volume found along Lake Michigan's eastern coast (Arbogast 2009). Significant northerly onshore sand drift potential occurs along the southern shore of Manitoulin Island in northern Lake Huron and significant westerly sand drift potential occurs along the eastern shore of the lake (Fig. 3.8b). Foredune plain and transitional dune complexes occur along a fairly large proportion of both of these coasts. Large transverse dune ridges occur at the inland edges of foredune plain complexes (Martini 1975; Davidson-Arnott and Pyskir 1988) on Manitoulin Island, at Wasaga Beach at the south end of Georgian Bay, at Sauble Beach and the southern Bruce Peninsula and the Pinery dune field at the southern end of the east shore of Lake Huron. These have been attributed to dune building during the high lake levels at the end of the Nipissing transgression (Sect. 3.7).

**Lake Erie** Relatively large onshore sand drift potential occurs from the north and northwest (Fig. 3.8b) along the southern shore of Lake Erie. However, sand dunes are relatively rare along the southern shore (Fig. 3.9b). This may be, in part, a consequence of the fine-grained nature of the till and glaciolacustrine deposits along this shore. Extensive modification of this coast by human activity may have also destroyed dune complexes. A few small dune complexes occur in Ohio, most of which survive or exist because of human structures like jetties (Jones 2017). The most extensive dune complex along the southern shore is a foredune plain on the spit Presque Isle near Erie, Pennsylvania (Delano 1991). The most prominent dune complexes are found along the Canadian (northern) shore. These are foredune plain complexes somewhat modified by the development of blowouts on large cuspate forelands and spits at (from east to west) Long Point (Martini 1981), Pointe-aux-Pins (Coakley 1989), and Point Pelee (Coakley 1976). Smaller dune fields are present just to the west of the Niagara River at Erie Beach and Point Abino.

**Lake Ontario** There is significant westerly sand drift potential along the eastern shore of Lake Ontario (Fig. 3.8b). Dune fields have developed along a 25 km stretch of the southeastern shore where deposits of lacustrine sand are exposed at or near the shore (Fig. 3.9b), but disappear northwards where thin mantles of clay and silt lie above bedrock (Cadwell and Pair 1991). Collectively this region is sometimes referred to as the Eastern Lake Ontario Dune and Wetland area (Earnest et al. 2002). Relatively large dune fields are also present on west-facing baymouth bars in Prince Edward County on the northern shore. These include Ontario's Sandbanks Provincial Park, Weller Bay, and Prequ'ile Provincial Park (Martini 1981; Davidson 1991), Toronto Island (the spit enclosing Toronto Harbor) and the baymouth barrier enclosing Hamilton Harbor at the southwest end of the lake. Dunes occur only sporadically along the eastern and southern shores of Lake Ontario, much of which has been heavily modified by human structures.

## 3.6 Environment and Processes

### *3.6.1 Impact of Storms on Great Lakes Dunes*

Most of the strongest storm winds in the Great Lakes region are associated with extratropical cyclones (Angel 1996). The most intense cyclones in the Great Lakes region occur between October and April (Angel and Isard 1998). During October through April in 2010–2011 and 2011–2012, 40 out of 55 (73%) strong wind events (wind speed greater than 2σ beyond the mean speed) were associated with extratropical cyclones (Yurk et al. 2014). As aeolian sand transport and, correspondingly, sand drift potential (Fryberger and Dean 1979) increase approximately with the cube of the wind speed (Bagnold 1954; Lettau and Lettau 1978; Pye and Tsoar 2009), the strongest wind events can have a disproportionate effect on dune surfaces. Thus, it has been hypothesized that strong winds associated with extratropical cyclones are important drivers of surface change in Great Lakes dune complexes (Yurk et al. 2014).

The impact of a cyclone on a particular dune complex depends largely on the strength of the cyclone (e.g., its pressure gradient), its track relative to the complex, and the positions of its fronts. The winds associated with cyclones in the Great Lakes region rotate in a counterclockwise direction about a low pressure center. Wind speeds increase toward the center and near the weather fronts moving along with the cyclone. Precipitation also tends to be concentrated along the fronts and can suppress aeolian sand transportation. However, strong sand transporting winds also occur in the "dry" portions of these storms. An example of this is given in Yurk et al. (2014). In this study surface change was measured with an array of erosion pins in a trough blowout on the southeastern shore of Lake Michigan. Roughly 20% of the total annual surface change in this dune was associated with the passage of an extratropical cyclone from 26 to 28 October. Sustained wind speeds of over 15 m/s with gusts of up to 25 m/s occurred over a 37-h period. Precipitation only occurred during 6 h towards the end of this period.

A cyclone's local impact often lasts for several days. During this time, local wind speeds and directions change as the storm center passes through the region (Fig. 3.10). This variation depends on the location of the complex relative to the storm track. Hence, the effects of a particular cyclone can vary greatly at different places along the Great Lakes shores (Yurk et al. 2014). The impacts of cyclones also vary with the intensity of the cyclone and its timing (e.g., relative to snow and ice cover (van Dijk 2014)). Given the impact of extratropical cyclone winds on Great Lakes dune complexes, changes in cyclone tracks, frequency, timing, or intensity associated with climate change (e.g., Bengtsson et al. 2006) may have important impacts on the region's dune systems.

The impact of storms on the foredune complex is generally similar to other midlatitude coastal dune systems and is an important element of beach/dune interaction (Houser and Ellis 2013). Large waves and water level set-up due to storm surge lead to erosion of the embryo dunes and scarping of established foredunes on both mainland and barrier beaches at all but the lowest levels in the decadal lake level cycle. However, the impact is greatest where intense storms coincide with lake level highs, as recognized by Olson (1958c). A number of studies at Long Point on Lake Erie's northern shore have documented the effects of an intense storm on 2 December 1985 that occurred during a period of record high lake levels. The foredune was scarped along more than 40 km of the spit shoreline (Stewart and Davidson-Arnott 1988) and overwash or inlet creation occurred over extensive areas of the proximal and central section of the spit (Davidson-Arnott and Fisher 1992; Davidson-Arnott

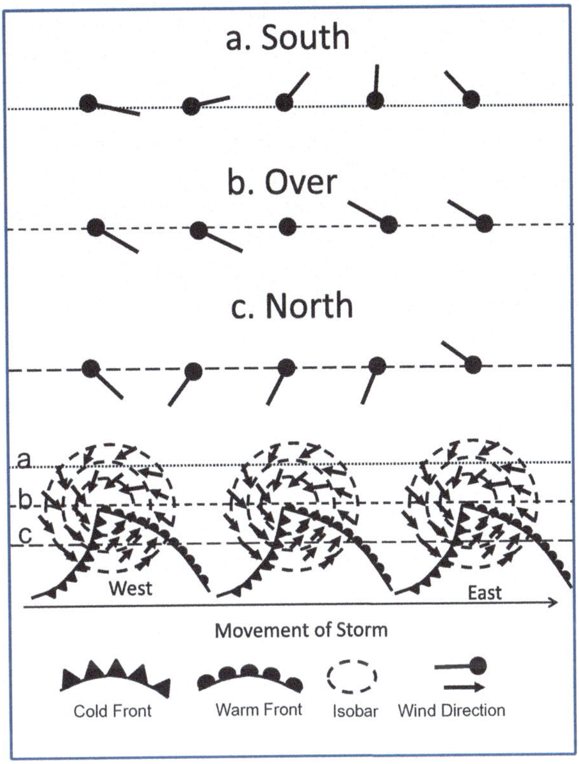


**Fig. 3.10** Wind directions as the center of an extratropical cyclone passes to the north (**a**), over (**b**), and to the south (**c**) of a dune complex. Wind directions are given by the arrows with the rounded heads along the three lines at the top of the diagram. The tails of the arrows indicate the direction from which the wind is coming. In the lower part of the diagram, a schematic wave cyclone is shown in three different positions: one to the west of the dune complex, one along a north south line passing through the complex, and one to the east of the complex

and Conliffe Reid 1994). The recovery of the foredune and associated vegetation is documented in Saunders and Davidson-Arnott (1990). Recovery coincided with a fall in lake levels after 1986. In areas where the foredunes had been completely breached, vegetation quickly colonized the outwash fans, trapping windblown sand. By 1990, a continuous foredune ridge had developed, trapping most of the sediment blown inland from the beach in these areas. Vegetation was slower to colonize places in which there was severe scarping of the foredune ridge. By 1990, there had been only limited new foredune growth in these areas.

The geometry and orientation of dunes relative to wind direction determine the dunes response to storm winds, though the relationship may be complicated. The compression of wind flow over a dune ridge creates a topographic acceleration which leads to maximum velocities at the dune crest. This effect was observed in early field studies of foredunes along Lake Michigan by Landsberg and Riley (1943) and Olson (1958a). Since that time, wind flow over foredunes has been modelled theoretically and studied experimentally in many parts of the world (Walker et al. 2017 and references therein). Flow deceleration occurs at the windward edges of the dunes. In places where the slope at the bottom of the dune is steep, for example after scarping by wave erosion, this deceleration can lead to the deposition of sand (Hesp et al. 2009). Flow acceleration occurs further up along the stoss slope. Winds that approach the foredune at an oblique angle tend to turn towards the direction perpendicular to the crest as they move up the stoss slope. This effect is most pronounced for winds approaching at angles between 30° and 60° (Hesp et al. 2015 and references therein).

Within "permanent" (transgressive, foredune plain, and transitional complexes) dune fields, wind patterns will also be modified by local topography. Windflow in parabolic dunes and blowouts along the Great Lakes has been studied by Fraser et al. (1998), Hansen et al. (2009) and Yurk et al. (2014) and in coastal dunes outside the Great Lakes by Gares and Nordstrom (1995), Hesp and Hyde (1996), Hesp and Pringle (2001), and Smyth et al. (2012). In trough blowouts and parabolic dunes, the topography steers local wind flow along the dune axis when winds are not at too great an angle relative to the axis (Hansen et al. 2009) (Fig. 3.11). Strong regional winds directed along the dune axis tend to result in the greatest surface change. But off-axis regional winds may still result in substantial sand transport. In some cases, regional winds directed down dune axes have been shown to result in a counter flow moving up the dune axes within the blowout trough (Fig. 3.11). These flows can be associated with considerable surface change (Fraser et al. 1998; Hesp and Walker 2012; Yurk et al. 2014). Wind directions will change as a cyclone passes, resulting in changes in wind steering, local wind patterns and sediment transport (Yurk et al. 2014; DeVries-Zimmerman et al. 2018). Hence, the contemporary orientations of large dune features within a complex can be expected to reflect the history of cyclone tracks within the region over some unknown time interval. It has been proposed that changes in the directions of axes in Great Lakes dune complexes might indicate historical changes in storm tracks and intensities within the region (Kilibarda et al. 2014).

Strong winds can blow sand beyond the brink of a dune where some of it may be temporarily suspended by updrafts associated with turbulence along the lee slope (Nickling et al. 2002). The sand carried in temporary suspension will eventually settle out as grainfall. Anderson and Walker (2006) found significant deposition by grainfall in a parabolic dune plain in British Columbia, most of which was associated with a blowout and an active parabolic dune. Bodenbender et al. (2018) measured grainfall in the lee of a large parabolic dune along the eastern shore of Lake Michigan. Although the amount of grainfall decreased exponentially with distance from the dune, sand from grainfall was still detected in a trap 215 m beyond the base of the lee slope. Roughly two thirds of the grainfall in a 5-month period (27 November 2017–23 April 2018) was associated with a single storm. DeVries-Zimmerman et al. (2014) have suggested that aeolian grainfall tied to sand suspension in the lee of large dunes is responsible for the accumulation of sand in lakes and bogs beyond these dunes. They further suggest that the amount of sand in these sediments of these lakes and bogs can be used as a proxy for the amount of aeolian activity. Björck and Clemmensen (2004) and deJong et al. (2006) suggest that the amount of aeolian sand in bogs can also be used as a proxy for storminess.

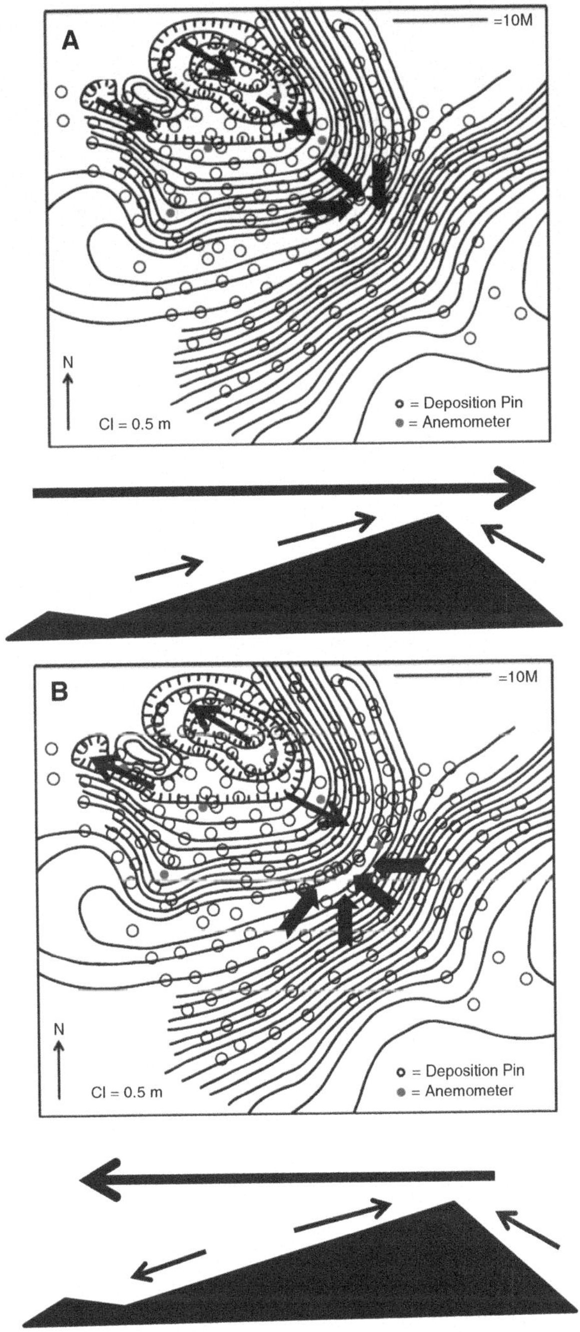


**Fig. 3.11** Wind directions within a trough blowout in a transgressive dune complex along the southeastern shore of Lake Michigan. The topographic map summarizes the results of experiments within the trough reported by Yurk et al. (2014) and DeVries-Zimmerman et al. (2018). The patterns are shown schematically in the silhouettes representing cross-sections along the dune axes below the topographic maps. Regional wind directions are shown by thick arrows. Local wind directions are depicted by thin arrows. (**a**) Steering of winds up the axis of the trough. This pattern occurs when regional winds approach towards the mouth of the trough. (**b**) Bifurcated flow in which winds closer to the crest blow up the axis of the trough while winds closer to the mouth blow in the opposite direction down the axis of the trough. This pattern occurs when regional winds approach towards the crest of the dune

### 3.6.2 Seasonal Processes

Seasonal variables and changes are distinct aspects of Great Lakes dune processes and landforms, adding complexity and interest to these environments. Key variables having seasonal patterns include wind, temperature, precipitation, vegetation, surface conditions, and lake levels. The variables, processes and morphologies of Great Lakes dunes are described seasonally through a complete annual cycle below. Figure 3.12 illustrates seasonal processes with the emphasis on Great Lakes foredunes and is based largely on the nearly two decade long study of van Dijk (2014). Studies of seasonal processes in other types of dunes have been less extensive and have largely concentrated on blowouts and parabolic dunes (Byrne 1997; Hansen et al. 2009; Kilibarda and Kilibarda 2016). Results from these studies are also included in Fig. 3.12.

**Summer** Great Lakes dune processes most resemble coastal dune processes in other parts of the world at this time. Surfaces dry quickly because of seasonally warmer temperatures (Ensign et al. 2006; van Dijk 2014). Although exposed sediments have the highest susceptibility to entrainment by wind at this time, wind energy is lower, reducing sand transport potential (Hansen et al. 2009; van Dijk 2014; Kilibarda and Kilibarda 2016). The spatial extent of exposed surfaces is also low (Fig. 3.13a) because vegetation height and leafiness are at their maximum (Byrne 1997; van Dijk 2014). This both shelters surfaces from wind action and reduces the availability of sand to the transport process. Average lake levels are typically highest in the summer, reducing beach widths and sand exposure as well as promoting wave erosion and scarping of the lakemost dune (van Dijk 2014).

**Autumn** Increasing storminess and strong wind events during autumn months provide the energy for putting exposed sand into motion even when moisture is present (van Dijk 2014; Kilibarda and Kilibarda 2016). Away from the beach, moisture depends on precipitation and proximity to the water table. On the beach, storm surges and higher waves cover large areas of the beach with water. At all locations, evaporation reduces surface moisture and observations of sand movement from very wet surfaces suggest that evaporation can be successful at freeing grains (Yurk et al. 2014; Kilibarda and Kilibarda 2016). The seasonal die-back of deciduous vegetation decreases plant height, leafiness, and percent cover (van Dijk 2014). Additionally, some vegetation is buried by windblown sand. Hence, sand tends to move further into vegetated areas (Fig. 3.13b) as the season progresses (Law and Davidson-Arnott 1990; Yurk et al. 2014; van Dijk 2014).

**Winter** Winter is comparable to autumn in strong wind events and storminess, but other factors can have significant effects on net winter sand transport (van Dijk 2014; Kilibarda and Kilibarda 2016). Lake levels are low in the winter and coastal ice (Fig. 3.13c) can fix the total beach width for periods of weeks to several months (Yurk et al. 2014; van Dijk 2014). Coastal ice also protects foredunes from wave erosion. Snow (Fig. 3.13c) also protects surfaces from wind erosion. However, this

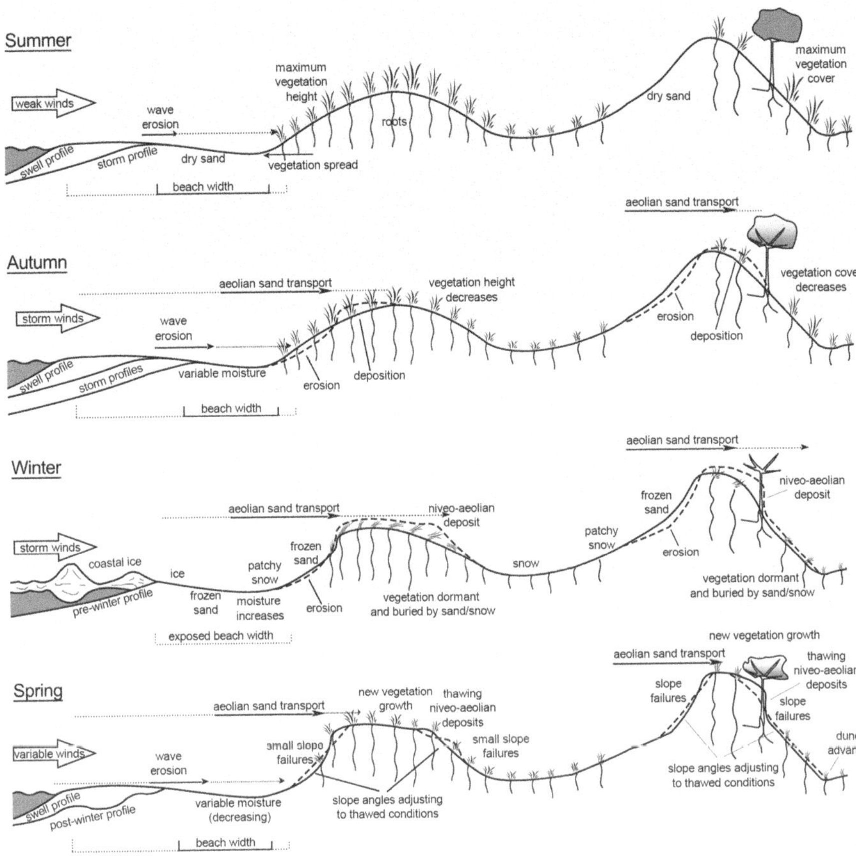


**Fig. 3.12** Examples of seasonal features, processes and conditions of Great Lakes dunes, as shown for a beach-foredune-blowout system. (Modified from van Dijk 2014)

protection is temporary as snow itself is susceptible to movement by wind. Thicker snow cover, wetter snow, more compact snow and older, denser, more metamorphosed snow are more effective at shielding surfaces from wind erosion (van Dijk 2014). Snowfalls and windblown snow deposits can bury vegetation and fill in depressions, enabling sand to travel further downwind.

Niveo-aeolian transport or deposition occurs when there is some combination, either sequential or simultaneous, of snow and sand transport by wind. The resulting deposits (Fig. 3.13c) contain temporal and spatial records of aeolian activity that can be accessed by shallow "coring" or digging a snow pit (Yurk et al. 2014). Layer thicknesses and ratios of sand to snow represent aeolian events (van Dijk 2004; Fisher et al. 2012) that can be traced back to specific time periods by comparison with local wind and precipitation records. Niveo-aeolian deposits tend to be thickest downwind of dune crests (Kilibarda and Kilibarda 2016), where deposits on large blowouts may exceed several meters in total depth (sand plus snow). Overall sand

**Fig. 3.13** Seasonal contrasts from a Lake Michigan coastal dune system. (**a**) In summer, vegetation is at its maximum cover and coastal processes (waves and changing lake levels) change beach width to increase or decrease sediment supply to the dunes. (**b**) In autumn, waves and precipitation add surface moisture, but windblown sand builds deposits in the edge zone of dune vegetation. (**c**) Winter conditions include coastal ice, surface ice, exposed frozen sand, snow cover, and niveo-aeolian deposits. The inset shows the niveo-aeolian layers in a snow pit. (**d**) In spring, melting snow and ice create temporary moist areas, and ephemeral features formed by sand deposits lowering in place are quickly erased by slumping and other slope processes

content and sand-to-snow ratios are greatest close to the source areas for sand transport (Yurk et al. 2014). During the winter, the deposits often maintain steeper slope angles than dry sand because freezing cements sand and snow together (Hansen et al. 2009; Yurk et al. 2014). These oversteepened slopes are prone to slope failure (Kilibarda and Kilibarda 2016).

Surface ice is a more substantial and persistent barrier to sand movement than snow because of the stronger bonding between the water molecules. Sources for ice include freezing rain (Kilibarda and Kilibarda 2016) or rainwater which later freezes at the surface, melting snow, and water from waves freezing onto the beach (van Dijk 2014). The nature of the ground is important as water will drain into sand as long as it is unsaturated. Frozen saturated ground is a barrier which keeps additional water at the surface (van Dijk 2014).

Ground freezing and pore ice can also be important to sand entrainment and transport by wind. If there is more than 0.5% moisture in frozen ground, the sand grains are cemented to each other. Increasing cementation occurs as moisture content increases (van Dijk and Law 2003). Pore-ice sublimation, in which water goes directly from solid to vapor, is effective at releasing sand grains to movement by wind (McKenna Neuman 1990; Law and van Dijk 1994; van Dijk and Law 2003). Grain-release rates by sublimation are increased by higher wind speeds, warmer temperatures, and lower relative humidity (van Dijk and Law 2003). The impacts of grains in motion may help to detach other grains (McKenna Neuman 1989; Kilibarda and Kilibarda 2016). A cemented surface causes saltating grains to bounce higher and travel further (McKenna Neuman 1989; Yurk et al. 2013) as the kinetic energy of the saltating grain is retained instead of being transferred to small movements of loose surface grains.

**Spring** Warming temperatures thaw the frozen ground and niveo-aeolian deposits, triggering mass movements that eventually adjust oversteepened slope angles to those more typical of dunes (Hansen et al. 2009). The 1–2+ m advances of large parabolic dune slipface edges in the spring are often associated with the gravity-driven downslope movement of sand originally deposited further upslope in vegetation or niveo-aeolian deposits in autumn and winter (Hansen et al. 2009; Yurk et al. 2014; Kilibarda and Kilibarda 2016). Early in the spring, frozen ground prevents meltwater from draining into the ground and the water-saturated surface sediments tend to flow or slide downslope on top of the frozen layer. The wet sand is less susceptible to movement by spring winds, which tend to greater variability in direction and an average decrease in energy compared to autumn and winter winds (van Dijk 2014; Kilibarda and Kilibarda 2016).

When snow and ice thaws in niveo-aeolian deposits, sand is concentrated and lowered to the surface below. Local differences in thawing rates result from variations in the albedo of the snow and sand surfaces and variations in the thermal gradient of the surface (a thicker surface sand layer insulates underlying snow from warming) (van Dijk 2014). Some buried snow/ice can persist well into April or May, providing a source of moisture for vegetation as it thaws (Ensign et al. 2006). As snow and ice disappear, the sand can take on unexpected, but ephemeral, forms (Fig. 3.13d), including wet pellets, layers suspended above the surface on vegetation, and layers with cracks and dimples (van Dijk 2004, 2014; Yurk et al. 2013).

### 3.6.3 Sedimentology of Great Lakes Dunes

Sources of sediment to the shores of the Great Lakes include the erosion of bedrock outcrops and bluffs made up of glacigenic sediments along the shoreline, erosion of modern and ancient lake bottom sediments, erosion and reworking of dune sediments, and sediment input from rivers and streams (Cressey 1922; Trask 1976; Loope and Arbogast 2000; Cioppa et al. 2010). Much of the Great Lakes basin is covered by glacial drift which is presumed to be the major source of stream sediment and the largest ultimate contributor of sediment overall (Hansen et al. 2011).

Regardless of source, as the sediment is eroded and transported, clay and silt fractions begin to be winnowed away. Once larger sediments reach the shoreline, they are redistributed by longshore drift. Sorting increases and grain size decreases in the direction of longshore drift (Trask 1976). At any one beach, grain size and mineralogy can vary seasonally if longshore currents change direction (Trask 1976) or as sand is moved onshore or offshore by storms or fair weather waves. Textural sorting continues within individual dune systems as sediments undergo aeolian transport from the beach to more distal parts of the dune complex. In a study of sediment textures in the dune complexes along the southern shore of Lake Michigan, Cressey (1922) found that sediments became progressively finer grained with distance away from the shoreline. He also found that sand in the dunes was more frosted and rounded in the dune complexes compared to the beach. However, studies in other parts of the world have shown that aeolian sediments are typically more angular than littoral or fluvial sediments. This is due to aeolian sediments experiencing greater amounts of fracturing during grain impacts in the air compared to water (Eamer et al. 2017). Dune sediments also vary in texture along the Great Lakes. On the southeastern shore of Lake Huron, individual samples from the surface of a blowout on an established dune ridge have coarse to medium sand as the modal grain size, accompanied by up to 30% fine sand (Byrne 1997). Medium sand is the modal grain size along northeastern (Lewis 1975) and southeastern Lake Michigan (Pettijohn 1931; Colgan et al. 2017) while fine sand is prevalent along southwestern Lake Michigan and can be predominant at other sites as well (Fig. 3.14) (Pettijohn 1931; Rimsnider 1958; Hansen et al. 2011).

One exception to the trend of decreasing grain size with distance from the beach is the occurrence of granules and gravel in dune settings. In high-perched dunes

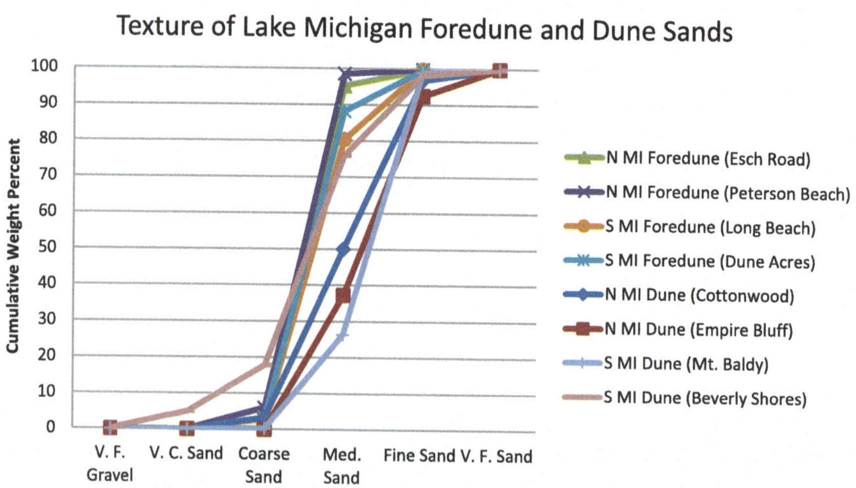


**Fig. 3.14** Particle size cumulative curve illustrating consistently finer texture of inland dune sands than foredune sands. *N* northern, *S* southern, *MI* Michigan, *VF* very fine, *VC* very coarse, *Med* medium

these can result from deflation of aeolian sediments exposing gravel in underlying glacial sediments. For example, lag deposits of coarse gravels to pebbles >2.5 cm in diameter are found in deflation basins and exposed bluff slopes within the Grand Sable Dunes along Lake Superior (Bergquist 1936; Marsh and Marsh 1987). Granules and gravels can also occur, however, on the upper stoss slopes of low-perched transgressive dunes. Kilibarda and Kilibarda (2016) attribute these coarser sediments to transportation during winter. Byrne (1997) describes seasonal variation in the coarseness of sand transported in a dunes complex on Lake Huron, where shifts in wind direction lead to changes in sediment supply and wind strength. Gentler summer winds winnow finer sands to leave a lag of coarser sands low on the dune slope. Stronger fall and winter storms then move this coarse sand to the upper slope.

The mineralogy of Great Lakes dune sediments partially reflects the mineralogy of their glacial sources. Tills throughout the Great Lakes region vary in mineralogy as a result of weathering, bedrock differences among source areas of different ice lobes, and differences in bedrock underlying the various routes of ice transport (Szabo 2006). Quartz, feldspar, and carbonate are usually the most abundant minerals. Heavy mineral studies demonstrate the presence of as many as 24 different mineral species depending on the source material (Gwyn and Dreimanis 1979; Dworkin et al. 1985). The mineral composition changes as sediment is moved to beaches and dunes (Pettijohn 1931). Trask (1976), for example, noted changes in composition down-drift in eastern Lake Ontario, and attributed a decrease in heavy mineral abundances to hydraulic lag. Dunes have orders of magnitude greater concentrations of heavy minerals in fine and very fine sand ranges, than source sediments (Schaetzl and Loope 2008; Hansen et al. 2011), a difference attributed to winnowing of less dense minerals from the smaller sand size classes.

Great Lakes dunes feature sedimentary structures common to many dune environments, such as climbing translatent strata, ripple-form laminae, ripple-foreset cross-laminae, sandflow cross strata, and planar grainfall laminae (Hunter 1977), as well as tabular planar, wedge planar, and trough cross-beds. Other structures are more prominent in some Great Lakes dunes. Intercalated with climbing translatent strata and sandflows are pin stripe laminations (Fryberger and Schenk 1988), thin, usually 1–3 mm, layers of very fine sand to silt-sized particles (Fig. 3.15a). These finer-grained fractions are enriched in denser, darker minerals, giving the pinstripes a noticeably dark color. The average spacing between pin stripe laminations varies from 4 to 18 cm, with a maximum spacing up to 100 cm. Closely-spaced pin stripes with even thicknesses are associated with climbing translatent strata, while more widely-spaced pin stripes that thin or pinch out laterally delineate sandflow strata. In very cold weather when some moisture is retained within the dune, pin stripe laminations freeze and stand out in relief while coarser-grained laminae crumble and remain flat or concave (Hansen et al. 2011; Kilibarda et al. 2014).

Several other sedimentary structures, more typical of cold climate dunes, develop in Great Lakes dunes during the late fall and winter. A variety of granule ripples (Fig. 3.15b), reaching heights of 3–15 cm and spacings of 20–160 cm, frequently develops near a dune's crest during winter. The coarse sand, granules, and pebbles

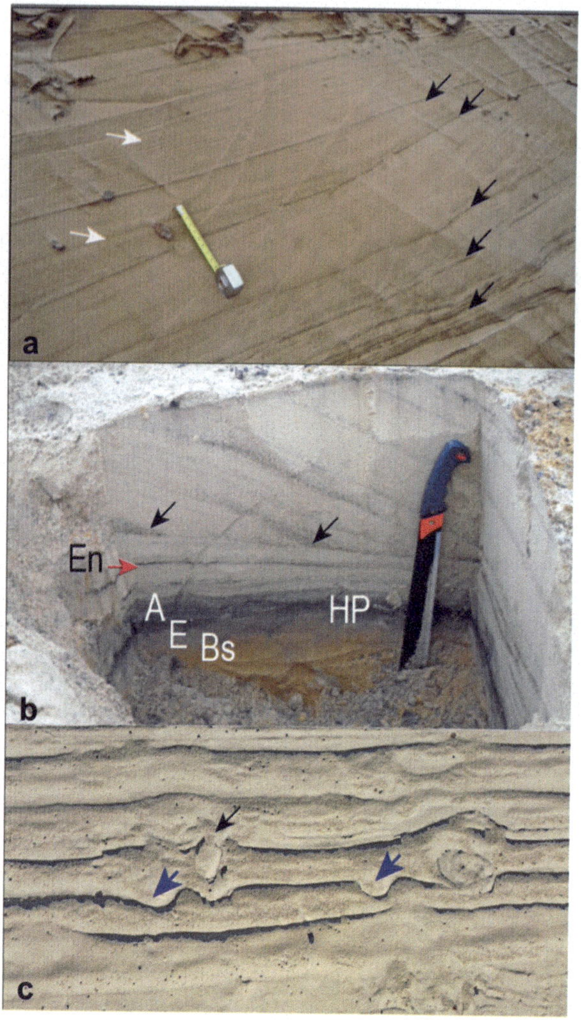


**Fig. 3.15** Sedimentary structures (**a**) White arrows point to wind ripple strata (climbing translatent strata) that are 3–5 mm thick and oriented perpendicular to the exposures. These occur within coset strata. Black arrows point to pinstripes that are parallel to wind ripple laminations and are made up of fine, mostly dark, grains. Tape measure is 25 cm long. (**b**) Darker colored sand (black arrows) is composed of very coarse sand and granules (predominantly lithic fragments of mudflows). They represent the bases (pinch-outs) of grainflows. They preserve a portion of granule ripples formed by strong winter winds. In this case, they overlie an Entisol (En) and the Holland Paleosol (HP), a spodosol containing dark A, ashy E, and orange $B_s$ horizons. (**c**) Deer tracks (blue arrows) preserved in wetted crust. Break-apart laminae (black arrow) in wetted crust

found in these granule ripples are derived from the frozen beach, because the dunes are predominantly comprised of medium to fine sand. Granules and gravel have a smaller relative surface area encased in ice and are more easily released by sublimation from the frozen surface than are finer particles. Frozen surfaces more readily facilitate saltation compared to loose sediment surfaces. Thus, coarse-grained laminae containing granules and pebbles are mainly the result of winter season niveo-aeolian deposition (Kilibarda and Kilibarda 2016). Crenulated lamination (Ruz and Allard 1995) is another niveo-aeolian sedimentary structure that develops during winter. Wind ripple laminations in a mixture of snow (60–90%) and sand (10–40%) can be buried by climbing translatent strata or sandflow strata. Subsequent thawing of trapped snow induces water saturation and the ensuing deformation results in crenulated lamination. Preservation potential of these niveo-aeolian structures is low.

Sand with wetted crusts (McKee et al. 1971) refers to laminated sand in which various layers have been wetted or dampened at the time of deposition or shortly after so that dry and wet laminae alternate. Wetted crusts may develop as a result of periodic dampening from dew, frost, or light rain, when silt-sized particles derived from the beach or exposed paleosols stick to the wet dune surface and make it very cohesive (Kilibarda and Kilibarda 2016). Cohesive wetted crusts are an excellent surface for preservation of bird, insect, lizard, and mammal tracks (Fig. 3.15c). Upon drying, the wetted crust will break up and potentially be preserved as break-apart laminae (Fig. 3.15c; McKee et al. 1971).

### 3.6.4 Ecological Communities

The varied Great Lakes dune geomorphologies host diverse ecological communities. The most abundant and well-studied ecosystem is the open dune community found on mobile foredunes, dune ridges, trough blowouts, and parabolic dunes. Cowles (1899) first studied this ecosystem in the area now preserved as the Indiana Dunes National Lakeshore and continued those studies along the eastern shore of Lake Michigan. Olson (1958b, d) and Maun (e.g., 1998, 2004) furthered Cowles' work, examining the interactions of these communities with sand transport/dune building. This ecosystem is characterized by the presence of migrating sand, dune-building grasses and herbaceous plants tolerant of sand burial (Peterson and Dersch 1981; Albert 1999). Two endemic plant species, Pitcher's thistle (*Cirsium pitcheri*) and Lake Huron tansy (*Tanacetum huronense*), grow in open dune and sandy/gravelly beach areas, respectively (Guire and Voss 1963). Table 3.2 provides an overview of the dominant pioneer plant species in the Great Lakes coastal dune complexes.

Older, stabilized dunes are often covered with a mesic forest comprised of *Acer rubrum* and *A. saccharum*, *Fagus grandifolia*, *Quercus rubra* and *Q. alba*, *Tsuga canadensis* and *Pinus strobus* (Olson 1958d; Cohen 2000, 2004). Unfortunately, insect pests such as the wooly adelgid (*Adelges tsugae*) and pathogens, including oak wilt (*Ceratocystis fagacearum*) and beech bark disease (*Cryptococcus fagisuga and Neonectria* spp.) have been noted at multiple locations along the Great Lakes coast, causing extensive tree fatalities which are anticipated to change the nature of this ecosystem over time. Excessive deer herbivory due to rising herd populations has decimated many of the herbaceous forbs along with seedling oak, hemlock, and maple populations (e.g., Parks Research Forum of Ontario 2001; Rooney and Waller 2003; Mudrak et al. 2009).

Interdunal wetlands, alternatively called wet pannes, palustrine sand plains, or slacks, are found within the dunes where the wind has scoured the sand to the water table. The hydrology of these areas can be complex as the water levels can be influenced by the Great Lakes and/or other hydrologic sources such as groundwater and precipitation and seasonal processes (evapotranspiration). Hence, the water levels in these features can fluctuate significantly within the growing season, and annually (Barko et al. 1977; Bissell 1993; Albert 2007; DeVries-Zimmerman et al. 2016).

**Table 3.2** Pioneer plants of the Great Lakes dunes

| |
|---|
| Lake Michigan – Indiana Dunes National Lakeshore (Hop et al. 2009); Wisconsin (Curtis 1959); Point Beach State Forest, Two Rivers, Wisconsin (van Denack 1961); (Hop et al. 2010a); Sleeping Bear (Thompson 1967; Hop et al. 2011); P.J. Hoffmaster State Park (Wells and Thompson 1983; Warners et al. 2005); Grand Mere (Wells and Thompson 1982) |
| *Ammophila breviligulata* |
| *Arctostaphylos uva-ursi* |
| *Artemisia campestris* |
| *Cakile edentula* |
| *Calamovilfa longifolia* |
| *Cirsium pitcheri* |
| *Elymus canadensis* |
| *Schizachyrium scoparium* |
| *Hudsonia tomentosa* |
| *Juniperus communis* |
| *Juniperus horizontalis* |
| *Lathyrus japonicus* var. *maritimus* |
| *Prunus pumila* |
| *Salix sp.* |
| Lake Superior – southern shore, Apostle Islands (Judziewicz and Koch 1993; Hop et al. 2010a, b); Grand Sable Dunes (Hop et al. 2010b) |
| *Ammophila breviligulata* |
| *Arctostaphylos uva-ursi* |
| *Artemisia campestris* |
| *Festuca saximontana* |
| *Juniperus communis* |
| *Juniperus horizontalis* |
| *Lathyrus japonicus* var. *maritimus* |
| *Lithospermum caroliniense* |
| *Oenothera oakesiana* |
| *Prunus pumila* |
| *Schizachyrium scoparium* |
| Lake Superior – northern shore (Bakowsky and Henson 2014) |
| *Ammophila breviligulata* |
| *Anemone multifida* |
| *Anthoxanthum hirtum* |
| *Arabidopsis lyrata* |
| *Artemisia campestris* ssp. *caudata* |
| *Elymus canadensis* |
| *Elymus trachycaulus* |
| *Lathyrus japonicus* |
| *Leymus mollis* |
| Northern Lake Huron – North Channel (Bakowsky and Henson 2014) |
| *Ammophila breviligulata* |

(continued)

**Table 3.2** (continued)

| |
|---|
| *Artemisia campestris* ssp. *caudata* |
| *Elymus canadensis* |
| *Elymus trachycaulus* |
| *Hesperostipa spartea* |
| *Lathyrus japonicus* |
| *Prunus pumila* |
| Southern Lake Huron – eastern shore and Manitoulin Island (Bakowsky and Henson 2014); Pinery Provincial Park, Grand Bend, Ontario (Baldwin and Maun 1983; Dech and Maun 2005)) |
| *Ammophila breviligulata* |
| *Andropogon gerardii* |
| *Artemisia campestris* ssp. *caudata* |
| *Calamovilfa longifolia* var. *magna* |
| *Elymus canadensis* |
| *Lithospermum caroliniense* |
| *Panicum virgatum* |
| *Schizachyrium scoparium* |
| Lake Erie – northern shore, Port Burwell (Iroquois Beach) Provincial Park (Zhang and Maun 1991) |
| *Cakile edentula* |
| *Cenchrus tribuloides* |
| *Elymus canadensis* |
| *Elymus lanceolatus* |
| *Melilotus alba* |
| *Panicum virgatum* |
| *Populus deltoides* |
| *Strophostyles helvola* |
| *Tussilago farfara* |
| Lake Erie – southern shore, Presque Isle, Pennsylvania (Bissell 1993); Headlands Dunes State Nature Preserve (Hicks 1933; ODNR 2017) |
| *Ammophila breviligulata* |
| *Artemisia campestris* ssp. *caudata* |
| *Cakile edentula* |
| *Elymus canadensis* |
| *Lathyrus japonicus* |
| *Panicum virgatum* |
| *Populus deltoides* |
| *Ptelea trifoliata* |
| *Sporobolus cryptandrus* |
| *Triplasis purpurea* |
| Lake Ontario – eastern shore, Sandy Pond Beach, New York (Bonanno 1992; Edinger et al. 2014) |
| *Ammophila breviligulata* |
| *Ammophila champlainensis* |

(continued)

**Table 3.2** (continued)

| |
|---|
| *Artemisia campestris* ssp. *caudata* |
| *Avenella flexuosa* |
| *Cornus amomum* |
| *Cornus sericea* |
| *Elymus canadensis* |
| *Euphorbia polygonifolia* |
| *Lathyrus japonicus* var. *maritimus* |
| *Populus deltoides* |
| *Prunus pumila* |
| *Salix cordata* |
| *Sporobolus cryptandrus* |

These areas tend to have a high biodiversity due to the many different species of sedges, rushes, and other wetland plants growing within them. However, the species present can vary markedly between growing seasons due to the changing water levels within the wetlands, sometimes shifting to open dune species when lower lake levels dry the wetlands (Keddy and Reznicek 1986; Bissell 1993; DeVries-Zimmerman et al. 2016).

Wooded dune and swale complexes which grow on the foredune plains often have different ecological communities across them due to the range in ridge ages. The ridges closest to the lake, and therefore the youngest ridges, tend to have early pioneer species similar to those found in the open dune community while older ridges further from the lake may be forested with *Pinus strobus*, *Quercus* sp., *Betula* sp., and other species (Lichter 1998; Comer and Albert 1993). The swales also show a range of ecological diversity from open wetlands in the younger swales near the lake edge to bogs to forested wetlands further from the lake in the older swales (Lichter 1998; Comer and Albert 1993).

## 3.7 Geomorphic History of the Great Lakes Dunes

### *3.7.1 Holocene Lake Level Change in the Laurentian Great Lakes*

The development and evolution of Great Lakes coastal dunes have taken place within the context of the changes in lake levels reviewed here. The late Pleistocene to modern lake level history of the Great Lakes can be divided into an early period dominated by the advance and retreat of the Laurentide Ice Sheet, a late Pleistocene to middle Holocene period of extremely low lake levels punctuated by the pass through of late proglacial lake floods, and a middle Holocene to present period influenced by glacial isostatic rebound and climate change. It is this later period that is relevant to the history of the coastal dunes.

The retreat of glacial ice from the Lake Huron basin exposed an outlet through the Nipissing-Mattawa lowland in Ontario, known as the North Bay outlet. Isostatic rebound of this sill caused water to backflow into Lake Superior, to rise in the confluent three lake basins, and to ultimately spillover at the Chicago, Illinois, and Port Huron, Michigan, outlets. Thompson et al. (2014) studied relict shorelines in embayments along the Lake Huron, Michigan, and Superior coasts to reconstruct mid to late Holocene lake levels in the combined lake basins. Recovered basal foreshore elevations from individual beach ridges and age models derived from radiocarbon and optically-stimulated luminescence (OSL) age determinations were used to produce a series of relative paleohydrographs (Thompson 1992; Thompson and Baedke 1997; Baedke and Thompson 2000; Thompson et al. 2011; Johnston et al. 2012). Combined and adjusted for isostatic rebound to a basal foreshore elevation of a shoreline at Fort Gratiot, Michigan, the data show the rise and fall of lake level from the last high stand of the upper Great Lakes, known as the Nipissing Phase of ancestral Lakes Huron, Michigan, and Superior.

The lake level reconstruction for Lakes Michigan and Huron shows lake level rising at a rate of ~0.7 cm/year to ~6.1 ka when the rate slowed (Fig. 3.16), apparently due to water debouching through the southern outlets of Lakes Huron and Michigan (Thompson et al. 2011). Erosional transgression was prevalent along the upper Great Lakes coastlines at this early rapid rate of rise. Lake levels continued to rise after 6.1 ka at a rate of ~0.3 cm/year. Depending on the sediment supply to reaches of the coast, shorelines at this time, such as the southern Lake Michigan shore, underwent a variety of shoreline behaviors from erosional and depositional transgression to aggradation and depositional regression (Fraser et al. 2012). The peak elevation of the Nipissing Phase was short-lived and lake level began to fall soon after the high was reached (Thompson et al. 2011, 2014). Over the next 800 years, lake level fell ~ 4 m at a rate of ~0.7 cm/year (Fig. 3.16). Along the coasts of the upper Great Lakes and depending on the available sediment supply, shorelines experiencing this fall underwent forced regression, developing beach ridges, to erosional regression, creating bluff-fronted terraces.

The fall from the Nipissing high established current outlet and shoreline configurations, long-term patterns of littoral drift and shoreline behavior, and groundwater and surface water flow patterns. Throughout the late Holocene, numerous embayments in the shorelines filled with foredune plain complexes (strand-plains) or developed a barrier beach across smaller reentrants. Relict shoreline studies have synthesized lake level change over the past 3.5 ka, showing continually falling lake levels at a rate to 5–7 cm/century (Thompson et al. 2009). However, several water level patterns emerge (Fig. 3.16), including high stands from 3.6 to 2.4 ka and from 1.8 to 1.2 ka. Superimposed on the overall fall and high stands are two quasi-periodic lake level fluctuations having periodicities of ~160 and 32 years in duration. The decadal fluctuation was instrumental in producing individual beach ridges, whereas the centennial fluctuation produced groups of beach ridges, showing an elevation rise and fall in each group. These patterns of lake level change are reflected in the historical record (Baedke and Thompson 2000).

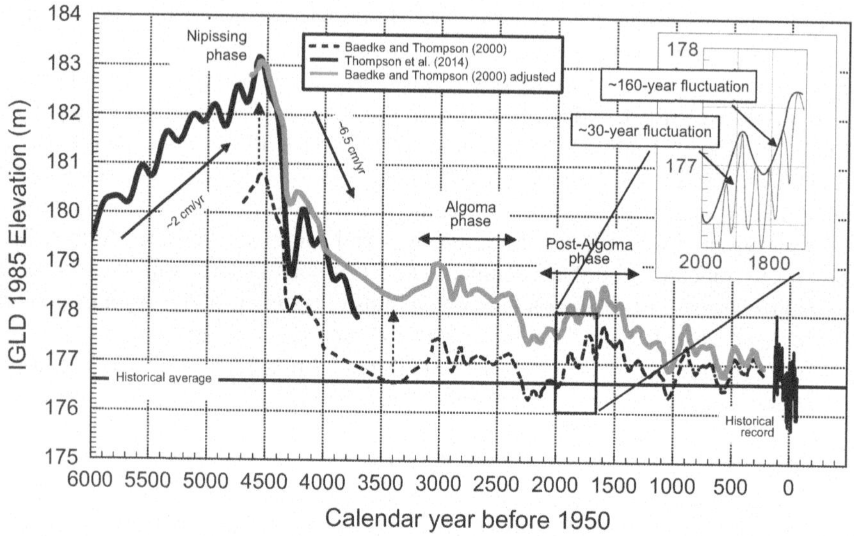


**Fig. 3.16** Paleohydrographs of the upper lake level limit for Lakes Michigan and Huron during the mid to late Holocene (Argyilan et al. 2018), utilizing data from Baedke and Thompson (2000) and Thompson et al. (2014)

### 3.7.2 *Chronology of Dune Growth and Migration*

In early work on the geomorphic history of Great Lakes coastal dunes, epitomized by Scott (1942) and his students (Tague 1946), the proximity of dunes to relict shoreline features, superposition and crosscutting relationships were used to construct relative chronologies of dune growth and migration. Radiocarbon analyses of organic material in paleosols exposed in eroding dune faces (Snyder 1985) and OSL dating of dune sands (e.g., Arbogast 2000; Hansen et al. 2010) enabled the construction of absolute chronologies. These chronologies were further refined by studies of variations in the amount of aeolian sand in the sediments in small lakes and wetlands in the lee of coastal dune fields (Fisher and Loope 2005; Timmons et al. 2007). These methods led to a resurgence of research on geomorphic dune history in the upper Great Lakes. With the exception of a few older ages on sinuous backdune ridges inland of the transgressive dune fields (Hansen et al. 2010), absolute ages from upper Great Lakes dune complexes tend to be younger than 6 ka and most are younger than the Nipissing high, ~4.5 ka (Fig. 3.17). Although dune growth may have been initiated during the rise to Nipissing high lake levels in many areas, the main coastal dune development in the upper Great Lakes occurred after the Nipissing high (Fig. 3.17).

In some dune complexes along the northern and eastern shores of Lake Huron, the inland edge is marked by a high dune ridge. Both Martini (1975) and Davidson-Arnott and Pyskir (1988) have suggested that the initial growth of this ridge occurred during the peak lake levels of the Nipissing transgression. Lakeward of this ridge are a series of younger dune ridges. In the complexes at Manitoulin Island, Wasaga and Sauble Beaches, it has been suggested that these younger ridges formed partly as the result of a relative decrease in lake level due to a combination of ongoing isostatic uplift relative to the Sarnia/Port Huron outlet and erosion of the St. Clair River channel (Martini 1975; Davidson-Arnott and Pyskir 1988). In these complexes, progradation due to sand supply from longshore sediment transport may have also played a role in the development of the younger dune ridges. Progradation was significant at the Pinery dune field at the southern end of the east shore of Lake Huron. The oldest dunes in this complex are dated at 4.8 ka (Lewis 1969) and occur at the eastern inland-margin of a foredune plain complex. The subsequent geomorphic history of this complex has been described by Eyles and Meulendyk (2012). The northern part of the Pinery dune field consists of a series of linear shore-parallel dune ridges 10 m or less in height. Dune ridges to the south are up to 20 m high and are interrupted by numerous blowouts and parabolic dunes. Dune stratigraphy revealed by ground penetrating radar suggests that this north-south contrast in dune form has occurred throughout the entire history of the complex. Eyles and Meulendyk (2012) attribute the differences in dune form to a southward increase in sand supply. In the foredune plain (i.e., strand-plain) complexes along Lake Michigan and Lake Superior studied by Baedke and Thompson (2000) and Johnston et al. (2012), the development of dune ridges after the Nipissing transgression was more or less continuous with new dune ridges forming during high stands of the quasi-periodic lake level cycles (Sect. 3.7).

The development of transgressive dune complexes occurred in distinct phases with millennial long periods of active dune growth punctuated by periods of relatively low aeolian activity. As Fig. 3.17 illustrates, in many places around the upper Great Lakes, a major period of dune growth and migration occurred during the drop in lake levels from the Nipissing high. Based on their studies along the southeastern shore of Lake Michigan, Hansen et al. (2010) suggested that broad dune fields developed at this time as the zone of active dune growth followed the migration of the shore lakeward. In southern Lake Michigan, dunes developed and migrated sub-parallel to the coastline, traversing through and over Nipissing foredunes and beach ridges and into landward wetland and fluvial areas (Argyilan et al. 2014). Lovis et al. (2012) found a somewhat different pattern in their studies of dunes along the northern and northeastern shores of Lake Michigan. In these areas, the first major period of dune building, beginning ~3.5 ka, lagged behind the drop from the Nipissing high. Lovis et al. (2012) proposed that sediment released during the erosion of shoreline features during the rise to the Nipissing high was stored offshore and was moved onshore and blown inland to form dunes only after the fall from the Nipissing high.

Throughout the upper Great Lakes, the initial period of dune growth in transgressive complexes was followed by a period of dune stability. Studies of some transgressive dune complexes along Lake Michigan (Fig. 3.17), western Lake Huron and southern Lake Superior have shown a resurgence of dune activity in the period between ~3–2 ka. By 1.8 ka, most transgressive dune complexes along the upper Great Lakes appear to have been stable (Fig. 3.17). Along the southern and

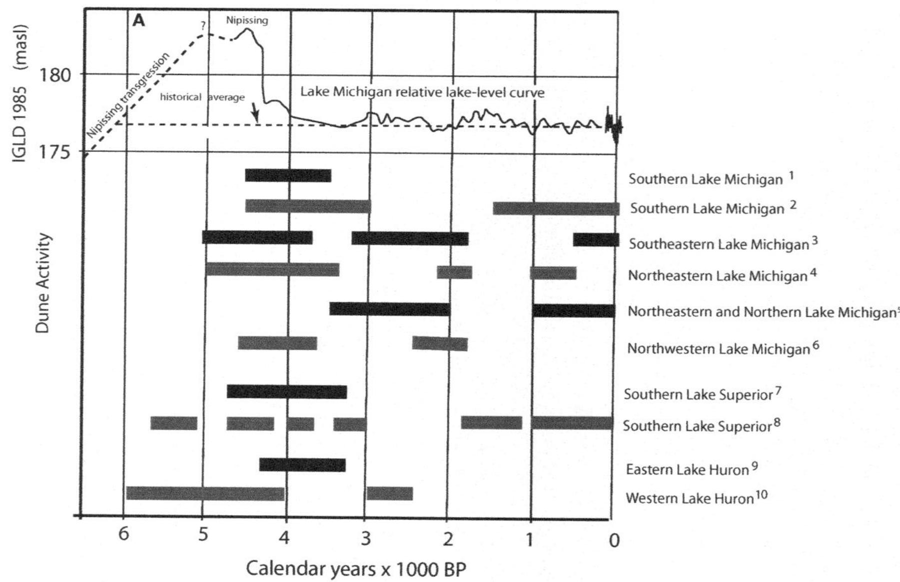


**Fig. 3.17** Periods of dune growth and mobility for transgressive and transitional dune fields at different localities along the upper Great Lakes (Superior, Michigan and Huron). The Lake Michigan lake level curve from Baedke and Thompson (2000) is plotted at the top of the figure. The bars in the lower portion of the figure give the age ranges for periods of dune growth and migration. Sources (keyed to the superscript following the location at the right of the figure): 1. Dune sand OSL ages from Argyilan et al. 2014; 2. Dune sand OSL ages from Kilibarda et al. 2014; 3. Dune sand OSL ages, radiocarbon paleosol ages, and sand concentrations in small lakes from Hansen et al. 2010; 4. Dune sand OSL ages from Blumer et al. 2012; 5. Dune sand OSL ages from Lovis et al. 2012; 6. Dune sand OSL ages from Rawling and Hanson 2014; 7. Dune sand OSL ages and radiocarbon paleosol ages from Arbogast 2000; 8. Radiocarbon paleosol ages from Anderton and Loope 1995; 10. Dune sand OSL ages from Arbogast et al. 2010

southeastern shore of Lake Michigan, this period of stability is marked by the development of a prominent spodic inceptisol known as the Holland Paleosol (Arbogast et al. 2004). There has been widespread dune mobility along the shores of the Great Lakes for at least the last 150 years as recorded in historical records, aerial photographs, and satellite images. It is tempting to attribute this renewed dune mobility to anthropogenic factors such as European settlement. However, in the northern portions of the Michigan basin, dune activity increased as early as 1 ka (Lovis et al. 2012) and subsequently decreased into the European settlement period. Along the southeastern shore of Lake Michigan, burial ages of the Holland Paleosol (Arbogast et al. 2004; Kilibarda et al. 2014) and the sedimentological record from small lakes (Timmons et al. 2007; DeVries-Zimmerman et al. 2014) also suggest that the increase in dune mobility predates the arrival of significant numbers of Europeans. The reason for the increase in dune activity within the last 1000 years remains one of the unsolved problems in the geomorphic history of Great Lakes coastal dunes.

Paleosol stratigraphy (Anderton and Loope 1995; Loope and Arbogast 2000), sand peaks in small lakes (Fisher and Loope 2005; Timmons et al. 2007), and OSL ages (Hansen et al. 2010) all suggest that longer periods of enhanced dune activity were broken up into smaller-scale periods of dune stability and mobility. Various attempts have been made to correlate these higher frequency periods of mobility and stability with the lake level curves determined from foredune plain complexes along the Great Lakes. These attempts are based on correlations between independent chronologies from radiocarbon or OSL ages with all of their inherent uncertainties. Despite this, a number of studies (Loope and Arbogast 2000; Loope et al. 2004; Fisher and Loope 2005; Fisher et al. 2007, 2012; Hanes et al. 2014) have concluded that alternating periods of dune mobility and stability coincide with the quasi-periodic lake level cycles recognized by Baedke and Thompson (2000). Working in a high-perched dune complex along the southern shore of Lake Superior, Anderton and Loope (1995) suggested that erosion of the bluffs during high lake levels exposed sand that was picked up by the wind and moved inland, initiating a period of dune building and mobility. This model was applied to dunes along Lake Michigan in general by Loope and Arbogast (2000) and Fisher and Loope (2005). Studies of the geomorphic history of dunes along European coasts (Wilson et al. 2001, 2004; Clarke et al. 2002; Clarke and Rendell 2009; Clemmensen et al. 2009; Costas et al. 2012, 2016; González-Villanueva et al. 2013), South America (Björck et al. 2012), and Alaska (Mason and Jordan 1993; Mason et al. 1997) have suggested that an increase in storminess is a major factor in increased aeolian activity. In the Great Lakes, periods of enhanced storminess correlate with periods of higher lake levels (Meadows et al. 1997), suggesting that both could be at play in triggering periods of enhanced dune mobility.

Most of what we know about the geomorphic history of foredune plain complexes in the lower Great Lakes (Erie and Ontario) originates from studies of the development and evolution of the cuspate forelands and spits on which they occur, such as Long Point (Davidson-Arnott and Conliffe Reid 1994), Pointe-aux-Pins (Coakley 1989), Long Point (Coakley 1976), and Presque Isle (Foyle and Norton 2006) on Lake Erie and Sandbanks on Lake Ontario (Martini 1981). All of these structures have antecedents in features that formed during the glacial or early post glacial history of the region. However, the consensus appears to be that their modern development began with the fall of lake levels (Coakley 1976, 1989; Davidson-Arnott and Conliffe Reid 1994; Foyle and Norton 2006) from a peak around 4 ka (Holcombe et al. 2003). The parallel dune ridges in the foredune plains mark the successive shoreline positions in the last 3–4 ka years during the growth and migration of the sedimentary platforms on which they sit.

## 3.8 Dune Management in the Great Lakes

### *3.8.1 Pressures*

Coastal dunes and ecosystems face a number of pressures due to human impact. These areas are very desirable for residential and commercial development, and the type and extent of development has spurred many lawsuits between builders/developers and regulatory agencies. Although controversial, commercial sand mining of the dunes has been done in some areas since the early 1900s. Along many coastal areas, a reduced alongshore movement of sand due to the construction of seawalls, harbor control structures, groins, or other structures, has changed the rates of dune growth and migration and increased coastal erosion (e.g., Bissell 1993; Kilibarda et al. 2014). Dams on streams can significantly impact the sediment supply to the Great Lakes and, hence, to dunes on these lakes. For example, dams in the St. Joseph River watershed have reduced the sediment flow from this watershed to Lake Michigan by as much as 80% (Nairn et al. 2006). Beach nourishment efforts in areas of high coastal erosion have had limited success as the sediments used are often not the same as the dune-building sands (Bissell 1993; Yurk et al. 2014). Heavy recreational use, including both foot traffic and off road vehicles, can destroy habitat, preventing it from becoming re-established (Bowles and Maun 1982; Peach 2006). Invasive species have successfully colonized areas of the dunes, often outcompeting the native species and replacing them (e.g., Swearingen and Bargeron 2016). Lastly, increasing deer herds have significantly overgrazed vegetation in many coastal dune ecosystems, sometimes eliminating new seedlings and other plants (e.g., Bissell 1993; Parks Research Forum of Ontario 2001; Rooney and Waller 2003; Mudrak et al. 2009). Deer culls have been implemented in selected areas to reduce the population and to allow the recruitment of new seedlings (e.g., Parks Research Forum of Ontario 2001).

### *3.8.2 U.S. Coastal Dune Management*

Despite the common issues facing the Great Lakes coastal dunes, there is no bilateral United States/Canadian agreement or approach to managing these features. U.S. management can be divided into two general categories: legislative/regulatory and conservation/preservation by either a governmental entity (federal, state or local park, etc.) or non-governmental organization (NGOs). No federal laws directly regulate coastal dunes and/or activities associated with them. However, coastal dune wetlands can be subject to Section 404, Dredge and Fill Permitting under the Federal Water Pollution Control Act of 1972, commonly called the Clean Water Act. Michigan is the only Great Lakes state with legislation directly regulating coastal sand dunes, Part 353 Sand Dunes Protection and Management, and Part 637 Sand Dune Mining of the Natural Resources and Environmental Protection Action, Public Act 451 of 1994. This legislation was originally enacted as the Sand Dunes Protection and Management Act, Public Act 222 of 1976 to regulate coastal sand dune mining. The Act was amended in 1989 to include a definition of "Critical Dune Areas" (CDAs) within which development would be regulated through a permit process. The statute was amended in 2012 to facilitate certain development activities within CDAs. New York regulates dunes through the Coastal Erosion Hazard Act to ensure these areas are not destabilized, thereby decreasing their ability to provide shoreline erosion and wildlife habitat/ecosystem protection. The Act defines two types of coastal erosion hazard areas (CEHA): natural protection feature areas and structural hazard areas. CEHA have been mapped and activities within these areas are regulated through a permitting process.

A patchwork of governmental agencies owns and preserves coastal dune areas along the U.S. side of the Great Lakes. The federal government, through the National Park Service, maintains four National Lakeshore areas with dune systems, Indiana Dunes National Lakeshore and Sleeping Bear Dunes National Lakeshore on the southern and northeastern shores of Lake Michigan, respectively, and Pictured Rocks and Apostle Islands National Lakeshores on Lake Superior's southern and southwestern shores, respectively. Coastal dune areas are also managed as state parks. Many of the dune systems in Wisconsin along Lake Michigan have been destroyed by urban and residential development. Two of the most extensive remaining dune systems are preserved in the Point Beach and Kohler-Andrae Dunes State Parks where they face heavy recreational pressures. The Headlands Dunes State Nature Preserve in Ohio contains one of the few remnants of the Lake Erie dunes community within that state. Michigan has the largest acreage of coastal dunes owned and managed as state, county, and local parks as it has the longest coastline containing coastal dunes. Historically, there has been no overall management approach for conserving ecosystems in these parks as the focus has been more on recreation opportunities. However, during the 1930s–1940s, drought conditions resulted in large sand advances damaging cropland and threatening infrastructure. Dune fixation projects, including the planting of non-native species on the open dunes, were undertaken to stabilize many dune systems (Kroodsma 1937; Lehotsky 1941). Similar programs were also implemented in Ontario (Davidson 1991). The result of these efforts was a decrease in native open dune ecosystems and biodiversity (Leege and Murphy 2001). Consequently, efforts to remove these non-native species from public lands have been undertaken (e.g., Leege and Kilgore 2014). However, removal and/or management of invasive species, both intentionally planted or inadvertently introduced, remain an ongoing problem for many coastal dune systems.

Dune areas are also owned and managed, primarily for conservation purposes, by a number of NGOs, including local land conservancies, land trusts and international conservation groups such as The Nature Conservancy. Partnerships between governmental agencies and NGOs are also used to conserve coastal dune areas such as the Eastern Lake Ontario Dunes in New York State.

### *3.8.3 Canadian Coastal Dune Management*

The Canadian shoreline of the Great Lakes is entirely contained within the Province of Ontario, simplifying the regulatory regime. While the Canadian Federal Government has jurisdiction over the lakebed, responsibility for shoreline management is provincial. Following the extensive shoreline losses associated with high lake levels in 1985–1987, the Ontario Government enacted a Shoreline Management Policy primarily geared towards addressing hazards associated with coastal erosion, flooding, and dynamic beaches. This policy is administered by 36 Conservation Authorities (CAs) whose boundaries were adjusted to accommodate this new task, and to account for natural boundaries associated with littoral cells. In areas lacking CAs, primarily on the east coast of Georgian Bay and large areas of Lake Superior, the Policy is administered by regional offices of the Ontario Ministry of Natural Resources and Forests (MNRF). An updated Provincial Policy Statement with new policy directions pertaining directly to public shoreline access, mitigating existing hazards, and preventing new hazards was enacted in 2014. It also recognized the need to address the environmental impacts of shoreline development and to account for the potential impacts of climate change.

The identification of "dynamic beaches" as a separate category explicitly recognized the processes associated with beach/dune interaction and sought to direct all development away from the zone where this occurs. Demarcation of the dynamic beach is based on the 100-year-flood event (combination of still water level and dynamic increase due to storm surge), a 15 m allowance for wave run-up and other related hazards such as ice piling, and a 30 m allowance for the dynamic zone associated with wave erosion during storms and dune accretion in intervening periods (Fig. 3.18). Mapping of the hazard zone and setbacks is performed by the CAs and incorporated in their Shoreline Management Plan. Individual CAs can adjust the landward boundary and the implementation guidelines encourage the setback to be beyond the base of the lee slope of the foredune. The Implementation of the Provincial Policy is guided by a detailed set of Technical Guidelines (OMNR 2001a) and by a more general guide on the nature of shoreline hazards (OMNR 2001b). Additional support for coastal dune protection is provided by a section of the Policy Statement which prohibits development and site alteration in significant coastal wetlands in addition to the dynamic beach zone. This is especially pertinent to coastal dune systems associated with baymouth barriers and spits.

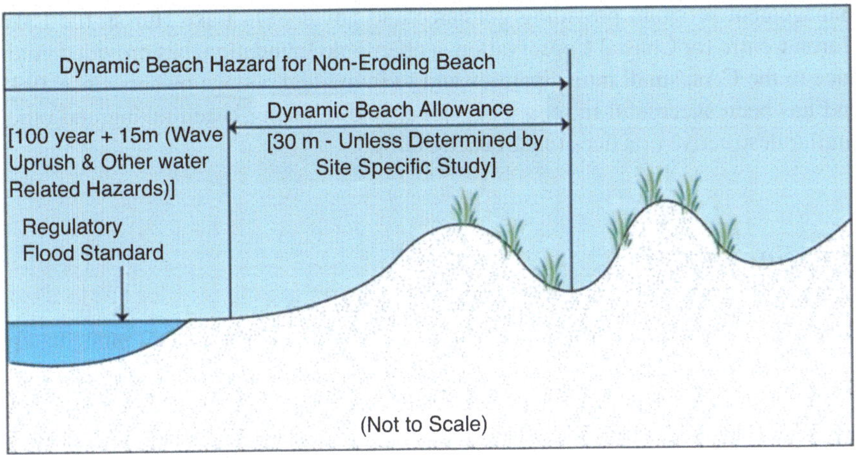


**Fig. 3.18** Schematic showing the setback associated with the Dynamic Beach Allowance for a non-eroding beach in Ontario. Where there is long-term erosion, an additional setback equal to 100 × the average annual recession rate is applied (OMNR 2001b)

While the policy applies to all sandy beaches with foredunes along the Great Lakes, in practice there are a number of limitations to the protection provided to dunes. Enforcement of the Policy varies with the resources of individual CAs, in particular the availability of staff and the degree to which municipalities within the CA have 'signed on' to the significance of this. In larger urban areas, the front line work is carried out by planners and by city staff. In areas where there is no CA, staff in the MNRF regional offices seldom have the training or time to enforce regulations. Most important, where there is existing encroachment on the foredune, much of it is now grandfathered as the regulations apply primarily to new development. Finally, the primary focus of the early policy was on hazard management and enforcement of regulations related to buildings. However, the recent updates provide more guidance in terms of environmental protection and there is now greater concern about managing activities within the foredune area.

The Provincial Policy applies only to the active foredune. Dune fields landward of this are not protected. However, portions of many important dune fields are incorporated within areas managed by various government agencies. Point Pelee on Lake Erie is a National Park. Long Point spit, the largest dune system in Ontario, is a UN Biosphere Reserve. Most of the central and distal zone of the spit is managed privately and by the Canadian Wildlife Service and is not accessible to the general public. Other smaller areas of the spit are managed by the Canadian Wildlife Service and Long Point Provincial Park. Provincial Parks manage some or all of other major dune areas such as Wasaga Beach on Georgian Bay, Pinery on Lake Huron, Rondeau on Lake Erie, and Presqu′ile and Sandbanks on Lake Ontario. Other smaller foredune systems are found within provincial and municipal parks and have varying levels of management. A major limitation for dune management within towns and cottage communities has been the absence of any coordinated conservation and management program providing guidance and advice. On Lake Huron, the Lake Huron Centre for Coastal Conservation, a charitable foundation, has provided guidance to the CAs, small municipalities and to individual cottage owners since 1995 and has been successful in rebuilding dunes and reducing beach raking and other similar destructive practices (e.g., Peach 2008).

## 3.9 Summary

Great Lakes coastal dune complexes can be broadly divided into dunes immediately adjacent to the shore and dunes further inland. The dunes adjacent to the shore are foredune complexes consisting of incipient foredunes which merge into incipient foredune ridges. Incipient foredunes and foredune ridges develop and grow during periods of lower lake levels when beaches are wide and sand supply is high. They are eroded, and sometimes completely destroyed by wave erosion during periods of high lake levels. Incipient foredunes and foredune ridges occur at the inland margins of beaches throughout the Great Lakes. In some foredune complexes an established foredune ridge occurs along the inland edge of the foredune dune complex. These ridges generally persist through periods of high lake levels, and are taller than incipient foredune ridges.

In contrast to the ephemeral foredune complexes, dune complexes somewhat further inland are not subject to destruction by wave erosion and, hence, have been characterized as permanent dune complexes. These can be further subdivided into foredune plain complexes, transgressive complexes and transitional complexes. Foredune plain complexes consist of multiple sets of parallel dune ridges separated by swales. They form on prograding sedimentary platforms in embayments that capture sediment, cuspate forelands, and spits. Each ridge in these complexes marks the position of the shore at different stages in the growth of the sedimentary platform. The formation of foredune plain complexes requires a positive sedimentary budget combined with some onshore aeolian sand drift potential. These conditions occur sporadically along many Great Lakes shores and, as a consequence, foredune plain complexes can be found in isolated patches along most of the shores of the Great Lakes. Transgressive dune complexes migrate inland over older geomorphic surfaces and typically consist of blowouts, parabolic dunes and dune ridges. In the Great Lakes region, these are generally subdivided into high-perched and low-perched transgressive dune complexes. High-perched transgressive dune complexes sit on top of coastal bluffs usually composed of glacial sediments. High-perched transgressive dune fields appear to form and grow during periods of high lake levels as wave erosion undercuts the shoreward edge of the bluffs on which they sit. This undercutting triggers slumps, exposing sediment along the bluff which is then picked up by wind and moved inland. Low-perched transgressive complexes lie on geomorphic surfaces close to current lake levels. Many dunes in low-perched transgressive complexes appear to have originated by the migration of blowouts that developed in established foredune ridges or ridges in foredune plain complexes.

Transitional dune complexes are in the earlier stages of this process and, hence, combine the features of transgressive complexes with the features of foredune or foredune plain complexes. Transgressive and transitional dune complexes occur along shores with both abundant sand supply and strong onshore aeolian sand drift potential. They are particularly well developed along the eastern shore of Lake Michigan, but also occur along the shores of Lake Huron, the southern shore of Lake Superior and the eastern shore of Lake Ontario.

Tidal range in the Great Lakes is typically less than 10 cm. Hence, tides have a negligible effect on coastal dunes. However, lake level changes linked to other processes have had a significant influence on the development of coastal dunes. Over time scales of centuries to millennia, changes in the elevations of the lake surface of many meters can be caused by changes in the elevations of their outlets caused by isostatic uplift or erosion. The formation of most of the transgressive and foredune plain complexes along the upper Great Lakes (Lakes Superior, Michigan, and Huron) began with the fall in lake levels from the Nipissing high from 5 to 3.5 ka. In a similar fashion, the growth of dune complexes along Lake Ontario and Erie began with a fall in lake levels from a peak approximately 4 ka. Variations in the balance between precipitation and evaporation drive changes in lake level on time scales that range from seasons to centuries. Paleohydrographs for the upper Great Lakes indicate meter-scale quasi-periodic lake level cycles with periods of ~160 and 32 years. Transgressive dune complexes along the upper Great Lakes have undergone centuries-long periods of stability punctuated by centuries-long periods of growth and migration. These longer periods of enhanced dune activity were broken up into smaller-scale periods of dune stability and mobility which some studies suggest are correlated with the quasi-periodic lake level cycles. The decadal scale lake level cycles are responsible for the cyclic growth and destruction of foredunes.

The Great Lakes region has a continental climate with strong seasonal variations in weather. As a result of seasonal changes in the balance between precipitation and evaporation, lake levels in the summer tend to be from 15 to 60 cm higher than lake levels in the winter. However, changes in lake levels are not the only seasonal factors that affect the dunes. The extent of vegetation on dunes is greater in the summer while storminess, together with wind and water energy, is lower than in other seasons. Although the base of foredunes may be scarped by wave erosion, aeolian transport is relatively low in the summer. Increasing storminess and dieback of vegetation cause an increase in aeolian transport during the autumn. During a typical winter, the buildup of ice along the shore prevents waves from reaching the beach. Surface and pore ice limit the amount of aeolian transport from bare sand surfaces. When sand is transported and deposited, it can be mixed with snow, forming niveo-aeolian deposits. Where they accumulate near the top of dunes, these niveo-aeolian surfaces often have slopes greater than the angle of repose for pure dry sand. During the spring thaw, these niveo-aeolian deposits collapse, triggering mass movements which deliver sand to the lower dune slopes. During the late spring, the extent of vegetation increases while storminess decreases, bringing the dunes back to their summer condition.

The majority of strong wind events in the Great Lakes regions are associated with the passages of extratropical cyclones. Aeolian transportation, especially during the dry portions of these storms, can be high and has a large impact on the coastal dunes. Wind directions depend on the position of a dune complex relative to the position of the storm and will change during the passage of the cyclone. When combined with the effects of local topographic steering, this can cause the transient direction of local sand transporting winds to depart significantly from the regional prevailing wind direction. Wave erosion at the base of cliffs and established foredune ridges can expose sediment to aeolian transport and trigger dune mobility. Higher lake levels and enhanced wave erosion are correlated with increased storminess in the Great Lakes. Increased wind energy during periods of high storminess may also play a role in increasing dune mobility. Thus, dune mobility may be a proxy for relative levels of storminess in the Great Lakes region.

Mobile and recently mobile dune systems in the Great Lakes host open dune ecological communities consisting of dune-building grasses and herbaceous plants tolerant of sand burial. Mesic forests develop on dune surfaces that have been stable for longer periods of time. Wetland communities occur in depressions in low-perched transgressive as well as swales in foredune plains. Changes in groundwater elevations tied to changes in lake levels cause quasi-periodic changes in the makeup of these wetland communities. The presence of different ecological communities in close proximity to each other leads to a diverse flora and fauna in Great Lakes coastal dune complexes. These communities are threatened by a spectrum of human activities, including residential development, sand mining, interruption of sand supply by coastal and inland structures and the introduction of invasive species. Attempts to manage these threats are complicated by the range of governmental entities involved in Great Lakes coastal management. The Great Lakes straddle the international border between Canada and the United States. All of the Canadian Great Lakes shoreline occurs in the province of Ontario which is responsible for coastal management. However, U.S. Great Lakes shorelines occur in seven different states. The federal government has only a limited role in Great Lakes coastal management, leaving most of the responsibility to the individual states. At this time, there is no coherent strategy for Great Lakes coastal dune management that crosses state or international boundaries.

## References

Albert DA (1999) Natural community abstract for open dunes. In: Species and community abstracts. Michigan Natural Features Inventory. http://mnfi.anr.msu.edu/abstracts/ecology/open_dunes.pdf. Accessed June 2017

Albert DA (2007) Natural community abstract for interdunal wetland. In: Species and community abstracts. Michigan Natural Features Inventory. https://mnfi.anr.msu.edu/abstracts/ecology/Interdunal_wetland.pdf. Accessed June 2017

Anderson JL, Walker IJ (2006) Airflow and sand transport variations within a backshore-parabolic dune plain complex: NE Graham Island, British Columbia, Canada. Geomorphology 77:17–34

Anderton JB, Loope WL (1995) Buried soils in a perched dunefield as indicators of Late Holocene lake-level change in the Lake Superior Basin. Quat Res 44:190–199

Angel JR (1996) Cyclone climatology of the Great Lakes. Illinois State Water Survey, Miscellaneous Publication 172, 122 pp

Angel JR, Isard SA (1998) The frequency and intensity of Great Lake cyclones. J Clim 11:61–71

Arbogast AF (2000) Estimating the time since final stabilization of a perched dune field along Lake Superior. Prof Geogr 52:594–606

Arbogast AF (2009) Sand dunes. In: Schaetzl RJ, Darden JT, Brandt D (eds) Michigan geography and geology. Pearson Custom Publishing, Boston, pp 274–287

Arbogast AF, Schaetzl RJ, Hupy JP, Hansen EC (2004) The Holland Paleosol: an informal pedostratigraphic unit in the coastal dunes of southeastern Lake Michigan. Can J Earth Sci 41:1385–1400

Arbogast AF, Shortridge AM, Bigsby M (2009) Volumetric estimates of coastal sand dunes in lower Michigan: explaining the geography of dune fields. Phys Geogr 30:479–500

Arbogast AF, Bigsby ME, DeVisser ME, Langley SA, Hanson PR, Daly TA, Young AR (2010) Reconstructing the age of coastal sand dunes along the northwestern shore of Lake Huron in Lower Michigan: Paleoenvironmental implications and regional comparisons. Aeolian Res 2:83–92

Argyilan EP, Lepper K, Thompson TA (2014) Late Holocene coastal development along the southern shore of Lake Michigan determined by strategic dating of stabilized parabolic dunes and wetlands of the Tolleston Beach. In: Fisher TG, Hansen EC (eds) Coastline and dune evolution along the Great Lakes, Geological Society of America special paper 508. The Geological Society of America, Boulder, pp 31–46

Argyilan EP, Johnston JW, Lepper K, Monaghan GW, Thompson TA (2018) Lake level and dune behavior along the Indiana souther shore of Lake Michigan. In: Florea LJ (ed) Ancient oceans, orogenic uplifts and glacial ice: geological crossroads in America's Heartland, Geological Society of America Field Guide 51. Geological Society of America, Boulder, pp 181–203

Baedke S, Thompson TA (2000) A 4,700 year record of lake level and isostasy for Lake Michigan. J Great Lakes Res 26:416–426

Bagnold RA (1954) The physics of blown sand and desert dunes. Methuen, London

Bakowsky WD, Henson BL (2014) Rare communities of Ontario: freshwater coastal dunes. Natural Heritage Information Centre, Ontario Ministry of Natural Resources

Baldwin KA, Maun MA (1983) Microenvironment of Lake Huron sand dunes. Can J Bot 61:241–255

Barko JW, Murphy PG, Wetzel RG (1977) An investigation of primary production and ecosystem metabolism in a Lake Michigan dune pond. Arch Hydrobiol 81:155–187

Bengtsson L, Hodges KI, Roeckner E (2006) Storm tracks and climate change. J Clim 19:3518–3543

Bergquist SG (1936) The Grand Sable Dunes on Lake Superior, Alger County, Michigan. Mich Acad Sci Arts Lett Pap 21:429–438

Bissell JK (1993) Rare plants and rare plant communities of Presque Isle. Bartonia 57(Supplement):2–8

Björck S, Clemmensen LB (2004) Aeolian sediment in raised bogs deposits, Halland SW Sweden: a new proxy of Holocene winter storminess variation in southern Scandinavia? The Holocene 14:677–688

Björck S, Rundgren M, Ljung K, Unkel I, Wallin A (2012) Multiproxy analyses of a peat bog on Isla de los Estados, earternmost Tierra del Fuego: a unique record of the variable Southern Hemisphere Westerlies since the last deglaciation. Quat Sci Rev 42:1–14

Blumer BE, Arbogast AF, Forman SL (2012) The OSL chronology of eolian sand deposition in a perched dune field along the northwestern shore of Lower Lake Michigan. Quat Res 77:445–455

Bodenbender BE, Hansen EC, Yurk BP, DeVries-Zimmerman S, Stid JT (2018) Sand deposition by grainfall beyond the lee slope of a large parabolic dune. Poster presented at the Geological Society of America annual meeting, Indianapolis, Indiana, USA, 4–7 2018

Boerboom T Minnesota geological survey, personal communication 2 August 2017
Bonanno SE (1992) Vegetation of a Lake Ontario dune barrier, Oswego and Jefferson Counties, NY, under high and low recreation pressure. MS Thesis, SUNY College of Environmental Science and Forestry, Syracuse, New York
Bowles JM, Maun MA (1982) A study of the effects of trampling on the vegetation of Lake Huron sand dunes at Pinery Provincial Park. Biol Conserv 24:273–283
Buckler WR (1979) Dune type inventory and barrier dune classification study of Michigan's Lake Michigan shore, Michigan geological survey report of investigation 23. Michigan Department of Natural Resources, Lansing
Byrne M-L (1997) Seasonal sand transport through a trough blowout at Pinery Provincial Park, Ontario. Can J Earth Sci 34:1460–1466
Cadwell DH, Pair DL (1991) Surficial geologic map of New York: Adirondack sheet, Surficial geologic map of New York, map and chart series #40. New York State Museum, New York State Geological Survey, Albany, New York
Calver JL (1946) The glacial and post-glacial history of the Platte and Crystal Lake depressions, Benzie County, Michigan, Publication 45, geological series 38, part II. Michigan Geological Survey, Lansing
Cioppa MT, Porter NJ, Trenhaile AS, Igokwe B, Vickers J (2010) Beach sediment magnetism and sources: Lake Erie, Ontario, Canada. J Great Lakes Res 36:674–685
Clarke ML, Rendell HM (2009) The impact of North Atlantic storminess on western European coasts: a review. Quat Int 195:31–34
Clarke M, Rendell H, Tastet JP, Clave B, Masse L (2002) Late-Holocene sand invasion and North Atlantic storminess along the Aquitaine coast, Southwest France. The Holocene 12:231–238
Clemmensen LB, Murray A, Heinemeier J, de Jong R (2009) The evolution of Holocene coastal dunefields, Jutland, Denmark: a record of climate change over the past 5000 years. Geomorphology 105:303–313
Coakley JP (1976) The formation and evolution of Point Pelee, Western Lake Erie. Can J Earth Sci 13:136–144
Coakley JP (1989) The origin and evolution of a complex cuspate foreland: Pointe-aux-Pins, Lake Erie, Ontario. Géog Phys Quatern 43:65–76
Cohen JG (2000) Natural community abstract for mesic northern forest. In: Species and community abstracts. Michigan Natural Features Inventory, http://mnfi.anr.msu.edu/abstracts/ecology/mesic_northern_forest.pdf. Accessed 8 June 2017
Cohen JG (2004) Natural community abstract for mesic southern forest. In: Species and community abstracts. Michigan Natural Features Inventory, https://mnfi.anr.msu.edu/abstracts/ecology/Mesic_southern_forest.pdf. Accessed 8 June 2017
Colgan PM, Amidon WH, Thurkettle SA (2017) Inland dunes on the abandoned bed of Glacial Lake Chicago indicate eolian activity during the Pleistocene-Holocene transition, southwestern Michigan, USA. Quat Res 87:66–81
Comer PJ, Albert DA (1993) A survey of wooded dune and swale complexes in Michigan. Michigan Natural Features Inventory report to Michigan department of natural resources, Land and Water Management Division, Coastal Zone Management Program. 159 pp
Costas S, Jerez S, Trigo RM, Goble R, Rebêlo L (2012) Sand invasion along the Portuguese coast forced by westerly shifts during cold climate events. Quat Sci Rev 42:15–28
Costas S, Naughton F, Goble R, Renssen H (2016) Windiness spells in SW Europe since the last glacial maximum. Earth Planet Sci Lett 436:82–92
Cowles HC (1899) The ecological relations of the vegetation on the sand dunes of Lake Michigan. Part I. Geographical relations of the dune floras. Bot Gaz 27:95–117, 167–202, 281–308, 361–391
Cressey GB (1922) Notes on the sand dunes of northwestern Indiana. J Geol 30:248–251
Curtis JT (1959) The vegetation of Wisconsin; an ordination of plant communities. University of Wisconsin Press, Madison

Davidson RJ (1991) Protecting and managing Great Lakes coastal dunes in Ontario. In: Davidson-Arnott RGD (ed) Proceedings of the symposium on coastal sand dunes. National Research Council of Canada, pp 455–471

Davidson-Arnott RGD, Conliffe Reid HE (1994) Sedimentary processes and the evolution of the distal bayside of Long Point, Lake Erie. Can J Earth Sci 31:1461–1473

Davidson-Arnott RGD, Fisher JD (1992) Spatial and temporal controls on overwash occurrence on a Great Lakes barrier spit. Can J Earth Sci 29:102–117

Davidson-Arnott RGD, Law MN (1996) Measurement and prediction of long-term sediment supply to coastal foredunes. J Coast Res 12:654–663

Davidson-Arnott RGD, Pyskir NM (1988) Morphology and formation of an Holocene coastal dune field, Bruce Peninsula, Ontario. Géog Phys Quatern 42:163–170

Davies JL (1980) Geographical variation in coastal development. Longman, London

De Jong R, Björck S, Björckman L, Clemmensen LB (2006) Storminess variation during the last 6500 years as reconstructed from an ombrotrophic peat bog in Halland, Southwest Sweden. J Quat Sci 21:905–919

Dech JP, Maun MA (2005) Zonation of vegetation along a burial gradient on the leeward slopes of Lake Huron sand dunes. Can J Bot 83:227–236

Delano HL (1991) Presque Isle State Park, Erie County: a dynamic interface of water and land, Pennsylvania Trail of geology, park guide 21. Pennsylvania Department of Conservation and Natural Resources, Bureau of Topographic and Geologic Survey, Harrisburg

DeVries-Zimmerman S, Fisher TG, Hansen EC, Dean S, Björck S (2014) Sand in lakes and bogs in Allegan County, Michigan, as a proxy for eolian sand transport. In: Fisher TG, Hansen EC (eds) Coastline and dune evolution along the Great Lakes, Geological Society of America special paper 508. The Geological Society of America, Boulder, pp 111–131

DeVries-Zimmerman S, Fuller JL, Peterson Jr DC, Watts AM, Van Gorp BCT, Hansen EC (2016) Wet and dry slacks: The relationship between Lake Michigan water levels and the hydrology/ecology of an interdunal wetland/slack on the southeast coast of Lake Michigan. Poster presented at the Geological Society of America annual meeting, Denver, Colorado, USA, 25–28 September 2016

DeVries-Zimmerman SJ, Hansen EC, Fisher TG, Bodenbender BE, Yurk BP, van Dijk D (2018) Coastal dune environments of southeastern Lake Michigan: geomorphic histories and contemporary processes. In: Florea LJ (ed) Ancient oceans, orogenic uplifts and glacial ice: geological crossroads in America's Heartland, Geological Society of America Field Guide 51. Geological Society of America, Boulder, pp 205–235

Dow K (1937) The origin of perched dunes on the Manistee Moraine, Michigan. Mich Acad Sci Arts Lett 23:427–440

Dworkin SI, Larson GJ, Monaghan GW (1985) Late Wisconsinan ice-flow reconstruction for the central Great Lakes region. Can J Earth Sci 22:935–940

Eamer JRB, Shugar D, Walker IJ, Lian O, Neudorf CM (2017) Distinguishing depositional setting for sandy deposits in coastal landscapes using grain shape. J Sediment Res 87:1–11

Earnest G, Kuehn D, Thompson M (2002) Sand wind & water: a recreational guide to eastern Lake Ontario's dunes and wetlands. New York Sea Grant

Edinger GJ, Evans DJ, Gebauer S, Howard TG, Hunt DM, Olivero AM (eds) (2014) Ecological communities of New York State, A revised and expanded edition of Carol Reschke's Ecological communities of New York State, 2nd edn. New York Natural Heritage Program, New York State Department of Environmental Conservation, Albany

Eichenlaub V (1979) Weather and climate of the Great Lakes region. University of Notre Dame Press, Notre Dame

Ensign KL, Webb EA, Longstaffe FJ (2006) Microenvironmental and seasonal variations in soil water content of the unsaturated zone of a sand dune system at Pinery Provincial Park, Ontario, Canada. Geoderma 136:788–802

Eyles N, Meulendyk T (2012) Ground-penetrating radar stratigraphy and depositional model for evolving Late Holocene aeolian dunes on the Lake Huron coast, Ontario. J Great Lakes Res 38:708–719

Fisher TG, Loope WL (2005) Aeolian sand preserved in Silver Lake: a reliable signal of Holocene high stands of Lake Michigan. The Holocene 15:1072–1078

Fisher TG, Loope WL, Pierce WC, Jol HM (2007) Big Lake records preserved in a little lake's sediment: an example from Silver Lake, Michigan, USA. J Paleolimnol 37:365–382

Fisher TG, Weyer KA, Boudreau AM, Martin-Hayden JM, Krantz DE, Breckenridge A (2012) Constraining Holocene lake levels and coastal dune activity in the Lake Michigan basin. J Paleolimnol 47:373–390

Foyle AM, Norton KP (2006) Late Holocene nearshore change at a nontidal transgressive systems tract strandplain complex, Presque Isle, Pennsylvania, USA. J Coast Res 22:406–423

Fraser GS, Bennett SW, Olyphant GA, Bauch NJ, Ferguson V, Gellasch CA, Millard CL, Mueller B, O'Malley PJ, Way JN, Woodfield MC (1998) Windflow circulation patterns in a coastal dune blowout, south coast of Lake Michigan. J Coast Res 14:451–460

Fraser GS, Thompson TA, Atkinson JC (2012) Sedimentary processes and sequence stratigraphy of Lake Michigan, United States. In: Baganz OW, Bartov Y, Bohacs K, Nummedal D (eds) Lacustrine sandstone reservoirs and hydrocarbon systems, AAPG Memoir 95. American Association of Petroleum Geologists, Tulsa, pp 385–415

Fryberger SG, Dean G (1979) Dune forms and wind regime. In: McKee ED (ed) A study of global sand seas, USGS prof paper 1052. United States Government Printing Office, Washington, pp 137–169

Fryberger SG, Schenk CJ (1988) Pin stripe lamination: a distinctive feature of modern and ancient eolian sediments. Sediment Geol 55:1–15

Gares PA, Nordstrom KF (1995) A cyclic model of foredune blowout evolution for a leeward coast: Island Beach, New Jersey. Ann Assoc Am Geogr 85:1–20

González-Villanueva R, Costas S, Pérez-Arlucea M, Jerez S, Trigo RM (2013) Impact of atmospheric circulation patterns on coastal dune dynamics, NW Spain. Geomorphology 185:96–109

Great Lakes Environmental Research Laboratory (GLERL) (2017a). https://www.glerl.noaa.gov/education/ourlakes/lakes.html. Accessed July 2017

Great Lakes Environmental Research Laboratory (GLERL) (2017b). https://www.glerl.noaa.gov/data/dashboard/data/. Accessed July 2017

Guire KE, Voss EG (1963) Distribution of distinctive plants in the Great Lakes region. Mich Bot 2:99–114

Gwyn QHJ, Dreimanis A (1979) Heavy mineral assemblages in tills and their use in distinguishing glacial lobes in the Great Lakes region. Can J Earth Sci 16:2219–2235

Hanes BE, Fisher TG, Becker RH, Martin-Hayden JM (2014) Elucidating paleo dune activity and timing from wetlands in the lee of coastal sand dunes, Grand Mere Lakes, Michigan, USA. In: Fisher TG, Hansen EC (eds) Coastline and dune evolution along the Great Lakes, Geological Society of America special paper 508. The Geological Society of America, Boulder, pp 133–149

Hansen EC, Arbogast A, Packman SC, Hansen B (2002) Post-Nipissing origin of a backdune complex along the southeastern shore of Lake Michigan. Phys Geogr 23:233–244

Hansen E, DeVries-Zimmerman S, van Dijk D, Yurk B (2009) Patterns of wind flow and aeolian deposition on a parabolic dune on the southeastern shore of Lake Michigan. Geomorphology 105:147–157

Hansen EC, Fisher TG, Arbogast AF, Bateman MD (2010) Geomorphic history of low-perched transgressive dune complexes along the southeastern shore of Lake Michigan. Aeolian Res 1:111–127

Hansen EC, Bodenbender BE, Johnson BG, Kito K, Davis AK, Havholm KG, Peaslee GF (2011) The origin of dark sand in eolian deposits along the southeastern shore of Lake Michigan. J Geol 119:487–503

Haslett SK, Davies P, Curr RHF (2000) Geomorphic and paleoenvironmental development of Holocene perched coastal dune systems in Brittany, France. Geogr Ann 82A(1):79–88

Hesp PA (2002) Foredunes and blowouts, initiation, geomorphology and dynamics. Geomorphology 48:245–268
Hesp PA (2013) Conceptual models of the evolution of transgressive dune fields. Geomorphology 199:38–149
Hesp PA, Hyde R (1996) Flow dynamics and geomorphology of a trough blowout. Sedimentology 43:505–525
Hesp PA, Pringle A (2001) Wind flow and topographic steering within a trough blowout. J Coast Res 34:97–601
Hesp PA, Thom BG (1990) Geomorphology and evolution of active transgressive dunefields. In: Nordstrom KF, Psuty NP, Carter RWG (eds) Coastal dunes: form and process. Wiley, Chichester, pp 253–287
Hesp PA, Walker IJ (2012) Three dimensional aeolian dynamics within a bowl blowout during offshore winds: Greenwich dunes, Prince Edward Island Canada. Aeolian Res 3:389–399
Hesp PA, Walker IJ (2013) Coastal dunes. In: Shroder J, Lancaster N, Sherman DJ, ACW B (eds) Treatise on geomorphology, CA. 11, Aeolian Geomorphology. Academic, San Diego, pp 328–355
Hesp PA, Illenberger W, Rust I, McLachlan A, Hyde R (1989) Some aspects of transgressive dunefield and transverse dune geomorphology and dynamics, south coast, South Africa. Z Geomorphol Suppl 73:111–123
Hesp PA, Walker IJ, Namikas SL, Davidson-Arnott R, Bauer BO, Ollerhead J (2009) Storm wind flow over a foredune, Prince Edward Island, Canada. J Coast Res Spec Issue 56:312–316
Hesp PA, Martinez M, Miot da Silva G, Rodríguez-Revelo N, Guitierrez E, Humanes A, Laínez D, Montaño I, Palacios V, Quesada A, Storero L, Trilla G, Trochine C (2011) Trangressive dunefield landforms and vegetation associations, Doña Juana, Veracruz, Mexico. Earth Surf Process Landf 36:285–295
Hesp PA, Smyth T, Nielsen P, Walker IJ, Bauer BO, Davidson-Arnott R. (2015). Flow deflection over a foredune. Geomorphology 230:64–74
Hicks LE (1933) The original forest vegetation and the vascular flora of Ashtabula County, Ohio. Dissertation, Ohio State University
Holcombe TL, Taylor LA, Reid DF, Warren JS, Vincent PA, Herdendorf CE (2003) Revised Lake Erie postglacial lake level history based on new detailed bathymetry. J Great Lakes Res 29:681–704
Hop K, Drake J, Lubinski S, Dieck J, Menard S (2009) National Park Service vegetation inventory program: Indiana Dunes National Lakeshore, Indiana. USGS, Upper Midwest Environmental Sciences Center, La Crosse
Hop K, Menard S, Drake J, Lubinski S, Dieck J (2010a) National Park Service vegetation inventory program: Apostle Islands National Lakeshore, Wisconsin, Natural resource report NPS/GLKN/NRR—2010/199. National Park Service, Fort Collins
Hop K, Menard S, Drake J, Lubinski S, Dieck J (2010b) National Park Service vegetation inventory program: Pictured Rocks National Lakeshore, Michigan, Natural resource report NPS/GLKN/NRR—2010/201. National Park Service, Fort Collins
Hop K, Drake J, Lubinski S, Menard S, Dieck J (2011) National Park Service vegetation inventory program: Sleeping Bear Dunes National Lakeshore, Michigan, Natural resource report NPS/GLKN/NRR—2011/395. National Park Service, Fort Collins
Houser C, Ellis J (2013) Beach and dune interaction. In: Sherman DJ, Shroder JF (eds) Treatise on geomorphology, Coastal Geomorphology, vol 10. Elsevier, Oxford, pp 267–288
Hunter RE (1977) Basic types of stratification in small eolian dunes. Sedimentology 24:361–387
Jennings JN (1967) Cliff-top dunes. Geogr Res 5:40–49
Johnston JW, Argyilan EP, Thompson TA, Baedke SJ, Lepper K, Wilcox DA, Forman SL (2012) A Sault-outlet-referenced mid- to late-Holocene paleohydrograph for Lake Superior constructed from strandplains of beach ridges. Can J Earth Sci 49:1–17
Jones M, Ohio Geological Survey, personal communication 1 August 2017

Judziewicz EJ, Koch RG (1993) Flora and vegetation of the Apostle Islands National Lakeshore and Madeline Island, Ashland and Bayfield Counties, Wisconsin. Mich Bot 32:43–189

Keddy PA, Reznicek AA (1986) Great Lakes vegetation dynamics: the role of fluctuating water-levels and buried seeds. J Great Lakes Res 12:25–36

Kilibarda Z, Kilibarda V (2016) Seasonal geomorphic processes and rates of sand movement at Mount Baldy dune in Indiana, USA. Aeolian Res 23:103–114

Kilibarda Z, Venturelli R, Goble R (2014) Late Holocene dune development and shift in dune-building winds along southern Lake Michigan. In: Fisher TG, Hansen EC (eds) Coastline and dune evolution along the Great Lakes, Geological Society of America special paper 508. The Geological Society of America, Boulder, pp 47–64

Kroodsma RF (1937) The permanent fixation of sand dunes in Michigan. J For 35:365–371

Landsberg H, Riley NA (1943) Wind influences on the transportation of sand over a Michigan sand dune. Proceedings of Second hydraulics Conference Bulletin 27, Univesity of Iowa studies in engineering

Larson G, Schaetzl R (2001) Evolution and origin of the Great Lakes. J Great Lakes Res 27:518–545

Law MN, Davidson-Arnott RGD (1990) Seasonal controls on aeolian processes on the beach and foredune. In: Davidson-Arnott RGD (ed) Proceedings of the symposium on coastal sand dunes. National Research Council of Canada, pp 49–68

Law J, van Dijk D (1994) Sublimation as a geomorphic process: a review. Permafr Periglac Process 5:237–249

Leege LM, Kilgore JS (2014) Recovery of foredune and blowout habitats in a freshwater dune following removal of invasive Austrian Pine (*Pinus nigra*). Restor Ecol 22:641–648

Leege LM, Murphy PG (2001) Ecological effects of the non-native *Pinus nigra* on sand dune communities. Can J Bot 79:429–437

Lehotsky K (1941) Sand dune fixation in Michigan. J For 39:998–1004

Lettau K, Lettau H (1978) Experimental and micrometeorological field studies of dune migration. In: Lettau HH, Letta K (eds) Exploring the world's driest climate. Madison, Center for Climatic Research, University of Wisconsin, pp 110–147

Leverett F (1911) Map of the surface formations of the southern Peninsula of Michigan. Geological Survey of Michigan, Lansing

Lewis CFM (1969) Late Quaternary history of lake levels in the Huron and Erie basins. Proceedings of 12th international association for Great Lakes research, pp 250–270

Lewis JD (1975) Michigan's industrial sand resources, Circular 11. Michigan Department of Natural Resources, Geological Survey Division, Lansing

Lichter J (1998) Primary succession and forest development on coastal Lake Michigan sand dunes. Ecol Monogr 68:487–510

Loope WL, Arbogast AF (2000) Dominance of an 150-year cycle of sand-supply change in late Holocene dune-building along the eastern shore of Lake Michigan. Quat Res 54:414–422

Loope WL, Fisher TG, Jol HM, Goble R, Anderton JB, Blewett WL (2004) A Holocene history of dune-mediated landscape change along the southeastern shore of Lake Superior. Geomophology 61:303–322

Lovis WA, Arbogast AF, Monaghan CW (2012) The geoarcheology of Lake Michigan coastal dunes, Environmental research series, vol 2. Michigan State Press, East Lansing

Mainville A, Craymer MR (2005) Present day tilting of the Great Lakes region based on water level gauges. Geol Soc Am Bull 117:1070–1080

Marsh WM, Marsh BD (1987) Wind erosion and sand dune formation on high Lake Superior bluffs. Geogr Ann Ser A Phys Geogr 69(3/4):379–391

Martinho CT, Hesp PA, Dillenburg SR (2010) Morphological and temporal variations of transgressive dunefields of the northern and mid-littoral Rio Grande do Sul coast, Southern Brazil. Geomophology 17:14–32

Martini IP (1975) Sedimentology of a lacustrine barrier system at Wasaga Beach, Ontario, Canada. Sediment Geol 14:169–190

Martini P (1981) Coastal dunes of Ontario: distribution and geomorphology. Géog Phys Quatern 35:219–229
Mason OK, Jordan JW (1993) Heightened North Pacific storminess during synchronous late Holocene erosion of Northwest Alaska beach ridges. Quat Res 40:55–69
Mason OK, Hopkins DM, Plug L (1997) Chronology and paleoclimate of storm-induced erosion and episodic dune growth across Cape Espenber Spit, Alaska, U.S.A. J Coast Res 13:770–797
Maun MA (1998) Adaptations of plants to burial in coastal sand dunes. Can J Bot 76:713–738
Maun MA (2004) Burial of plants as a selective force in sand dunes. In: Martinez ML, Psuty NP (eds) Ecol Studies 171:119–135
McKee ED, Douglas JR, Rittenhouse SS (1971) Deformation of leeside laminae in eolian dunes. Geol Soc Am Bull 82:359–378
McKenna Neuman C (1989) Kinetic energy transfer through impact and its role in entrainment by wind of particles from frozen surfaces. Sedimentology 36:007–1015
McKenna Neuman C (1990) Observations of winter aeolian transport and niveo-aeolian deposition at Crater Lake, Pangnirtung Pass, N.W.T., Canada. Permafr Periglac Process 1:235–247
Meadows GA, Meadows LA, Wood WL, Hubertz JM, Perlin M (1997) The relationship between Great Lakes water levels, wave energies and shoreline damage. Bull Am Meteorol Soc 78:675–682
Michigan Department of Environmental Quality (2017) Critical Dunes GIS Open Data, State of Michigan. http://gis-michigan.opendata.arcgis.com/datasets/ae5112507e3a40f1ae570a4966821002_6. Accessed June 2017
Michigan Sea Grant (2018) About the Great Lakes. http://www.miseagrant.umich.edu/explore/about-the-great-lakes/. Accessed Dec 2018
Mortimer CH (2004) Lake Michigan in motion: responses of an inland sea to weather, earth-spin, and human activities. University of Wisconsin Press, Madison
Mudrak EL, Johnson SE, Waller DM (2009) Forty-seven year changes in vegetation at the Apostle Islands: effects of deer on the forest understory. Nat Areas J 29:167–176
Nairn R, Brunton A, Selegean J (2006) Multiple approaches to assessing the impact of dams on sediment delivery in the St. Joseph River Watershed, Michigan/Illinois. Proceedings of the eighth federal interagency sedimentation conference, pp 515–522
National Ocean Service (1986) Great Lakes water levels, 1860–1985. U.S. Department of Commerce, National Oceanic and Atmospheric Administration, National Ocean Service, Rockville
National Oceanic and Atmospheric Administration (2017) Realtime Great Lakes weather data and marine observations. https://coastwatch.glerl.noaa.gov/marobs/marobs.html. Accessed June 2017
Neff BP, Day SM, Piggott AR, Fuller LM (2005) Base flow in the Great Lakes basin. US Geol Surv Sci Investig Rep:2005–5217, 23 p
Nickling WG, McKenna Neuman C, Lancaster N (2002) Grainfall processes in the lee of transverse dunes, Silver Peak, Nevada. Sedimentology 49:191–209
Notaro M, Holman K, Zarrin A, Fluck E, Varus S, Bennington V (2013) Influence of the Laurentian Great Lakes on regional climate. J Clim 26:789–804
Ohio Department of Natural Resources (2017) Headlands Dunes State Nature Preserve. http://naturepreserves.ohiodnr.gov/headlandsdunes. Accessed July 2017
Ohio Department of Natural Resources, Office of Coastal Management (2007) Ohio Coastal Atlas, 2nd edn
Olson JS (1958a) Lake Michigan dune development: 1. Wind-velocity profiles. J Geol 66:254–263
Olson JS (1958b) Lake Michigan dune development: 2. Plants as agents and tools in geomorphology. J Geol 66:345–351
Olson JS (1958c) Lake Michigan dune development 3: Lake level, beach, and dune oscillations. J Geol 66:473–483
Olson JS (1958d) Rates of succession and soil changes on southern Lake Michigan sand dunes. Bot Gaz 119:125–170

Ontario Ministry of Natural Resources (2001a) Great Lakes St. Lawrence River system and large inland lakes technical guides for flooding, erosion and dynamic beaches in support of natural hazards policies 3.1 of the Provincial Policy Statement (1997) of the Planning Act. Watershed Science Centre, Trent University Peterborough, Ontario, Canada

Ontario Ministry of Natural Resources (2001b) Understanding natural hazards. Great Lakes–St. Lawrence River system and large inland lakes, river and stream systems and hazardous site. An introductory guide for public health and safety policies 3.1—Provincial Policy Statement. Queen's Printer for Ontario, Ontario, Canada

Parks Research Forum of Ontario (2001) Ecological integrity and protected areas. Proceedings of the Parks Research Forum of Ontario annual general meeting, Map 9 and 10, 2001, North York, Ontario. Porter J and Nelson JG (eds)

Peach G (2006) Management of Lake Huron's beach and dune ecosystems: building up from the grassroots. Great Lakes Geogr 13:39–49

Peach GH (2008) The Port Franks Beach and dune stewardship guide. Lake Huron Centre for Coastal Conservation. http://lakehuron.ca/uploads/pdf/Beach.Stewardship.Guide.for.Port.Franks-2008.pdf

Peterson JM, Dersch E (1981) A guide to sand dune and coastal ecosystem functional relationships. Michigan State University Extension Bulletin E-1529, MICHU-SG-81–501, East Lansing

Pettijohn FJ (1931) Petrography of the beach sands of southern Lake Michigan. J Geol 39:432–455

Pye K, Tsoar H (2009) Aeolian sand and sand dunes. Springer, Berlin

Rawling E, Wisconsin geological and natural history survey, personal communication 1 August 2017

Rawling JE, Hanson PR (2014) Dune formation on late Holocene sandy barriers along Lake Michigan's Door Peninsula: the importance of increased sediment supply following the Nipissing and Algoma high lake-level phases. In: Fisher TG, Hansen EC (eds) Coastline and dune evolution along the Great Lakes, Geological Society of America special paper 508. The Geological Society of America, Boulder, pp 65–83

Rimsnider DO (1958) Texture and mineralogy of beach and dune sands, southeastern shore of Lake Michigan. Master's thesis, University of Illinois

Rooney TP, Waller DM (2003) Direct and indirect effects of white-tailed deer in forest ecosystems. For Ecol Manag 181:165–176

Ruz MH, Allard M (1995) Sedimentary structures of cold-climate coastal dunes, Eastern Hudson Bay, Canada. Sedimentology 42:725–734

Saunders KE, Davidson-Arnott RGD (1990) Coastal dune response to natural disturbances. In: Davidson-Arnott RGD (ed) Proceedings of the symposium on coastal sand dunes. National Research Council of Canada, pp 321–346

Saye SE, Pye K, Clemmensen LB (2006) Development of a cliff-top dune indicated by particle size and geochemical characteristics: Rubjerg Knude, Denmark. Sedimentology 53:1–21

Schaetzl RJ, Loope WL (2008) Evidence for an eolian origin for the silt-enriched soil mantles on the glaciated uplands of eastern Upper Michigan, USA. Geomorphology 100:285–295

Scott ID (1942) The dunes of Lake Michigan and correlated problems. In: 44th Annual report Michigan academy of science arts and letters, pp 53–61

Scott RW, Huff FA (1996) Impacts of the Great Lakes on regional climate conditions. J Great Lakes Res 22:845–863

Simiu E, Scanlan RH (1996) Wind effects on structures, 3rd edn. Wiley-Interscience, New York

Smyth TAG, Jackson DWT, Cooper AG (2012) High resolution measured and modelled three-dimensional airflow over a coastal bowl blowout. Geomorphology 177–178:62–73

Snyder FS (1985) A spatial and temporal analysis of Sleeping Bear Dunes Complex, Michigan (A contribution to the geomorphology of perched dunes in humid continental regions). Dissertation, University of Pittsburgh

Stewart CJ, Davidson-Arnott RGD (1988) Morphology, formation and migration of longshore sandwaves; long point, Lake Erie, Canada. Mar Geol 81:63–77

Swearingen J, Bargeron C (2016) Invasive plant atlas of the United States. Indiana Dunes National Lakeshore (Indiana) University of Georgia Center for Invasive Species and Ecosystem Health. https://www.invasiveplantatlas.org/park.html?id=INDU. Accessed July 2017

Szabo JP (2006) Textural and mineralogical characteristics of tills of northeastern and north-central Ohio. Ohio J Sci 106(2):9–16

Tague GC (1946) The post-glacial geology of the Grand Marais embayment, Berrien County, Michigan. Michigan Geology Survey Publication 45. Geological series, 38 (Part I)

Thompson PW (1967) Vegetation and common plants of Sleeping Bear, Cranbrook Institute of Science bulletin 52. Cranbrook Institute of Science, Bloomfield Hills

Thompson TA (1992) Beach-ridge development and lake-level variation in southern Lake Michigan. In: Donoghue JF, Davis RA, Fletcher CH, Suter JR (eds) Quaternary coastal evolution, Sedimentary Geology 80. Elsevier, Amsterdam, pp 305–318

Thompson TA, Baedke SJ (1995) Beach-ridge development in Lake Michigan shoreline behavior in response to quasi-periodic lake-level events. Mar Geol 129:163–174

Thompson TA, Baedke SJ (1997) Strandplain evidence for late Holocene lake-level variations in Lake Michigan. Geol Soc Am Bull 109:666–682

Thompson TA, Baedke SJ, Johnston JW (2004) Geomorphic expression of late Holocene lake levels and paleowinds in the Upper Great Lakes. Mich Acad 35:355–371

Thompson TA, Lepper K, Baedke SJ, Argyilan EP, Wilcox DA (2009) Strandplain evidence for late Holocene lake level and isostatic rebound in the Lake Huron basin (abs.). Abstracts for the International Association for Great Lakes Research 52nd annual meeting, p 218

Thompson TA, Lepper K, Endres AL, Johnston JW, Baedke SJ, Argyilan EP, Booth RK, Wilcox DA (2011) Mid Holocene lake level and shoreline behavior during the Nipissing phase of the upper Great Lakes at Alpena, Michigan, USA. J Great Lakes Res 37:567–576

Thompson TA, Johnston JW, Lepper K (2014) The contemporary elevation of the peak Nipissing phase at outlets of the upper Great Lakes. In: Fisher TG, Hansen EC (eds) Coastline and dune evolution along the Great Lakes, Geological Society of America special paper 508. The Geological Society of America, Boulder, pp 15–29

Timmons EA, Fisher TG, Hansen EC, Eiasman EC, Daly T, Kashgarian M (2007) Elucidating eolian dune history from lacustrine sand records in the Lake Michigan coastal zone, USA. The Holocene 17:789–801

Trask CB (1976) Mineralogy, texture, and longshore transport of beach sand, eastern shore of Lake Ontario. Dissertation, Syracuse University

van Denack JM (1961) An ecological analysis of the sand dune complex in Point Beach State Forest, Two Rivers, Wisconsin. Bot Gaz 122:155–174

van Dijk D (2004) Contemporary geomorphic processes and change on Lake Michigan coastal dunes: an example from Hoffmaster State Park, Michigan. Mich Acad 35:425–453

van Dijk D (2014) Short and long-term perspectives on the evolution of a Lake Michigan foredune. In: Fisher TG, Hansen EC (eds) Coastline and dune evolution along the Great Lakes, Geological Society of America special paper 508. The Geological Society of America, Boulder, pp 195–216

van Dijk D, Law J (2003) The rate of grain release by pore-ice sublimation in cold-aeolian environments. Geogr Ann Ser A 85:99–113

Walker IJ, Davidson-Arnott RGD, Bauer BO, Hesp PA, Delgado-Fernandez I, Ollerhead J, Smyth TAG (2017) Scale-dependent perspectives on the geomorphology and evolution of beach-dune systems. Earth-Sci Rev 171:220–253

Wang J, Bai X, Hu H, Clites A, Colton M, Lofgren B (2012) Temporal and spatial variability of Great Lakes ice cover, 1973–2010. J Clim 25:1318–1329

Warners D, Gosselink A, VanBrandt L (2005) Botanical inventory, floristic quality assessment and habitat delineation of P.J. Hoffmaster State Park, Michigan: report to P.J. Hoffmaster State Park and the Michigan Department of Natural Resources. Calvin College Department of Biology, Grand Rapids

Weather Underground (2017) Weather forecasts and reports. https://www.wunderground.com. Accessed June 2017

Wells JR, Thompson PW (1982) Plant communities of the sand dune region of Berrien County, Michigan. Mich Bot 21:3–38

Wells JR, Thompson PW (1983) Ecological and floristic survey of P.J. Hoffmaster State Park, Ottawa and Muskegon counties, Michigan. Contract No. SDPMA 83-A, Cranbrook Institute of Science, Bloomfield Hills, Michigan and State of Michigan – Department of Natural Resources

Wilson P, Orford JD, Knight J, Braley SM, Wintle AG (2001) Late-Holocene (post-4000 years B.P.) coastal dune development in Northumberland, Northeast England. The Holocene 11:215–229

Wilson P, McGourty J, Bateman MD (2004) Mid- to late-Holocene coastal dune event stratigraphy for the north coast of Northern Ireland. The Holocene 14:406–416

Wisconsin Department of Natural Resources (2017a) Apostle Islands Sandscapes (No. 268). http://dnr.wi.gov/topic/Lands/naturalareas/index.asp?SNA=268. Accessed July 2017

Wisconsin Department of Natural Resources (2017b) Woodland Dunes (No. 252). http://dnr.wi.gov/topic/Lands/naturalareas/index.asp?SNA=252. Accessed July 2017

Yurk BP, DeVries-Zimmerman S, Hansen E, Bodenbender BE, Kilibarda Z, Fisher TG, van Dijk D (2013) Dune complexes along the southeastern shore of Lake Michigan: geomorphic history and contemporary processes. In: Gillespie R (ed) Insights into the Michigan Basin: salt deposits, impact structure, youngest basin bedrock, glacial geomorphology, dune complexes, and coastal bluff stability, Geological Society of America field guide, vol 31. The Geological Society of America, Boulder, pp 57–102

Yurk B, Hansen EC, DeVries-Zimmerman S, Kilibarda Z, van Dijk D, Bodenbender B, Krehel A, Pennings T (2014) The role of extratropical cyclones in shaping dunes along southern and southeastern Lake Michigan. In: Fisher TG, Hansen EC (eds) Coastline and dune evolution along the Great Lakes, Geological Society of America special paper 508. The Geological Society of America, Boulder, pp 167–194

Zhang J, Maun MA (1991) Establishment and growth of *Panicum virgatum* L. seedlings on a Lake Erie sand dune. B Torrey Bot Club 118:141–153

# Chapter 4
# The Central and Southern Great Plains

**William C. Johnson, Paul R. Hanson, Alan F. Halfen, and Aaron N. Koop**

**Abstract** An often-windy landscape with few major topographic features, poorly consolidated fine-grained geology, and limited and variable precipitation has endowed the Central and Southern Great Plains with the ideal environment for development and repeated reactivation of dune fields and sand sheets. Mapping efforts have documented large numbers, sizes, and a wide distribution of aeolian sand deposits, and the application of soil texture, geochemistry and other data have begun to enhance our perspective on these deposits. In recent decades, numerical dating techniques have spurred inquiry into the development of activation chronologies, which have defined periods of sediment flux and prehistoric droughts, including megadroughts commonly observed in other paleoclimatic records. Nearly thirty dune fields within the region have been investigated and dated with radiocarbon and luminescence techniques. Activation, which has usually been climatically forced, occurs when sediment becomes transportable under the prevailing wind regime. Currently, most dunes throughout the Great Plains are, however, inactive, but those that are active occur primarily along the Texas-New Mexico border region and in areas impacted by human activity such as cattle ranching and off-road vehicle recreation. With the major exception of the Nebraska Sand Hills, most dune fields of the Central and Southern Great Plains are associated geomorphically with and in some cases geochemically-linked to one or more fluvial systems. Dunes composed of fine sediments (silt, clay) also occur in the region—lunettes, or crescentic dunes associated with playa basins (Kansas through to Texas into New Mexico), and the parna dunes of the Oklahoma Panhandle. Adaptation of mineralogical and geochemical finger printing of aeolian sand deposits and potential sand sourcing has made possible the identification of provenances (e.g., the Miocene Ogallala

W. C. Johnson (✉) · A. N. Koop
Department of Geography and Atmospheric Sciences, University of Kansas, Lawrence, KS, USA
e-mail: wcj@ku.edu; aaronkoop@ku.edu

P. R. Hanson
CSD, School of Natural Resources, University of Nebraska, Lincoln, NE, USA
e-mail: phanson2@unl.edu

A. F. Halfen
Department of Geography, University of Wisconsin-Milwaukee, Milwaukee, WI, USA
e-mail: ahalfen@wiley.com



N. Lancaster, P. Hesp (eds.), *Inland Dunes of North America*, Dunes of the World, https://doi.org/10.1007/978-3-030-40498-7_4

Formation and late Pleistocene-modern river systems) and by implication the directions of formative paleowinds. Given the early indications of global warming and model scenarios for the future, the Central and Southern Great Plains may experience future intense and extended-duration droughts and the attendant reactivation of dune fields and sand sheets.

**Keywords** Aeolian landforms · Great Plains dunes · Dune activation histories · Dust Bowl · Aeolian sand provenance

## 4.1 Introduction

The Central and Southern Great Plains (CSGP), herein defined as Nebraska and eastern Wyoming south to southern Texas and adjacent New Mexico, comprise nearly 60% of the North American Great Plains. The latter is an expansive grassland biome extending from northeastern Mexico north into Canada, which is bound on the south by the Chihuahuan desert, the west by forests of the Rocky Mountains, the east by temperate forest, and the north by boreal forest. One can define the boundary of the Great Plains using many criteria, and, though a physiographical approach is usually adopted when landforms are the focus, the footprint of the Great Plains adopted for this review is that of the United States (US) Environmental Protection Agency Level 2 map (USEPA 2012) because it embodies a plant community approach, which is appropriate given the control that vegetative cover plays in stability of aeolian sand deposits (Fig. 4.1).

Floristically, short- and tall-grass communities covered the CSGP until European colonization and its resultant agricultural development. Original native vegetation included tall-grass prairie (e.g., Big Bluestem (*Andropogon gerardi*), Indian grass (*Sorghastrum nutans*), and Switchgrass (*Panicum virgatum*)) to the east, transitioning westward to short-grass prairie (e.g., Blue Grama (*Bouteloua gracilis*), Buffalo grass (*Buchloe dactyloides*)). Cool-season grasses ($C_3$) occur, but warm-season ($C_4$) grasses dominate regionally, though near the limits of the region other species become common, such as yucca (*Yucca glauca*) and sagebrush (*Artemisia* spp.) to the west and savanna-steppe to the far south in Texas. In larger dune fields, edaphically suited species such as sand sage (*Artemisia filifolia*) and sand bluestem (*Andropogon hallii*) occur (Küchler 1967). Notably, the relative importance of $C_4$ versus $C_3$ grasses within the region has not been static during the late Pleistocene and Holocene, but rather has, based on $\delta^{13}C$ values from paleosol organic matter, fluctuated significantly (Johnson and Willey 2000; Mason et al. 2008; Nordt et al. 2008).

The pronounced east-to-west transition in plant community composition is a function of the marked decrease in annual precipitation with increasing distance from Gulf-derived moisture and greater influence of the Rocky Mountain rain shadow (Fig. 4.2). Kansas City, Missouri (94.58° W), for example, receives 980 mm of annual precipitation, though Denver, Colorado to the west (104.98° W) receives

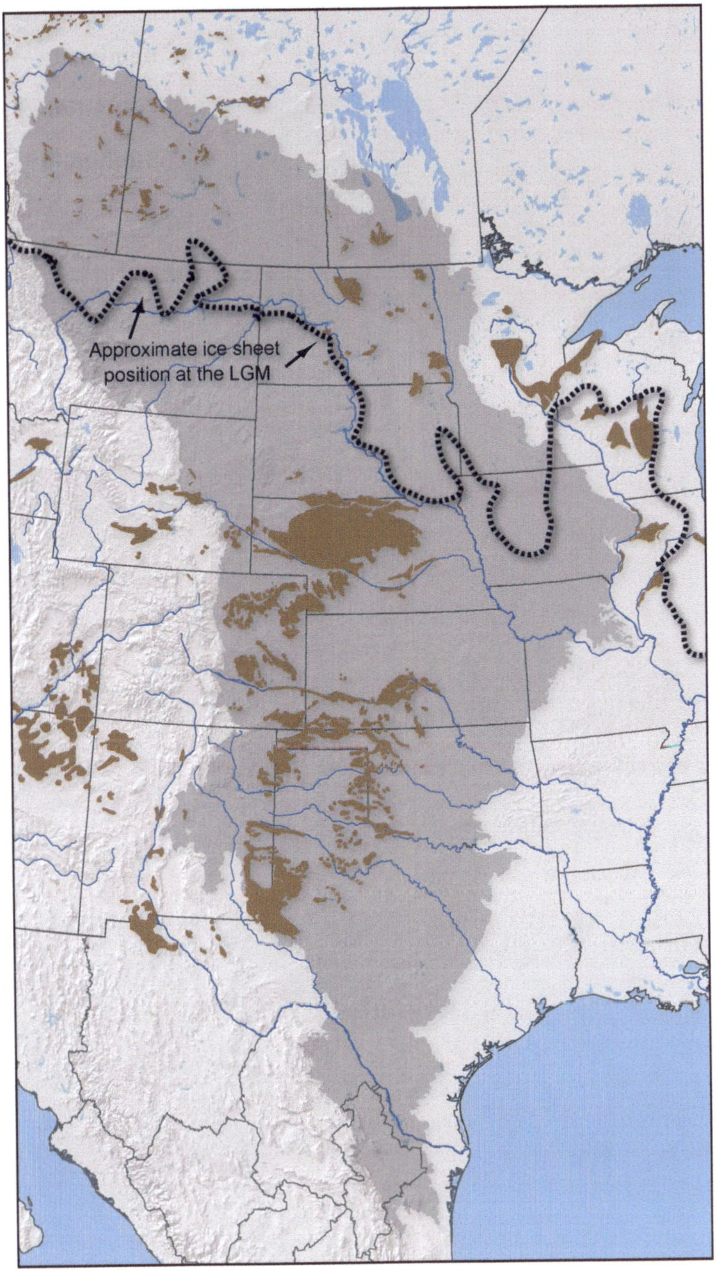


**Fig. 4.1** Definition of the North American Great Plains, location and extent of dune fields within and adjacent to the Great Plains, and the approximate last glacial maximum (LGM) positon of the Laurentide ice sheet. Figure is modified from Halfen and Johnson (2013), with the approximate position of the last glacial maximum ice sheet margin after Fullerton et al. (2003, 2004) and Martin et al. (2004), dune fields (brown shading) after Muhs and Holliday (1995), Wolfe et al. (2009) and sources cited therein, and the boundary of the Great Plains as defined by the U.S. Environmental Protection Agency Ecoregion Level 2 map (USEPA 2012)

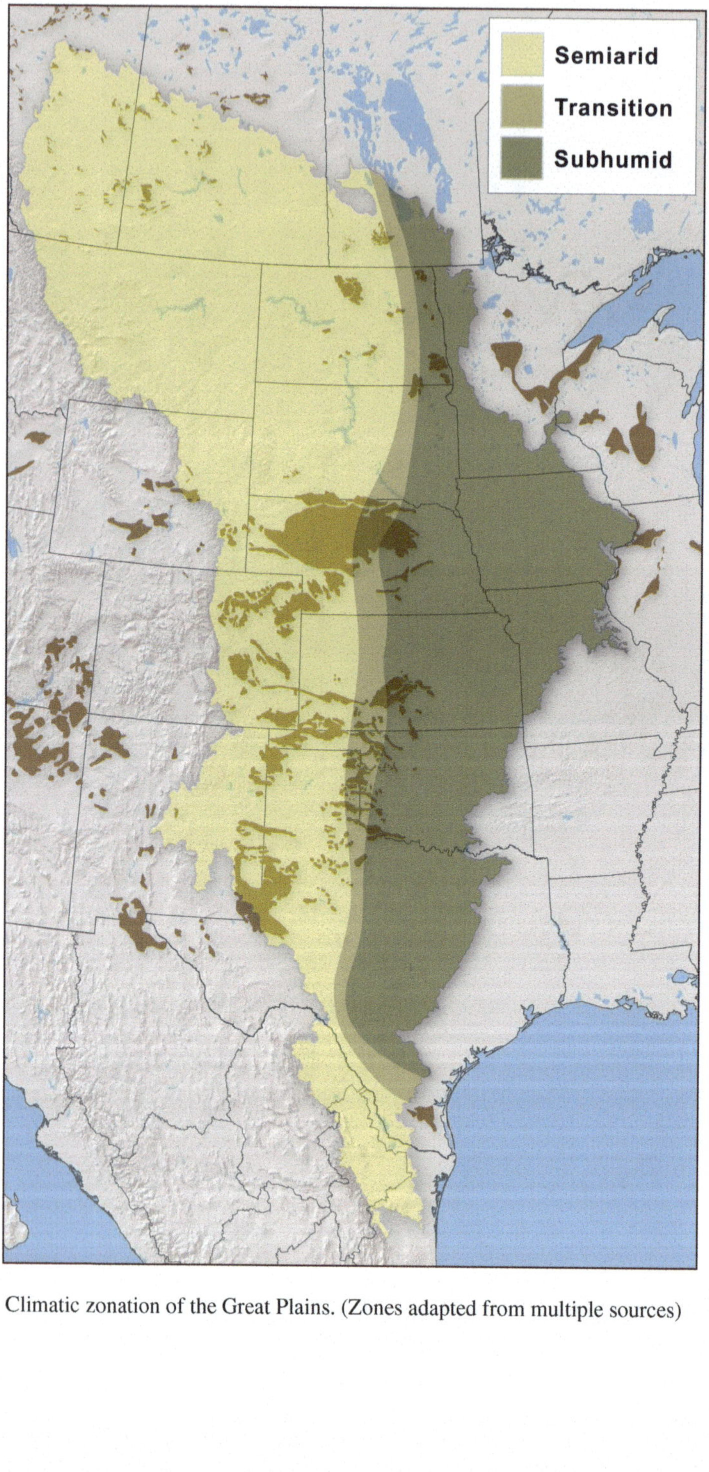


**Fig. 4.2** Climatic zonation of the Great Plains. (Zones adapted from multiple sources)

ca. 390 mm, resulting in a decrease of ca. 100 mm/150 km. Whereas the east-to-west gradient is primarily that of precipitation, the north-to-south gradient is expressed mostly in winter (January) temperature, with, for example, Omaha, Nebraska at ca. −6.4 °C and San Antonio, Texas at nearly 10 °C, but summer (July) temperatures vary little latitudinally, averaging ca. 24 °C (Table 4.1).

As with the entire Great Plains, the CSGP are known for strong and variable winds, resulting from the seasonal shifts in air-mass influence, pressure cells and wind patterns, in combination with the relatively flat terrain providing unimpeded flow (Fargione et al. 2012). Present winds in the CSGP tend to be seasonal, with northwesterly winds in winter and southerly winds in summer. Moreover, the characteristically strong winds have resulted in a high drift potential (Fryberger and Dean 1979), a measure of potential aeolian sand transport.

The CSGP contain the largest concentration of dune fields within the Great Plains (Table 4.2). Dunes occur in eastern Wyoming, Nebraska, eastern Colorado, Kansas, central and western Oklahoma, eastern New Mexico and west-central Texas, and the largest dune field in the Great Plains (and all of North America) is the Nebraska Sand Hills (NSH). A wide array of dune morphologies occurs among and within the various dune fields, resulting in often-complex dune landscapes, though the most common type of dune is the parabolic dune, with others including dome, barchan, barchanoid ridge, transverse, and linear. This complexity could be due to prehistoric wind shifts (e.g., Sridhar et al. 2006; Schmeisser et al. 2010), more recent episodes of full or partial activation (e.g., Forman et al. 2005; Halfen et al. 2012), and human activity (e.g., Lee and Gill 2015). Other lesser-known dunes

**Table 4.1** Climatological data for major cities in the Central and Southern Great Plains

| Location[a] | Lat. (°N) | Long. (°W) | Mean Jan. Temp (°C) | Mean July Temp (°C) | Mean annual precip. (mm) | Jan. Mean wind vector (°) | Jan. Mean wind speed (m $s^{-1}$) | July Mean wind vector (°) | July Mean wind speed (m $s^{-1}$) |
|---|---|---|---|---|---|---|---|---|---|
| Omaha, NE | 41.250 | 96.000 | –6.4 | 24.7 | 778.5 | 337 | 4.9 | 360 | 3.9 |
| Denver, CO | 39.733 | 104.983 | −0.1 | 23.3 | 394.2 | 180 | 3.9 | 180 | 3.7 |
| Kansas City, MO | 39.100 | 94.580 | −3.5 | 25.8 | 977.6 | 203 | 4.9 | 180 | 4.1 |
| Oklahoma city, OK | 35.467 | 97.533 | 2.2 | 27.8 | 931.2 | 360 | 5.6 | 157 | 4.8 |
| Dallas, TX | 32.767 | 96.800 | 6.3 | 29.6 | 953.8 | 180 | 4.9 | 180 | 4.3 |
| San Antonio, TX | 29.417 | 98.500 | 9.6 | 29.4 | 819.0 | 360 | 3.9 | 157 | 4.1 |

[a]Location of each city can be referenced in Fig. 4.7

Climatological data obtained from the High Plains Regional Climate Center (HPRCC 2012)

**Table 4.2** Mapped dune field areas of the Central and Southern Great Plains by state

| State | Total area[a] (Km$^2$) | Central & Southern Great Plains Dune area (km$^2$)[b] | Dunes % of land area |
|---|---|---|---|
| Colorado | 113,191 | 27,045 | 23.89 |
| Iowa | 136,579 | –[c] | – |
| Kansas | 212,592 | 16,352 | 7.69 |
| Missouri | 70,204 | –[c] | – |
| Nebraska | 200,001 | 70,818 | 35.41 |
| | *(w/o Sand Hills* | *20,850* | *10.43)*[d] |
| New Mexico | 103,635 | 15,252 | 14.71 |
| Oklahoma | 142,184 | 15,881 | 11.17 |
| Wyoming | 73,535 | 3163 | 4.30 |
| Texas | 408,256 | 12,112 | 2.97 |
| *Central and Southern Great Plains* **Total 1,460,177** | | **160,623** | **11.00** |
| *Entire North American Great Plains* **Total 2,735,067** | | **182,806** | **6.68** |

[a]Great Plains area was calculated in ArcGIS using the USEPA Ecoregion Level 2 Map boundary (USEPA 2012) and WGS 1984 Datum

[b]Dune area was calculated in ArcGIS using dune polygons modified from Wolfe et al. (2009) and Koop et al. (2012)

[c]States where no mapped dune fields are found within the boundaries of the Great Plains (isolated dunes may appear in county-level surficial geological maps)

[d]The Nebraska Sand Hill are discussed separately in this volume by Mason et al.

occur in the CSGP—the lunettes, associated with playas (Holliday 1997a; Bowen and Johnson 2012), and the recently recognized parna dunes (Fine et al. 2011; Johnson et al. 2012).

Inquiry into dune fields can originate from any number of perspectives, such as ecology, meteorology, physics of sand grain transport, geology, and past or present human experience or impact. This chapter on dunes of the CSGP takes a geological perspective, first describing the dune fields, then the climate and chronology of development and subsequent reactivations, and finally information on the provenance of the sand constituting the dune fields.

## 4.2 Aeolian Sand Deposits and Activation Histories

### *4.2.1 Mapping Aeolian Sand Deposits*

Aeolian sand deposits were first mapped region-wide by Muhs and Holliday (1995) using geologic data from a wide array of sources in combination with STATSGO (State Soil Geographic Database) from the US Soil Conservation Service (now the National Resources and Conservation Survey). Muhs and Wolfe (1999) later published a map of sand dunes on the Northern Great Plains of Canada and the US. A continental-scale map of aeolian deposits for the Great Plains of North America by Muhs and Holliday (1995) and Wolfe et al. (2009) has been presented in modified form by Halfen and Johnson (2013), Muhs (2017), Johnson et al. (2019) and others.

With SSURGO (Soil Survey Geographic Database; successor to STATSGO), widespread coverage of LiDAR imagery data, and increased availability of detailed surficial geologic mapping, aeolian sand deposits are mappable in relatively high resolution. For example, these data were used to depict an assemblage of map layers showing the distribution of aeolian sand deposits in Kansas (Koop et al. 2012) (Fig. 4.3a). Aeolian sand deposits can also be differentiated into sand dunes and sand sheets using the data sources (Fig. 4.3b) to display sand sheets such as the St.

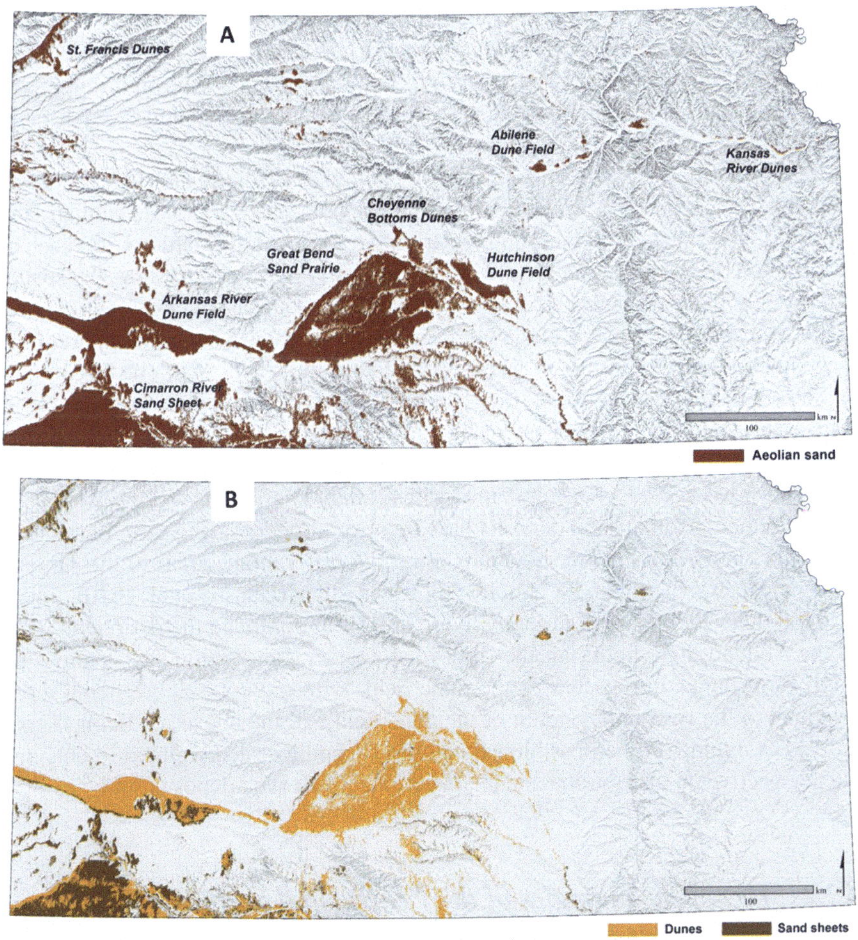


**Fig. 4.3** Distribution of aeolian sand deposits in the state of Kansas. (**a**) Mapped areas in sand dunes and sheets with major deposits indicated; (**b**) Mapped areas in aeolian sand differentiated into sand dunes and sand sheets

Francis aeolian sand along the South fork of the Republican River in extreme northwestern Kansas and the Cimarron River valley area of southwestern Kansas. Major dune formation appears along the Arkansas River of central and southwestern Kansas, the relief of which was rendered using terrain slope values derived from SSURGO and spot-verified with LiDAR data (Fig. 4.4). Relatively low-relief aeolian sand dunes (0–8% slope) occur across most of the Great Bend Sand Prairie (GBSP) but no more than half of the Hutchinson dune field (Fig. 4.4a). Depiction of the 5–20% slope dune terrain highlights the river valley-proximal dunes on the western side of the GBSP as well as those scattered from the southwestern part, and much of the Hutchinson dune field (Fig. 4.4b). Other than scattered dunes, the highest slope dunes (10–30%) occur river proximal in the southwestern part of the GBSP and the central core of the Hutchinson dune field (Fig. 4.4c). Similarly, combining field observations and stereoscopic aerial photograph interpretation, Arbogast (1998) and Arbogast and Johnson (1998) produced a landform map of the GBSP that included the categories of loess plain, low-relief sand sheet, high-relief sand sheet, compound subparabolic dunes, compound parabolic dunes, and parabolic dunes.

Relying on field observations, SSURGO, and high-resolution surficial geologic mapping (e.g., Johnson and Woodburn 2011), the relationship between aeolian sand deposits (sand sheets vs. dunes) and underlying terraces of the Garden City reach of the Arkansas River valley in southwestern Kansas is portrayed to show the spatial extents of both sand sheets and dunes (Fig. 4.5a): sand sheets and exposed underlying terrace and loess (intervening colorless areas) do occur in the eastern part of the reach, though dune mantling of terraces dominates the reach. Also, the discontinuous dunes of the Kansas River valley, northeastern Kansas (Johnson et al. 2019) (Fig. 4.5b) comprise a linear array of small patches of dunes resting on the Menoken Terrace (Fader 1974; Sorenson et al. 1987), consisting of glaciofluvial and ice-contact deposits. SSURGO, augmented with scattered field observations, were used to differentiate aeolian sand deposits based on overall texture (Fig.4.6). Fine sandy loam and loamy fine sand designations proved to differentiate areas of the Abilene dune field, associated with the Smoky Hill River valley (Hanson et al. 2010). Finer-textured dune sediments (fine sandy loam) appear to dominate the peripheral parts of the Abilene dune field, but the somewhat coarser dune sediments (loamy fine sand) form dunes in the core of the dune field, with the coarse core being more proximal to the river valley south of the dune field. All the above renderings serve only as examples of one possible expeditious mapping strategy aimed to enhance the perspective on distribution and character of aeolian sand deposits.

### 4.2.2 *Dune Chronologies of the Central and Southern Great Plains*

Dunes and sand sheets of the CSGP (Fig. 4.7) are indeed impressive in terms of ubiquity and stature and, as a consequence, have historically garnered a great deal of attention from an array of scientific disciplines. Geoscientists, specifically, have

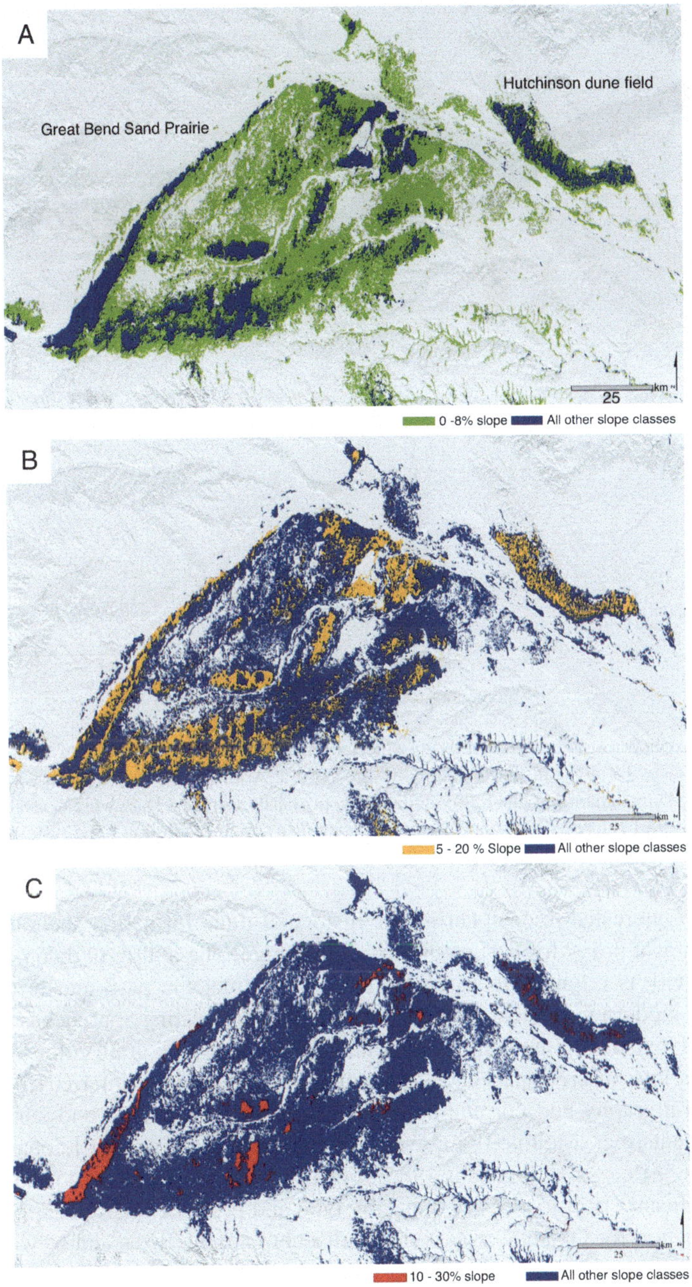


**Fig. 4.4** Slope (%) for the Great Bend Sand Prairie (GBSP) and Hutchinson dune field of the Arkansas River valley, central Kansas. (**a**) 0–8% slope dunes (green); (**b**) 5–20% slope dunes (orange); (**c**) 10–30% slope dunes (red). Areas within the GBSP that are not rendered in any color are relatively level areas of exposed loess or terrace treads. Slope data were derived from the USDA Web Soil Survey (2018) and spot-checked using LiDAR data

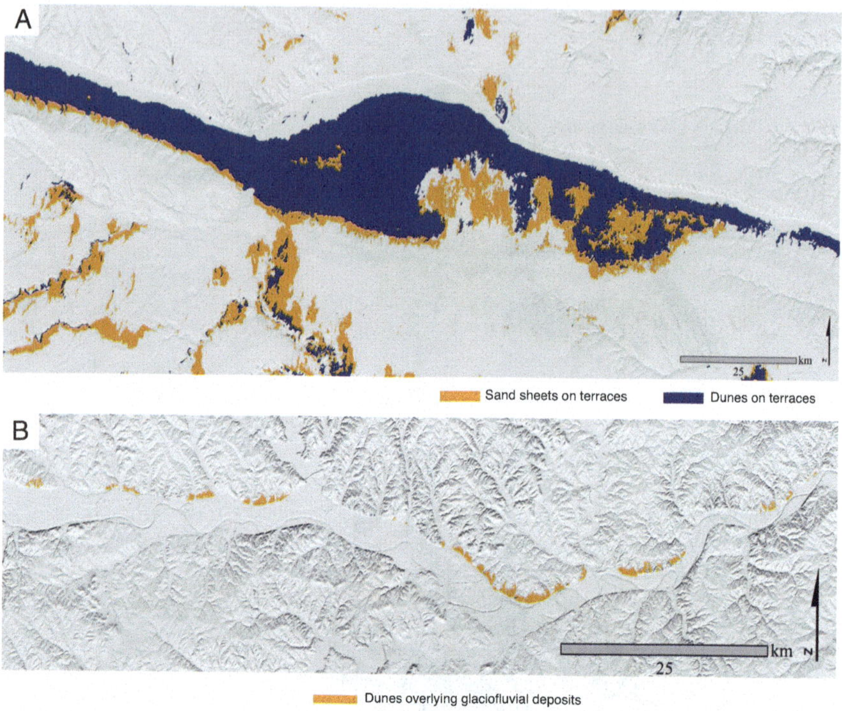


**Fig. 4.5** Aeolian sand deposits overlying alluvial terraces. (**a**) Dunes (blue) and sand sheets on alluvial terraces (orange) in the Garden City area of the Arkansas River valley in southwestern Kansas; (**b**) Dunes mantling glaciofluvial deposits, primarily terraces. Data were derived from the USDA Web Soil Survey (2018) and field-checked during high-resolution, county-level surficial geologic mapping

long been interested in deciphering the history of dune formation and subsequent activation as a proxy for past environmental change. The ability to date periods of dune activity is requisite to using them to reconstruct the past, and prior to the advent of modern numerical dating techniques, relative dating approaches were the norm. Relative dating methods have included the use of sand grain color to classify sediments into discrete packages, formative wind directions inferred from dune forms, stratigraphy and correlation with loess units or terraces, and surface soil development (e.g., Melton 1940; Wendorff et al. 1955; Frye and Leonard 1964; Smith 1965; Reeves 1976; Muhs 1985). For example, Melton (1940) produced the first significant effort to classify dunes by type and presumed ages, and associated wind direction for Oklahoma, as well as adjacent parts of Texas and New Mexico; he defined three categories of dunes based on morphology, interpretation of paleowind and his three estimated categories of age (<5 ka, 5–12 ka, >15 ka). Using information from decades of numerical dating of aeolian sand bodies, Holliday and Rawling (2006) calibrated relative dating using observations from numerically dated localities on the High Plains of Texas and New Mexico. They did this by relating the numbers and thicknesses of pedogenic lamellae (clay-enriched bands) to the

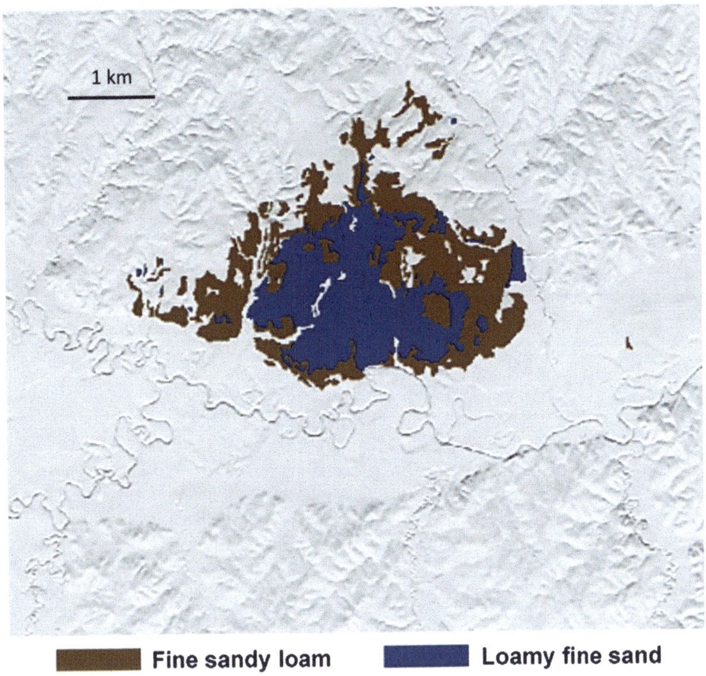


**Fig. 4.6** Textural differences within the Abilene dune field, Kansas. Distributions of the coarser aeolian sand component (loamy fine sand (blue)) and finer aeolian sand component (fine sandy loam (brown)). Data were derived from the USDA Web Soil Survey (2018) and field spot-checked

absolute ages of sand bodies; for example, late Pleistocene to early-Holocene sand deposits exhibit the greatest number and thickness (10–12 mm) of lamellae, whereas those in late Holocene deposits are fewer in number and thinner (ca. 3 mm).

Dune field numerical chronologies were based initially on radiocarbon dating of organic material (e.g., paleosols, bone, charcoal), which was used widely (e.g., David 1971; Ahlbrandt et al. 1983; Arbogast 1996b; Holliday 2001), but such materials are often difficult to locate and typically represent only dune stability. The advent of luminescence dating—thermoluminescence (TL), infrared stimulated luminescence (IRSL), and optically stimulated luminescence (OSL)—was a major advance because (1) organic carbon was no longer needed to obtain ages, (2) aeolian processes allow for ample exposure of grains to sunlight for the requisite resetting of the signal, and (3) luminescence dating techniques are usually used to date an aeolian sedimentation event, in contrast to radiocarbon dating that again typically provides ages indicating times of relative land-surface stability.

Of the luminescence approaches, OSL, developed by Huntley et al. (1985), has become the mainstream luminescence technique (Duller 2004; Rhodes 2011). Given that a major focus of aeolian sand deposit research for the last few decades has been the development of chronologies of their formation and particularly subsequent activation (Lancaster et al. 2016), the following discussion takes that approach,

**Fig. 4.7** Dune fields and sand sheets of the Central and Southern Great Plains. (Modified from Halfen and Johnson (2013) which was generated from Muhs and Holliday (1995), Wolfe et al. (2009) and sources cited therein)

proceeding generally north to south by state (Table 4.3). Radiocarbon ages in referenced sources have been calibrated where necessary using Calib 7.1 (Stuiver et al. 2019) and are usually cited as thousands of years ago (ka), as are the luminescence ages. Refer to Fig. 4.7 for the locations of the dune fields discussed in the following.

**Table 4.3** Chronological studies of sand dunes and sand sheets in the Central and Southern Great Plains

| Dune field or sand sheet[a] | Studies | $^{14}C$ ages | Lum. Ages[b] |
|---|---|---|---|
| *Wyoming* | | | |
| Casper dunes | Albanese (1974a, b) | 2 | – |
| | Halfen et al. (2010) | 6 | 12 |
| Ferris dunes[c] | Gaylord (1982, 1990) | 9 | – |
| | Stokes and Gaylord (1993) | – | 6 |
| *Colorado* | | | |
| Greeley dunes | Forman et al. (1992) | 1 | – |
| Fort Morgan dunes | Forman et al. (1992) | 4 | – |
| | Madole (1994) | 8 | – |
| | Forman et al. (1995) | 8 | 8 |
| | Madole (1995) | 13 | 2 |
| | Clark and Rendell (2003) | – | 8 |
| | Madole et al. (2005) | – | 3 |
| Hudson dunes | Forman and Maat (1990) | 2 | 2 |
| North Park dunes | Albrandt et al. (1983) | 4 | – |
| Wray dunes | Forman et al. (2005) | – | 4 |
| | Mason et al. (2011) | – | 4 |
| *Nebraska* | | | |
| Stanton dunes | Puta et al. (2013) | – | 24 |
| Duncan dunes | Hanson et al. (2009) | – | 17 |
| Imperial dunes | Mason et al. (2011) | – | 7 |
| *Kansas* | | | |
| Kansas River dunes | Johnson et al. (2019) | – | 8 |
| Abilene dunes | Hanson et al. (2010) | – | 17 |
| Cheyenne Bottoms dunes | Willey and Johnson (this chapter) | 1 | – |
| Great Bend Sand Prairie | Arbogast (1996b, 1998) | 27 | – |
| Hutchinson dunes | Halfen et al. (2012) | – | 60 |
| Arkansas River dunes | Forman et al. (2008) | – | 22 |
| | Halfen (2012) | 2 | 50 |
| Cimarron River Bend dunes | Olson and Porter (2002) | 9 | – |
| | Werner et al. (2011) | – | 8 |
| *Oklahoma* | | | |
| Cimarron River Valley dunes | Cordova et al. (2005) | 2 | 1 |
| | Lepper and Scott (2005) | – | 12 |
| | Brady (1989) | 3 | – |
| Dempsey Divide dunes | Thurmond and Wyckoff (1998) | 7 | – |
| *New Mexico* | | | |
| Mescalero dunes | Stokes (1994) | – | 2 |
| | Hall and Goble (2006, 2011) | 4 | 11 |
| *New Mexico and Texas* | | | |
| Muleshoe dunes | Holliday (1995, 1997a, b, 2001) | 16 | – |

(continued)

**Table 4.3** (continued)

| Dune field or sand sheet[a] | Studies | $^{14}C$ ages | Lum. Ages[b] |
|---|---|---|---|
| | Haynes (1995) | 4 | – |
| | Rich and Stokes (2001) | – | 1 |
| Lea–Yoakum dunes | Holliday (2001) | 1 | – |
| | Holliday and Meltzer (1995) | 3 | – |
| *Texas* | | | |
| Dune site | Holliday (1985) | 9 | – |
| Andrew dunes | Holliday (2001) | 1 | – |
| | Rich and Stokes (2011) | – | 6 |
| Monahans dunes | Stokes (1994) | – | 2 |
| | Rich and Stokes (2011) | – | 6 |
| Misc. dunes and sand sheets | Holliday (1989) | – | 4 |
| | Rich and Stokes (2011) | – | 8 |
| South Texas sand sheet[c] | Forman et al. (2009) | – | 12 |
| | **Total** | **146** | **327** |

[a]Listed approximately north to south
[b]Includes all luminescence data (OSL, IRSL, TL)
[c]Dune field included due to close proximity to the Great Plains as defined by the USEPA Ecoregion Level 2 map boundary (USEPA 2012)

#### 4.2.2.1 Wyoming

The Casper dunes form an elongate (west-east) but discontinuous 2000 km$^2$ dune field located on the western perimeter of the Great Plains and occupy several adjoining drainages of the Central Wyoming Basin. Dune morphologies vary: the eastern part of the dune field consists primarily of complex and hairpin parabolics with poorly developed surface soils, ranging to 15 km in length and 30 m in height, whereas the western part has less complex dunes with well-developed surface soils and a table-like morphology (Halfen et al. 2010). Notably, the Casper dune field lies within the 'Wyoming wind corridor,' a topographically defined southwest to northeast-oriented zone of southwest winds extending from the easternmost end of the Killpecker dune field (Ahlbrandt 1974; Mayer and Mahan 2004) to the North Platte River. Albanese (1974a, b) and Albanese and Frison (1995) were the first to report numerical ages from the Casper dune field and concluded that dune activation began after 10.3 ka and that alluvium was deposited in the basin prior to dune activity. Halfen et al. (2010), employing radiocarbon and OSL dating to develop an expanded chronology of dune activation in the dune field, reported widespread aeolian activity after ca. 10 ka, subsequent to the alluviation of the basin, which lasted until the middle Holocene at ca. 6.2 ka. Aeolian activity resumed at 4.1 ka and 1.0–0.40 ka. Whereas regional climate was a forcing agent, local factors including sediment supply and water table fluctuations also likely had a role in dune activation (Halfen et al. 2010).

Though not in the Great Plains as defined, two other dune fields are worthy of note due to their geologic and geomorphic relationship to the Casper dunes. The

Ferris dune field and its southern appendage, the Seminoe dune field (300 $km^2$; a.k.a. Ferris-Lost Soldier area) are also in the Wyoming wind corridor (Gaylord 1982). Dunes are primarily parabolic and display a dramatic converging trajectory as the dune field constricts to the northeast as it approaches Windy Gap. Stratigraphic and other geologic data suggest that the average wind velocity and direction varied little during the Holocene, and age data indicate aeolian activity began ca. 12 ka with periods of stability interpreted from calibrated radiocarbon ages of interdunal deposits at ca. 8.5, 8.2, 7.3–6.6, 5.2, and 2.2 ka (Gaylord 1982, 1990); Stokes and Gaylord (1993) subsequently OSL dated the intervening periods of instability, establishing 8.8–8.1 ka and 4.3–4 ka as the major periods of sand deposition. Age data from the far western end of the Killpecker dune field (Ahlbrandt et al. 1983; Mayer and Mahan 2004) are in approximate agreement with both the Ferris and Casper dune field records.

#### 4.2.2.2 Nebraska

The Nebraska Sand Hills (50,000 $km^2$) has produced a spectacular array of aeolian sand chronologies, one of the first being by Ahlbrandt et al. (1983) and more recently by Mason et al. (2011). Due to the large size and historical significance of the NSH, it is addressed in a separate chapter (Mason et al. this volume). However, with such opportunities for dune study in the NSH, only a few other dune fields in the state have been numerically dated: the Duncan dunes (Hanson et al. 2009), Stanton dunes (Puta et al. 2013), and the Imperial dunes (Mason et al. 2011). The Wray dunes (Forman et al. 2005; Mason et al. 2011) extend into southwesternmost Nebraska but are deferred to the Colorado discussion.

Hanson et al. (2009) investigated the Duncan dune field (ca. 170 $km^2$), which is in a riverine environment located between the Loup and Platte rivers immediately upstream from their confluence. Dune heights range 10–25 m, and slip faces indicate a formative northwest wind. Samples of dune sand and underlying terrace alluvium produced ages ranging from ca. 16.1–12.6 ka, and dune activation ages clustered at 4.4–3.4 ka and 0.8–0.5 ka, with relative stability for the last 500 years. At 97.7° west, this was the first study to address the recognized lack of dune activation chronologies from the eastern side of the CSGP, information needed to assess the prehistoric eastern penetration of drought-induced dune field activity. Another dune field on the eastern side of the CSGP, the Stanton dunes (ca. 162 $km^2$) are situated on the south side of the Elkhorn River in northeastern Nebraska, ca. 145 km east from the easternmost appendage of the NSH. In contrast to the high-relief barchanoid and other types of dunes in the NSH, Stanton dunes are mostly low-relief barchans perched on the Elkhorn River terrace (Puta et al. 2013). The underlying alluvial terraces were deposited 23.7 to 16.4 ka, and dune activation occurred at 15.8 and 9.8 ka, 5.8–3.8, 0.96–0.41 ka, and historically at 120 and 50 years ago. The Imperial dunes of southwestern Nebraska were dated at two locations by Mason et al. (2011), one producing three ages of 14.7, 3.0, and 0.54 ka, and the other four ages of 13.6, 13.7, 10.7, and 7.5 ka.

#### 4.2.2.3 Colorado

Dunes of northeastern Colorado comprise two broad groups, one consisting of linear dune fields along the north and south sides of the South Platte River, where they mantle the adjacent uplands and partially cover the river terraces, and another much larger dune field to the southeast. The assemblage of dune fields was initially differentiated into three discrete fields, designated by Muhs (1985) as the Greeley dunes, Fort Morgan dunes, and Wray dunes. Muhs (1985) derived geomorphic and pedologic information which he used to develop a preliminary model of evolution for the three dune fields: those dunes exhibiting surface soil development similar to that of the NSH were likewise probably active ca. 3–1.5 ka; parabolic dunes, the most common form, are probably underlain by older sand from an earlier episode of activity; based on dune morphology, formative winds during the late Holocene were northwest, as in the NSH; and dune field distribution was influenced by the pre-existing landscape. On the South Platte River terraces in the Greeley area, Holliday (1987) concluded that aeolian sand started to accumulate by 11.6 ka, though most dunes are late Holocene in age.

From a small (ca. 40 $km^2$) singular dune field due south of Greeley, Colorado and the South Platte River they informally named the Hudson dune field, Forman and Maat (1990) derived ages of 9.2–8.6 ka relatively deep within their profile, while surface soil development indicated dune stabilization after 3 ka. Forman et al. (1992), using Landsat imagery to better define the stabilized single and compound parabolic dunes of the Hudson dune field, proposed four possible periods of Holocene aeolian activity from the Kersey road section: 10.9–6.3 ka, 6.3–5.4 ka, 5.4 -> 1 ka, and < 1 ka, each of which was presumably separated by less than 2 k years of stability. Subsequently Forman et al. (1995) reported additional ages from the site, indicating a basal age of 13.5 ka, as well as ages from the Coors site, in the western extent of the Fort Morgan dunes, which captured ages of 143 and 137 ka at a depth over 16 m and ages 17.5 to 6.6 ka in the upper 12 m of aeolian sediment, thereby encompassing ca. 150 k years of aeolian deposition.

In an effort to reconstruct dune activity within the last 1000 years, Madole (1994) obtained ages from widely distributed sites north and south of the South Platte River, which showed that dune activity had been widespread and episodic in response to only slight variations in climate, given that the region is environmentally close to the threshold of sand reactivation. Madole (1995) provided preliminary age limits for his three recognized sand units (22.5–9.0, 8.0–1.0, and 1.0–0.15 ka) and also presented a detailed history of dune study in eastern Colorado in which he defined and described the major dune fields of the region: South Platte (collectively the Greeley, Fort Morgan, Hudson and Sterling dunes fields), Wray, Black Squirrel, Big Sandy, Baca, Apishapa, and Arkansas River valley dune fields. Clarke and Rendell (2003) subsequently provided eight IRSL ages to supplement previously reported radiocarbon age data (Madole 1994, 1995), which refined the late Holocene times of activity.

Muhs et al. (1996) considered the origin and to a lesser extent, the activation chronology of Late Quaternary dune fields in northeastern Colorado. Radiocarbon ages from two sites in the Fort Morgan dunes and one in the Greeley dunes were derived from carbonate nodules, organic coating on grains, and rhizoliths, suggesting three episodes of dune activity: 27–11 ka, 11–4 ka, and after 1.5 ka. They reasoned that paleowinds were similar to those of today, sand sources were the South Platte River for Fort Morgan and Wray dune fields and outcrops of the Laramie Formation for the Greeley dune field, and, like Madole (1995), concluded that dunes of the region are extremely sensitive to even minor fluctuations in climate. Madole et al. (2005) produced a pamphlet and companion map summarizing chronologically-based dune studies to date in eastern Colorado. Forman et al. (2005) later reported OSL ages ranging from 540 to 70 years ago from a borrow pit near Benkleman, Nebraska, in the eastern Wray dune field. Most recently, Mason et al. (2011) documented activation 9.6 ka deep within one of the largest dunes in the Wray dune field, which is younger than the Pleistocene ages obtained deep within the NSH to the north.

#### 4.2.2.4 Kansas

For Kansas, the first intensive studies of aeolian sand deposits occurred in the GBSP, the largest dune field in the state (ca. 4500 km$^2$) and defined by the 'great bend' of the Arkansas River in central Kansas as it flows west to east and then south across the state (Fig. 4.4). Johnson (1991) described the stratigraphy as a silty sand unit with multiple intercalated paleosols, overlain by varying thicknesses of aeolian sand. Radiocarbon ages from paleosols within the silty sand unit ranged from 16.6 to 0.76 ka, with spruce charcoal (*Picea* cf. *glauca*) within the unit dating to 21.6 ka. Expanded investigation of the GBSP stratigraphy by Arbogast (1996b, 1998) and Arbogast and Johnson (1996, 1998) consisted of radiocarbon dating of paleosols within the silty unit, overlying aeolian sand and probable isolated loess layers. The silty sand unit was subdivided into a lower unit yielding ages from 20 to 8 ka and an upper unit dating to 7.0 to 0.8 ka. Paleosols within dune and sand sheet produced radiocarbon ages indicating middle to late Holocene periods of stability. While parabolic dunes were the only identifiable dune type, they reportedly occur in two orientations, northwesterly and southwesterly, the oldest being formed by late Pleistocene northwesterly winds and the other by late Holocene southwesterly winds. To compare stratigraphy with the GBSP, a brief exploratory investigation was conducted in the Cheyenne Bottoms dune field located ca. 15 km north of the GBSP on the east-northeast side of Cheyenne Bottoms, a large (170 km$^2$) subsidence basin containing the largest US interior wetland. The silty sand unit again dated ca. 21.6 ka (ISGS-4485: 17,950 ± 310 years BP), whereas aeolian sand 1.6 m below an overlying dune crest dated 0.62 ka (ISGS-4475: 660 ± 70 years BP) (K.L. Willey and W.C. Johnson, unpublished data).

To the northeast of the GBSP lies the Hutchinson dune field, a linear dune field subparallel to the Arkansas River valley. Given the location of this dune field on the north side of the river and its northwest to southeast orientation, dune-building sediment was likely deflated from the Arkansas River by south-southwesterly winds (Bayne 1956), which is in contrast to the upstream Arkansas River valley, where dune fields were constructed on the south side of the river by north-northwesterly winds (Simonett 1960; Arbogast and Muhs 2000; Halfen 2012). This stabilized field has a core of dunes averaging 8 m high, surrounded by smaller, isolated dunes on its periphery. Halfen et al. (2012) documented only late Holocene dune activity for the Hutchinson dunes: 2.1–1.8, 1.0–0.9, and after 0.6 ka, indicating widespread dune field activation.

Northeast of the Hutchinson dunes lies the Abilene dunes, a relatively small, amoeba-shaped dune field (ca. 50 $km^2$) north of the Smoky Hill River in northeastern Kansas (Fig. 4.6) (Hanson et al. 2010). Two alluvial terraces occur on the north side of the Smoky Hill River: aeolian sand mantles the low terrace partially and the high terrace completely. Dunes are amorphous and tend to be of low relief, most under 6 m, but where steep dune faces do occur they are usually oriented southward, suggesting northwesterly activating paleowinds. As with the Hanson et al. (2009), Halfen et al. (2012), and Puta et al. (2013) studies, this investigation was an attempt to expand chronologically based dune studies beyond the larger, more concentrated dune fields well west of the 98th meridian. With the exception of one infinite age of >80 ka, the oldest ages from the dunes were 13.1 and 12.0 ka, but all others ranged 1.1–0.46 ka. The low-terrace fill radiocarbon dated at 30.5 ka and the high terrace was OSL- and radiocarbon dated to 44–50 ka.

Recent research continuing the quest to identify the eastward extent of megadroughts in the Central Great Plains (CGP) examined the small dunes perched on the high terrace of the lower Kansas River valley, northeastern Kansas (Johnson et al. 2019). These dunes are of relatively low relief (<10 m), occur in small patches, and are formed on terrace remnants on the north side of the river valley, reflecting likely southerly formative winds. OSL ages from dune crests were 40.2–33.7 ka, whereas those from the interdunal locations were 7.0–5.3 ka, indicating crest stabilization within late Marine Isotope Stage 3 (MIS 3), an interstadial of the last glacial period and a time with a relatively mild climate (Van Meerbeeck et al. 2009), a solar insolation maximum (Berger and Loutre 1991), proposed regional southerly winds (Johnson et al. 2019), and widespread upland soil development (e.g., Johnson et al. 2007; Muhs et al. 2008). The much younger middle Holocene interdunal ages suggest minor sediment contributions from the dune flanks and, given the high proportion of interdunal silt, from aeolian sedimentation as well. Persistence of the Kansas River dunes indicate that latest Pleistocene and Holocene droughts have not been of sufficient intensity to effectively extend their impact to this part of the eastern CGP.

Simonett (1960) conducted the first dune study focused solely on the Arkansas River of southwestern Kansas, where he reported a weak tendency for the dunes and interdunal depressions to orient south-southwest to north-northeast. He defined a discontinuous layer of loess within the aeolian sand body, a loess he proposed as the Peoria Loess of the last glacial maximum (LGM) and concluded, based on the pattern of dune lineation, that southwesterly winds were responsible for recent dune activity. However, given his observations of southeast trending bedding structures and the near-continuous distribution of late Pleistocene loess (Peoria Loess) above the old alluvium and below and within the dune sand effectively excluding the old alluvium as a possible dune sand source, he assumed the river, to the north, was the source of aeolian sand. In a subsequent study, Olson et al. (1997) conducted subsurface investigations in the same valley reach as Simonett (1960), and, although they encountered a loess zone within the aeolian sand, a paleosol immediately below the loess produced radiocarbon ages ranging from 7 to 7.6 ka, indicating that at least some of the loess was of Holocene age.

Forman et al. (2008) produced the first luminescence ages for dunes in the Arkansas River valley in a study focused on a reach west from Garden City to near the Colorado state line, where dune activation ages ranged from the Late Quaternary to the 1930s' Dust Bowl. Ages of 33.4 and 16.6 ka dated the terrace underlying the dunes, which in turn produced a range of 16.3–12.2 ka to 65–80 years ago. In a recent related investigation, Bolles et al. (2017) returned to the same study reach and obtained additional dune-derived ages dating to the historical period.

Halfen et al. (2012) focused on the entire reach of the Arkansas River valley from the GBSP to the Colorado state line, nearly 250 valley km. Ages from aeolian sand range from the late Pleistocene and early Holocene (15.7–10.3 ka) to those indicative of recent activation (34–30 years ago), although the bulk of the ages cluster during the middle Holocene and the last 2000 years, especially after ca. 700 years ago, all of which reinforce and add to those periods of activation identified by Forman et al. (2008). Terrace stratigraphy and fill ages indicate four terraces dating to ca. 60, 40, 30, and 15–13 ka, the latter two of which correspond to terrace fill ages identified by Forman et al. (2008). The most recent fill is historical and correlates to when the Arkansas River channel metamorphosed from a braided channel habit present as the valley underwent initial European settlement to the present single-thread, higher-sinuosity channel.

From its headwaters in southeastern Colorado and northeastern New Mexico, the Cimarron River flows through the panhandle of Oklahoma into southwestern Kansas where it forms a large bend in its course back southeastward toward Oklahoma. This bend is similar to that of the Arkansas River, in that its surficial geology consists of a complex pattern of loess, sand dunes and sheets, and alluvium (Johnson et al. 2009). In the western part of the Cimarron Bend, Olson and Porter (2002) reported alluvial ages as early as 38 and 10 ka, and noted late Pleistocene and early Holocene aeolian activity, which resumed during the middle Holocene as evidenced by sand sheets and dunes covering paleosols and other organic sediment dating 7.5–6 ka. The subsequent study in the Cimarron Bend area by Werner et al. (2011) investigated three localities, one in adjacent Oklahoma and two in Kansas, resulting in OSL ages ranging from a singular infinite age of >266 ka and late Pleistocene ages 40.6 and 12.4 ka, to Holocene ages of 6.4, 3.6–2.5, and 0.8–0.5 ka. A simple, but important observation resulting from the Werner et al. (2011) study was that the degree of surface-soil development has previously been largely overlooked in the interpretation of luminescence ages from aeolian sand deposits, in that a dune with a well-developed mature soil will resist reactivation by drought and wind significantly longer than a dune with little or no surface soil development, and the more frequently a dune is activated, the lower its threshold to reactivation due to depletion of the silt and clay components.

#### 4.2.2.5 Oklahoma

In the Cimarron River valley of northwestern Oklahoma, Brady (1989) described dune morphology, soil development and paleowinds, and produced radiocarbon ages from aeolian sand paleosols to 13.3, 8.5, 7.2, and 1.1 ka. Lepper and Scott (2005), incorporating and expanding upon earlier findings by Scott (1999), focused on the geomorphology and stratigraphy of a ridge dune on the second terrace above the Cimarron floodplain, dating the underlying terrace fill at 12.4 ka and buried dune soils at 3.3, 1.6 and 1.2 ka, as well as a cluster at 0.87–0.77 ka. From these data, they surmised that ridge dunes adjacent to shallow, sand-load rivers can potentially yield a record of activation over a 5 km reach. Cordova et al. (2005) assessed sand dune stability along the west-east bioclimatic gradient traversed by the Cimarron River towards its confluence with the Arkansas River in northeastern Oklahoma. They examined six sites situated from the westernmost Oklahoma panhandle (ca. 102.4° W) to north-central Oklahoma (ca. 97.5° W); that is, from High Plains sand-sage grassland to the oak cross timbers. Only four ages were obtained from two mid-transect sites: ca. 6.5, 3.1, and 0.2 ka, and post 1950 CE, but, based on historical aerial photography, the west-east trend in the sand mobility index (Lancaster 1988), and landscape observations, they concluded that most dune activity occurred west of the 98th meridian.

One dune study site not associated with the Cimarron River was that of Thurmond and Wyckoff (1998), which documented the existence of late Pleistocene dunes on the Demsey Divide, the interfluve between the North Fork of the Red River and the Washita River in extreme western Oklahoma. They attributed the dune sand source to the underlying Ogallala Formation (Ogallala Group in Nebraska), consisting primarily of Miocene–age, poorly consolidated Rocky Mountain-derived alluvium (Swinehart and Diffendal 1990; Ludvigson et al. 2009). Radiocarbon ages of ca. 31–17.4 ka obtained from paleosols within multiple dunes were far older than anticipated based on nearby late Paleoindian and younger archaeological sites and provide evidence of pre-31 ka activation.

#### 4.2.2.6 Texas and New Mexico

Discussion of aeolian deposit and activation research for this two-state region should begin with the Blackwater Draw Formation (BDF), named by Reeves (1976) and previously referred to as 'cover sands' (e.g., Frye and Leonard 1964). It is a sand- to clay-rich, cyclically deposited Quaternary aeolian deposit (>100 k km$^2$) including six major paleosols and mantling (<1 to 27 m thick) the Southern High Plains Plateau of northwestern Texas and eastern New Mexico. Its source was likely the Pecos River to the south, southwest and west (Holliday 1989, 1990). A maximum age of over 1.4 Ma was derived by Holliday (1989, 1990), based on TL dating (>270 ka) and volcanic ash ages, and subsequent OSL dating by Rich and Stokes (2011) indicated that the BDF was episodically accreting at ca. 204, 165, 100,

81–62, and 48–43 ka. Notably, dunes of the Southern High Plains, most of which are grass-covered and inactive, share an ancestry tied to the BDF.

Major aeolian sand deposits of the region include the Muleshoe (ca. 1850 $km^2$), Lea-Yoakum (ca. 1330 $km^2$), Mescalero (ca. 3000 $km^2$), and Monahans-Andrews (ca. 2570 $km^2$) dune fields and the Seminole sand sheet (ca. 13,000 $km^2$) situated between the Lea-Yoakum and Monahans-Andrews dunes (Holliday 2001). The Muleshoe dunes, consisting of barchanoid forms and blowouts associated with parabolic dunes, are a west-east, linear sand body located on the western side of the Southern High Plains escarpment and extend eastward into Blackwater Draw. Lea-Yoakum dunes occur in tributaries to the eastward flowing Mustang Draw and exhibit parabolic dunes with blowouts as well as historic fencerow dunes (Holliday and Meltzer 1995, Holliday 2001). The Mescalero and Monahans-Andrews dunes and sand sheets occur on the west and southwest side of the Southern High Plains Plateau. The Mescalero sands are 20–30 km wide and track the Southern High Plains escarpment for ca. 200 km and include blowout, parabolic, coppice and barchan dunes (Holliday 2001), whereas the Monahans field has developed between the plateau and the Pecos River, with a small dunal appendage, the Andrews dunes, jutting east into a valley cut into the plateau,. The Seminole sand sheet is an expansive but discontinuous body of thin (<1 m) sand but contains a few dunes locally on the western side of the Southern High Plains Plateau (Holliday 2001).

An early numerical dating study by Wells (1992) derived TL ages of 18–11 ka from a blowout and road cut in the Muleshoe dunes. Subsequent investigations of the Clovis archaeological site in the Muleshoe dunes by Haynes (1995) and Holliday (1995) produced late Holocene ages ranging from 4.2 to 0.65 ka and 5.6 ka, respectively. The most in-depth studies providing age control on the Muleshoe dunes were, however, those by Holliday (1997b, 2001), who reported ages from late Pleistocene, sub-dune sediments of 18–17 ka to early middle to late Holocene aeolian sand ages of 8.4–7.2, 5.6–4.1, and 1.3–0.5 ka. Rich and Stokes (2001) complemented Holliday's chronology with OSL ages of 12.9–12.5 ka and 0.7 ka. In OSL dating of Southern High Plains archaeological sites, Feathers et al. (2006) reported an age of 1.0 ka at the Clovis archaeological site. Most recently, Rich and Stokes (2011) produced OSL ages indicative of activation ca. 30 ka, again during the Holocene at 7.5, 4–3.6, 1.3–1.1 ka, and historically 1910, 1920, and 1930 CE.

The first luminescence ages derived for the Lea-Yoakum dune field were two OSL ages of 34 ka and 28 ka from dunes in the Milnesand, New Mexico area (Stokes 1994). Holliday et al. (1996), in an investigation of the lithostratigraphy and geochronology of fills in small playa basins, reported an age of 3.4 ka in lacustrine sediments immediately below aeolian sands, indicating deposition of the sand within the last 3000 years, but later obtained an older age of ca. 7 ka from a paleosol within a dune at another site (Holliday 2001). Feathers et al. (2006) produced older ages of 11.4–10.4 ka and a younger age of 120 years ago from a fence-line dune at the Ted Williamson and Milnesand archaeological sites in eastern New Mexico. Rich and Stokes (2011) OSL dated a sequence consisting of the BDF and two overlying sand units exposed in a road cut to 48, 3.6, and 0.11 ka, the youngest of which reflects activity in the late 1800s.

The Mescalero dune field, which rests on Permian-Triassic shale and sandstone, was determined by Hall and Goble (2006) to consist of a lower and upper aeolian sand unit that they OSL dated to 81.7–87.4 ka and 8.9–6.3 ka, respectively, and a near-surface weakly-developed paleosol dating to 370–150 years ago. Hall and Goble (2011) subsequently obtained additional OSL ages of 87.0 and 61.1 ka on the lower sand unit and 17.3, 13.9, 7.18, and 1.43 ka on the upper sand unit, concluding that the lower unit accumulated 90–50 ka and the upper, 18–5 ka. Rich and Stokes (2011) extracted OSL ages of 81 ka on the lower sand, 3.9, 3.7 and 2.3 ka from two coppice dunes overlying the BDF, 75 ka from a blowout (likely the BDF), and 47 ka from sands overlying the Ogallala Formation (again likely the BDF).

Monahans dunes, historically an active field, was first dated by Stokes (1994), who reported two OSL ages of 13 and 4.5 ka. Rich and Stokes (2011) dated dunes from the late Pleistocene (22 and 15 ka) and middle to latest Holocene (7.5 ka to 70 years ago). The adjoining Andrews dune field was dated 2.3 ka at the Bedford Ranch archaeological site by Holliday (2001), with Feathers et al. (2006) supplementing this site chronology with OSL ages of 7.5–7.3 ka, and also dating the Shifting Sands archaeological site to 10.6 ka.

Several other scattered sites have been dated, some of which are located on the southeastern end of the Southern High Plains Plateau in the lower reaches of Sulfur Springs, Mustang Spring, and Monahans draws (Holliday 1995). A wave-cut face exposing sand units at Red Lake dated to 9.5 ka and 1.8–1.6 ka (Holliday 2001), and an exposure at the Midland (a.k.a. Scharbauer) archaeological site produced ages of 12.1, 8.2, and 7.4 ka (Rich and Stokes 2001). At the Mustang Springs archaeological site, Rich and Stokes (2001) and Feathers et al. (2006) reported sand OSL ages of 5.8 ka and 6.4 and 3.8 ka, respectively. Elsewhere, Holliday (2001) dated aeolian sand along the margins of a playa basin at the Terry County site in the Seminole sand sheet to 360 years ago. OSL ages on aeolian sand at the Lubbock Lake archaeological site were reported by Rich and Stokes (2001) at 5.9 and 1.9 ka, and subsequently by Feathers et al. (2006) at 7.9, 4.9, and 2.4 ka.

Though not within the Southern Great Plains (SGP), the South Texas sand sheet on the coastal plain is sufficiently close that it warrants recognition, and more importantly, it has in the past been influenced by offshore continental winds. The South Texas sand sheet, consisting of a sequence of aeolian sand with intercalated paleosols, yielded ages representing two periods of deposition at 2.7 ka and 2 ka (Forman et al. 2009). Continentally derived northwesterly winds produced parabolic dunes, but the most recent activity ca. 200 years ago produced parabolic dunes from on-shore southeasterly winds. Dune activity since then was attributed to both overgrazing and minor climate fluctuations (Forman et al. 2009).

### 4.2.3 Fine-Grained Dunes of the Central and Southern Great Plains—Lunettes and Parna Dunes

Not all dunes consist of single-grain, sand-sized particles—some are comprised of silt with varying amounts of fine sand and clay, the lunettes, and others of sand-sized silt-clay aggregates, the parna dunes. Whereas lunettes are ubiquitous, ranging from Wyoming, Nebraska, Colorado, Kansas, and Texas to New Mexico, parna dunes are a recently recognized landform in the CSGP, occurring in the panhandle of Oklahoma. While the focus of this chapter is aeolian sand dunes and sheets, these fine-grained aeolian landforms merit recognition due to their regional ubiquity (lunettes) and unique nature (parna dunes).

#### 4.2.3.1 Lunettes

Lunettes are arcuate-shaped dunes that form in combination with playas, which in turn are endorheic and ephemeral basins: internally drained, closed depressions that are isolated above the water table and have infrequent short-term hydroperiods, becoming flooded only when runoff results from heavy rainfalls or unusually large, rapid snowmelts (Smith 2003). Lunettes were first named and studied in Australia by Hills (1940), where they were recognized in association with ephemeral or dry lakes; they have since been reported in semi-arid and arid environments globally (Goudie and Wells 1995). Lunettes of the SGP were the first in the Great Plains to be investigated (e.g., Judson 1950; Reeves 1965; Holliday 1997a, 1985). Lunette investigations in the CGP were first conducted on the Rainwater Basin area of southern Nebraska by Starks (1984), Krueger (1986), Kuzila and Lewis (1993) and Zanner and Kuzila (2001). Arbogast (1996a, 1998) was the first to investigate lunette stratigraphy and age in Kansas, at Wilson lunette (a.k.a. Wilson Ridge), located on the southwestern fringe of the GBSP. Wilson lunette is one of the best formed in Kansas, extending 2.4 km along the south and southeastern fringe of the playa and exhibiting its highest point along the southeastern edge (Fig. 4.8a, b). Subsequently, Bowen and Johnson et al. (2012) investigated age and stratigraphy of a lunette at the 1.5 km-long Ehmke playa of west-central Kansas.

Due to the inherent relationship between playas and lunettes, this combination has been designated to as the 'playa-lunette system' (Bowen and Johnson 2012). During times of extreme drought, airborne silt-dominated sediment deflated from the playa is deposited on its downwind side to form the lunette. Despite the tens of thousands of playas on the CGP and SGP, only a small percentage have lunettes. For example, of the greater than 20,000 playas mapped on the High Plains of western Kansas (Bowen et al. 2010), less than 200 lunettes have thus far been recognized (Bowen et al. 2018). It appears that large playa size, and perhaps depth and wind fetch are the major determining factors in the development of lunettes. Similarly, on the SGP of Texas and adjacent New Mexico, lunettes are associated with some

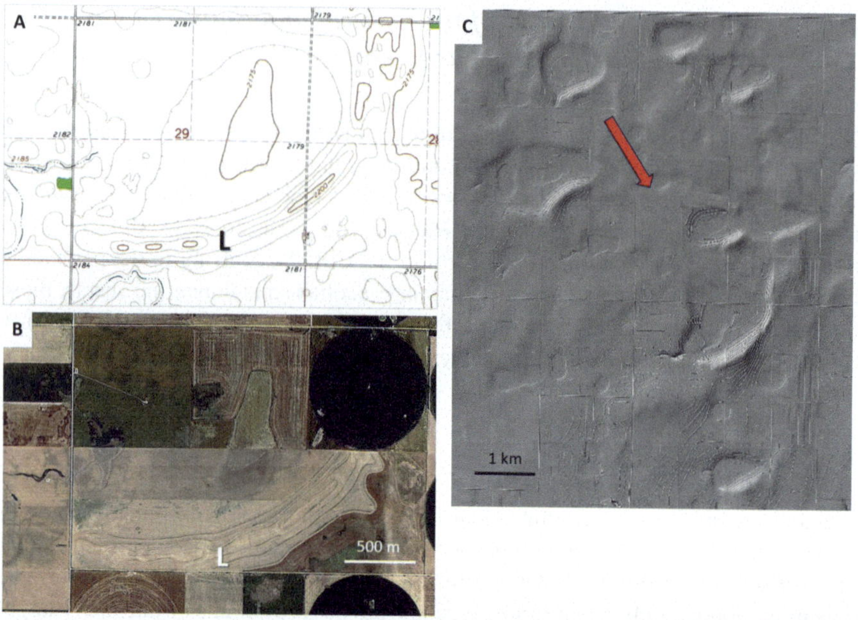


**Fig. 4.8** Lunettes within Kansas. (**a**) US Geological Survey topographic map of Wilson lunette (L) and playa on the southwest edge of the Great Bend Sand Prairie of the Arkansas River in central Kansas; (**b**) Aerial image of Wilson lunette and playa (USDA Farm Service Agency National Agriculture Imagery Program); (**c**) Lunette formation on the southeastern edge of playas on the High Plains of western Kansas, with the arrow representing the formative northwest glacial-age winds

larger, deeper playas (Holliday 1997a; Sabin and Holliday 1995), though lunette mapping has only covered part of the region (Sabin and Holliday 1995).

Timing of lunette formation has been documented through numerical dating using both radiocarbon and luminescence methods by investigators in Kansas and Texas into New Mexico (Table 4.4). In Kansas, Arbogast (1996a, 1998) produced radiocarbon ages of 20.9–2.9 ka from paleosols in Wilson lunette, whereas Bowen and Johnson et al. (2012) radiocarbon dated paleosols in the Ehmke and Bush lunettes to 42.1–8.3 ka. Lunettes in Kansas, however, may be much older in that Bowen and Johnson et al. (2012) documented through drilling that the Ehmke playa had it origin at least as early as the last interglacial; a similar observation was made of the playas comprising the Rainwater Basins of south-central Nebraska (Kuzila 1994). Therefore, the playa-lunette systems appear to be relict features that have migrated upward as successive layers of last glacial period loess had been deposited and intercalated soils formed.

The largest lunette age database has been reported from Texas and adjacent New Mexico: Holliday (1985, 1997a, b) derived an age range of 37.8–0.51 ka and Frederick (1993, 1998) 11.7, 9.9, 3.2, and 0.81 ka. Earliest luminescence dating by Rich et al. (1999) reported a range of 5.6–0.7 ka; subsequently, Wood (2002)

**Table 4.4** Silt dunes and chronological studies of the Central and Southern Great Plains

| Silt dunes | Lat.[a] (°N) | Long.[a] (°W) | Studies | $^{14}C$ ages | Lum. Ages[b] |
|---|---|---|---|---|---|
| *Lunettes* | | | | | |
| Kansas | | | | | |
| Wilson playa | 37.841 | 99.320 | Arbogast (1996a, 1998) | 12 | – |
| Bush playa | 38.352 | 100.458 | Bowen and Johnson (2012) | 6 | – |
| Ehmke playa | 38.437 | 100.598 | Bowen and Johnson (2012) | 6 | 5 |
| Texas | | | | | |
| Southern High Plains | widespread | widespread | Holliday (1985, 1997a, b) | 37 | – |
| Lubbock Lake | 33.619 | 101.889 | Feathers et al. (2006) | – | 2 |
| Red Lake | 32.168 | 101.666 | Frederick (1998) | 1 | – |
| Sulfur Springs draw | 32.330 | 101.738 | Frederick (1998) | 2 | – |
| Rich Lake | 33.266 | 102.170 | Rich et al. (1999) | – | 5 |
| Mound Lake | 33.219 | 102.068 | Rich et al. (1999) | – | 6 |
| Double Lakes | 33.115 | 101.810 | Rich et al. (1999) | – | 3 |
| Double Lakes | 33.115 | 101.810 | Wood (2002) | – | 20 |
| Double Lakes | 33.201–33.203 | 101.876–101.916 | Rich (2013) | – | 20 |
| Illusion Lake | 33.842 | 102.453 | Rich (2013) | – | 2 |
| White Lake | 33.947 | 102.763 | Rich (2013) | – | 2 |
| Jackson Lill | 34.662 | 101.540 | Rich (2013) | – | 3 |
| Julia Lake | 33.990 | 101.731 | Rich (2013) | – | 3 |
| *Parna dunes* | | | | | |
| Oklahoma | | | | | |
| Blue Mound | 36.551 | 100.773 | Fine et al. (2011, 2012) | 3 | – |
| Blue Mound | 36.551 | 100.773 | Johnson et al. (2010, 2012) | 3 | 3 |
| Gray Mound | 36.564 | 100.792 | Fine et al. (2011, 2012) | 1 | 3 |
| **Total** | | | | **70** | **75** |

[a]Approximate midpoint location of the lunette or parna dune, except for Double Lakes, Texas, where range is provided given the multiple lunettes

[b]Includes all luminescence-derived ages (OSL, IRSL, TL)

provided ages ranging between 122 and 0.5 ka. Rich (2013) dated a sampling of both large and small playa lunettes, which produced an age range on the former of 277–5.6 ka and on the latter 207–0.29 ka. Though the luminescence age data from Texas indicate that lunette development formation began as early as 277 ka (Rich 2013), the oldest ages were from relict playa-distal lunettes in contrast to the younger proximal lunettes, which began taking form since the last glacial period, supplemented by periodic inputs during the Holocene (Holliday 1985, 1997a).

The propensity of lunettes to occur on the south to southeast side of playas indicates formation by northwesterly winds, whereas present dominant winds for most of the region are southerly. In Kansas, the effectiveness of the northwesterly paleowinds is apparent in the greatest number of playa-lunette systems being located on a relatively broad and undissected upland in west-central Kansas, where wind fetches can exceed 40 km northwest to southeast, thus promoting relatively active deflation of playas during the LGM (Bowen et al. 2018)(Fig. 4.8c).

In addition to being another fascinating landform for geoscientists, lunettes have been a draw for archaeologists. Early Native American use of playas and lunettes has been well documented in the CGP and SGP, including locations in New Mexico (Hill et al. 1995), Texas (Holliday 1997b; Litwinionek et al. 2003), Oklahoma (Labelle et al. 2003), New Mexico (Feathers et al. 2006), Kansas (Witty 1989; Mandel and Hofman 2003), and Colorado (Stanford 1979). Ehmke lunette is the only lunette in Kansas with a documented coherent archaeological site, which consists of multiple cultural components (Paleoindian to Late Ceramic) and artifacts including lithics (e.g., Paleoindian projectile points) and faunal remains (e.g., mammoth, bison, horse) (Witty 1989). The highest probability of burial and preservation of archaeological remains is on the leeward side of the lunette crest (Arbogast 1996a; Holliday 1997a; Bowen and Johnson 2012) because compression of the wind streamlines on the windward side of the lunettes promotes sediment transport and erosion, whereas on the leeward side the wind flow streamlines diverge thereby reducing velocity and promoting deposition (Lancaster 1985). Historically, many lunettes have been cultivated, which has disturbed much of the cultural record, though such disturbance has also resulted in surface exposure of artifacts that has in several instances enticed archaeologists and amateur collectors.

#### 4.2.3.2 Parna Dunes

Working in central Australia, Hills (1939) was the first to recognize the type of dunes comprised of aggregated clay particles rather than individual sand grains. Butler and Hutton (1956) subsequently introduced the term 'parna' for this aeolian clay material, which is an aborigine word for 'dusty and sandy ground.' Parna has been documented in Beaver County in the Oklahoma panhandle, where upland dune forms are composed of sand-sized aggregates of clay and silt. These dunes occur in two swarms, range in height 10–15 m, and have asymmetrical dome morphologies with approximate north-south long-axis orientations (Johnson et al. 2012) (Fig. 4.9a–c). While morphologically similar to sand dunes, their origin and evolution are yet unknown. Stratigraphies of two parna dunes, Blue and Gray mounds (Fig. 4.9c), were characterized through coring, which documented a basal hydric paleosol dating to 25–21 ka at a depth equivalent to the present-day surrounding landscape (Table 4.4). Given that hydric soils are associated with many of the playas on the surrounding landscape today, the dune forms may have been anchored upon

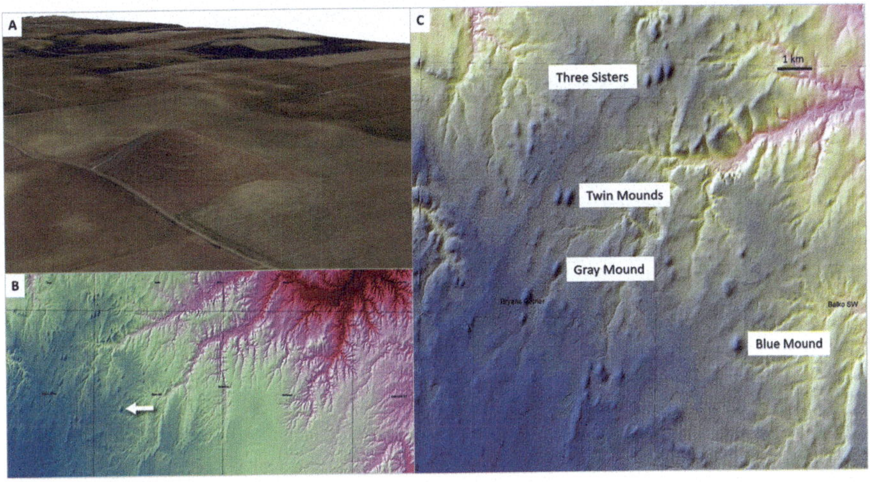


**Fig. 4.9** Parna dunes of the Oklahoma panhandle. (**a**) Oblique view of the parna dune Blue Mound as depicted by image-draped IFSAR elevation data (view to the northeast); (**b**) False-color hill-shade image of the western and eastern parna dune fields, with the arrow identifying Blue Mound in the western dune field (fine black grid line demarcate the 7.5-min topographic quadrangle boundaries); (**c**) Larger-scale view of named dunes to illustrate the approximate north-south orientations. (Images provided by S. McGowen, USDA Natural Resources Conservation Service)

encountering paleoplayas during the LGM (Johnson et al. 2010, 2012). Dune pedostratigraphy indicated up to ten major episodes of sediment flux (Fine et al. 2011, 2012), though near-surface luminescence (OSL) ages from Blue Mound are similar to calibrated radiocarbon ages from the basal paleosol, suggesting rapid dune construction, with little or no Holocene accumulation of sediment (Johnson et al. 2010, 2012). Based on chronological data and geomorphic observations, parna dune construction in the Oklahoma panhandle was driven by strong, northerly winds (Johnson et al. 2012). In that parna formation is initiated typically by the desiccation of saline water bodies, their development in this area was likely linked to the Permian bedrock exposed in the dissected landscape immediately to the north, which is a ready source of silt- and clay-sized particles (likely ancient loess deposits or 'loessite'; see Soreghan et al. 2008) and salt. With additional research, parna dunes, like lunettes, may prove to be a potential source of paleoenvironmental and archaeological information for the CSGP.

## 4.3 Dune Development and Subsequent Activation

When considering the history of aeolian sand systems, three basic factors converge to determine accumulation, reactivation, or stability: sediment availability, transport capacity, and sediment supply, (Kocurek and Lancaster 1999). Availability of sand can be a function of the condition of stabilizing vegetative cover or lack thereof, which relates to climate (drought) and attendant changes in land cover, or to changes in the amount or caliber of sediment carried by river systems or yielded by outcrops. The prevailing wind regime determines transport capacity, given favorable conditions of supply and availability, but only if the winds are from the requisite direction relative to the sand source and of sufficient velocity when sand is vulnerable (seasonality). This section considers the chronology of prehistoric dune activity and the climatic forcing of the activity, whereas sediment source is discussed in Sect. 4.4.

### *4.3.1 Chronology of Prehistoric Regional Dune Activity*

While the oldest preserved CSGP dune forms yet documented appear to be in northeastern Kansas (CGP), where stabilization occurred ca. 40–34 ka (Johnson et al. 2019), several other studies have reported early evidence of regional dune building or activation, such as the sand dune ages of ca. 140 ka from the Fort Morgan dune field (Forman et al. 1995), 87–61 ka from the Mescalero dune field (Hall and Goble 2006; Rich and Stokes 2011), and 41 ka from the Cimarron Bend (Werner et al. 2011), and also the late Pleistocene ages from lunettes and parna dunes (e.g., Holliday 1997a; Bowen and Johnson 2012; Johnson et al. 2012; Rich 2013). Latest Pleistocene activity has been recognized at some sites from ca. 25 ka through the LGM to the earliest Holocene, such as the Stanton dunes (Nebraska: Puta et al. 2013), South Platte River dunes (Colorado: Madole 1995), Arkansas River dunes (Kansas: Forman et al. 2008; Halfen 2012) and Demsey Divide dunes (Oklahoma: Thurmond and Wyckoff 1998), and sites in Texas and New Mexico (Holliday 2001). Mason et al. (2011) identify the interval 13–10 ka (i.e., the Pleistocene-Holocene transition) as a time of very little activity in the NSH, which they interpret as a time of stability (minimal wind transport) as expressed by the development of the Brady Soil in the upland loess (e.g., Johnson and Willey 2000; Mason et al. 2008; Muhs et al. 2008). Notwithstanding the disparate and low-resolution age data elsewhere, it does appear that less activity has been documented during this interval in the CGP, but not in the SGP.

Given the coarse age resolution and widely scattered nature of the documented sites, the early Holocene and middle Holocene periods cannot be split into two distinct periods. Though certainly not ubiquitous, indications of early Holocene aeolian activity occur widely distributed throughout the CSGP. These include, for example, Wyoming's Casper and Ferris dune fields at ca. 10 and 8.8 ka, respectively (Halfen et al. 2010; Stokes and Gaylord 1993), Stanton and Imperial dune fields 11–10 ka (Puta et al. 2013; Mason et al. 2011) of Nebraska, Colorado's Fort Morgan, Greeley, and Hudson dune fields ca. 11 ka (Muhs et al. 1996; Holliday 1987; Forman et al. 1992), Arkansas River dune field ca. 10 ka (Forman et al. 2008), and Muleshoe dune field ca. 8.4–7 ka (Holliday 1997b, 2001). Activity continued into the middle Holocene was reported also throughout the CSGP and appeared to have occurred at ca. 6.5 and 4 ka; for example, Wyoming's Casper and Ferris dune fields were active ca. 4 ka (Halfen et al. 2010; Stokes and Gaylord 1993), Nebraska's Duncan and

Stanton dune fields ca. 6 and 4 ka and 4.4–3.4 ka, respectively (Puta et al. 2013; Hanson et al. 2009), Colorado's Hudson dune field ca. 6.3–5.4 ka (Forman et al. 1995) and the Fort Morgan dunes ca. 8–4.9 ka (Madole 1995; Forman et al. 2005; Clarke and Rendell 2003), Kansas' Arkansas River and Cimarron River dune fields ca. 5.8 and 6.4 ka, respectively (Halfen 2012; Werner et al. 2011), and the Texas and New Mexico dune fields of the Southern High Plains Plateau ca. 5.6–4 ka (Holliday 1997b, 2001). If a hiatus or reduction in activity did occur, it possibly did so between ca. 6 and 4 ka, at least for the CGP. This is consistent with the CGP loess record, which shows extended early to middle Holocene activity (ca. 10–6 ka), followed by a period of soil formation (stability) from ca. 6–4 ka, with a distinct resumption in sedimentation after 4 ka and on (Miao et al. 2007).

Most all of the CSGP dune fields were active during the late Holocene, especially during the past 2000 years (e.g., Holliday 2001; Forman et al. 2005, 2008; Halfen et al. 2012), much of which occurred during the past few hundred years (see Halfen and Johnson 2013 for a summary). Of note is the degree of synchrony among sites within the CSGP for the period ca. 1100 and 700 years ago. For example, of the 60-plus OSL ages obtained from within the Hutchinson dune field of eastern Kansas, with only very few exceptions ages fell within the last 2000 years and many of those were within the above time interval (Halfen et al. 2012). Interestingly, aerial imagery, field observations and luminescence ages indicate that the extensive activity, rather than developing new dune forms, involved migration of existing dunes and modification of late Pleistocene dunes such as parabolic forms. Holocene activity has, for example, modified parabolic into subparabolic dunes in the GBSP (Arbogast and Johnson 1996) and reversed the direction of sand transport on transverse dunes in the Arkansas River dune field (Halfen 2012).

Given the vast number of ages indicative of dune building or subsequent activation, one could assume that such a database should make it possible to accurately interpret spatial and temporal patterns of prehistoric dune activity. There are, however, limitations that must be taken into consideration when interpreting patterns. A primary consideration is the fact that ages reported for the region are not distributed proportionally; for example, some dune fields have a high density of reported ages, whereas most have a low density. In addition, no standard research design for sampling has been adopted, which has resulted in unintentional biases and often disregard for the complexity of dune fields and individual dunes. Another consideration, or reality, is that some dune fields have been so thoroughly reworked, even to significant depths, during the late Holocene such that only sparse and often fragmentary records of earlier activity may have been preserved. Lastly, because aeolian sand bodies are typically void of regionally-expressed paleosols such as those expressed within the regional loess stratigraphy (e.g., Johnson et al. 2007; Miao et al. 2007; Muhs et al. 2008) or have only isolated paleosols occurring in segments of individual dunes, an inherent issue to be recognized when luminescence dating of aeolian sand bodies is that most resulting ages are representing stratigraphic hiatuses rather than capturing detailed activation histories.

### 4.3.2 Climate and Dune Activation

Despite the multiple issues surrounding the interpretation of prehistoric temporal and spatial patterns of dune activity, observations regarding the link between aeolian sand activity and climate can be gleaned from the CSGP database of ages reported to date. Chronologies of individual dune field activation throughout the entire CSGP do not neatly align with one another due in large part to the sheer size of the region, but dune chronologies appear asynchronous primarily due to the behavior of climate shifts, which can be synchronous but in opposite directions, or time-transgressive (Miao et al. 2007; Williams et al. 2010). Timing of reactivation in the Monahans dune field may not, for example, be synchronous with reactivation in the Hutchinson or Wray dune fields, given the variable geography of drought (e.g., Woodhouse and Overpeck 1998; Cook et al. 2007, 2016; Stahle et al. 2007). Broad temporal patterns do, however, emerge.

The causal connection between large-scale atmospheric changes and droughts has not yet been clearly established. Some studies have identified El Nino Southern Oscillation (ENSO) conditions as the primary driver of prehistoric droughts (Herweijer et al. 2006; Seager et al. 2009; Woodhouse et al. 2009), others concluded from models that indicated shifts in sea-surface temperatures (SSTs) brought drought condition to the CSGP (Feng et al. 2008; Forman et al. 2001), and others as well to changes in the North Atlantic Multidecadal Oscillation (AMO) (Enfield et al. 2001; Knight et al. 2006; Shin et al. 2010) or in the Pacific Decadal Oscillation (PDO) (McCabe et al. 2004; Tian et al. 2006). Human-induced land degradation may complicate the quest for climate-drought linkages: SSTs did not fully model the Dust Bowl drought, but the addition of human forcing (e.g., reduction of vegetation cover and increased atmospheric dust loading) did produce simulations similar to observational data (Cook et al. 2009).

While the first significant preserved activity in the dune fields of the CSGP has been reported during the late Pleistocene and early Holocene, there is a paucity of dune activity indications for the late Pleistocene overall, which was due at least in part to the cooler and moister climate as evidenced by stable isotope data from an area adjacent to the western limit of the dunes in central New Mexico (Hall and Penner 2012), a peak in relative humidity during the Younger Dryas (Voelker et al. 2015), and spruce macrofossils including charcoal, needles and cones from sites in Kansas (Fredlund and Jaumann 1987; Wells and Stewart 1987; Feng et al. 1994). Conversely, it can be argued that low $CO_2$ levels in the atmosphere may also have created plant stress sufficient to weaken anchoring vegetation (Monnin et al. 2001; Prentice et al. 2017). Further, Holliday (2000) reported evidence for episodic drought on the Southern High Plains during the Younger Dryas.

As noted above, early and middle Holocene activity did occur but the timing is unclear, though proxy data peripheral to the CSGP indicate more or less tenacious drought (e.g., Williams et al. 2010; Shuman and Serravezza 2017). Middle Holocene dune activity has been largely attributed to well-documented early to middle Holocene drought in the Great Plains (e.g., Dean et al. 1996; Fritz et al. 2001; Smith et al. 2002; Schwalb et al. 2010), which appeared to have abated somewhat ca. 4 ka (Booth et al. 2005; Carter et al. 2018).

Widespread activity documented for the late Holocene, particularly the past 2000 years, occurred during a period characterized by multiple long-duration droughts (e.g., Dean et al. 1996; Denniston et al. 1999; Schwalb et al. 2010). Of note is the relatively good region-wide synchronous dune activity documented during the Medieval Climatic Anomaly (MCA) (a.k.a. Medieval Warm Period), ca. 1100–700 years ago (e.g., Cook et al. 2004; Herweijer et al. 2007; Feng et al. 2008), with the warmest period ca. 950 years ago (Esper et al. 2002). Periodic drought during this period has been documented by multiple proxies, including lake sediments, tree-rings, alluvial sequences, and indications of wind shifts (e.g., Laird et al. 1998; Daniels and Knox 2005; Sridhar et al. 2006; Denniston et al. 2007; Cook et al. 2009; Schmeisser et al. 2010). The term 'megadroughts' has been used to characterize the long-duration (multi-decadal), intense and often spatially extensive droughts recorded during the MCA (Woodhouse and Overpeck 1998; Stahle et al. 2007; Cook et al. 2011, 2016), which is in contrast to the shorter historical droughts. Prehistoric megadroughts during and after the MCA occurred at irregular intervals, which adds to the difficulty of ascertaining the cause. Accordingly, several potential causes have been proposed, including reduced volcanic activity, increased solar irradiance, changes in SSTs, and land-atmosphere interactions, but the available data are too low in resolution (Cook et al. 2016).

Dune activity after ca. 600 years ago, which is common to dune fields in the western CSGP, has been correlated to several intervals in the Little Ice Age (LIA), a cool to cold period following the MCA. Due to an apparent time-transgressive nature, the LIA temporal boundaries are not fixed but are ca. 700–100 years ago, or 1300–1900 C.E. (Miller et al. 2012). Though seemingly not as widespread as the megadroughts of the MCA, the LIA megadroughts have been recognized in the tree-ring record (Stahle et al. 2000; Herweijer et al. 2006). Activity during the LIA has been documented in several dune chronologies in the CSGP, but LIA dune activity is particularly pronounced in the record from the Hutchinson dunes (Halfen et al. 2012), where it was most pronounced during the coolest times of the LIA (Mann et al. 2009). Overall, the record indicates, however, that though dune activity was most intense during the MCA, it continued to be chronic into the twentieth century.

Several detailed proxy records of late Holocene climate have been derived from lakes and wetlands in the Great Plains but have been from sites located within the northern US Great Plains or in the immediately adjacent prairie-forest ecotone (e.g., Dean et al. 1996; Denniston et al. 1999; Schwalb and Dean 2002; Schwalb et al. 2010; Grimm et al. 2011; Hobbs et al. 2011). One such study site relatively close to the CSGP, Pickerel Lake, northeastern South Dakota (Schwalb et al. 2010) documents the recurrence of windy and dry conditions ca. 3250, 2700, 2150, 1600–1400, 1000–600, and 150 cal years BP, with the two drought maxima occurring 1000–600 years ago (coincides with the MCA) and 150 years ago (coincides with the LIA). As with other proxy sites, Pickerel Lake data document the complexity of late Holocene climate and the prevalence of drought periods, but given the variability in spatial and temporal climatic patterns, it is unlikely that the record in South

Dakota would be the same or even similar to that of the entire CGP and certainly the SGP. Consequently, an increased understanding of the large-scale atmospheric mechanisms causing CSGP droughts would be extremely useful in deciphering the temporal and spatial relationships among dune field activation records within the CSGP.

While the generally accepted paradigm is that prehistoric aeolian sand activity is a function of drought, some activity was most certainly caused by localized disturbances including grass and shrub fires, bison traversing dunes, and Paleoindians and their successors (Loope 1986; Schlesinger et al. 1990; Forman et al. 2001; Mason et al. 2004).

## 4.4 Aeolian Sand Sources

When considering the history of aeolian sand systems, a fundamental issue is that of the sand source, or its provenance. Sediment supply is essential in that it creates the potential for dune or sand sheet development or activation and can be any sand-bearing source, such as easily eroded bedrock outcrops (e.g., Ogallala Formation), older unconsolidated aeolian sand deposits (e.g., BDF), and sand-load river channel systems (e.g., Platte, Arikaree, Arkansas, and Cimarron rivers). Further, sediment comprising a dune field or sand sheet may have originated from not a single, but multiple sources at a given time or throughout its history.

### *4.4.1 Fluvial Systems as a Sediment Source*

Most dune fields and sand sheets of the CGP, unlike those of the Northern Great Plains and to a lesser extent the SGP, are geomorphically or at least spatially linked to fluvial systems (a major exception being the NSH), which can be sources of sand especially when dry (Muhs and Holliday 1995; Muhs et al. 1996; Muhs 2004; Forman et al. 2008; Halfen and Johnson 2013; Halfen et al. 2016). For example, initial dune development along the Arkansas River, expressed as large transverse dunes, occurred 16–13 ka (Forman et al. 2008; Halfen 2012), which Halfen et al. (2012) and Halfen et al. (2016) argued was due to large volumes of sediment being transported by the river in response to collapse of the Pinedale glaciers in the Rocky Mountain headwaters (Briner 2009; Young et al. 2011). This coincides with aeolian sand activity in the GBSP (Arbogast 1996b; Arbogast and Johnson 1998) and with the intercalated loess deposits (Simonett 1960; Olson et al. 1997). The Cimarron River, draining the eastern side of the Sangre de Christo Mountains, may have similarly provided a late Pleistocene sand supply for the Baca, Cimarron Bend, Cimarron Valley, Little Sahara and other related dune fields (Halfen et al. 2016), but the Cimarron does not have the high-relief dunes of the Arkansas River and GBSP probably due to its smaller drainage area and attendant sediment load (Fig. 4.10).

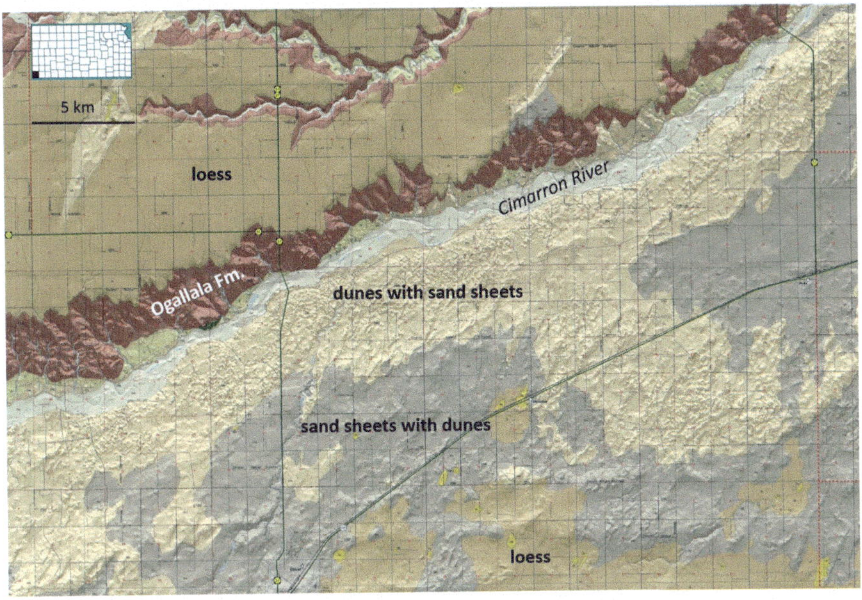


**Fig. 4.10** Surficial geologic map of the Cimarron River corridor in Morton County, extreme southwestern Kansas (Johnson et al. 2009). Aeolian sand is confined to the south side of the Cimarron River with the dunes (coarser sand and higher relief) proximal to the channel and the sand sheets more distal (finer sand and lower relief) and grading into loess

Age data suggest that dune fields proximal to large river systems were constructed during the high-sediment supply of the late Pleistocene, whereas the Holocene was characterized by activations largely related to drought-induced sediment availability (Halfen et al. 2016). Conversely, dune fields developed adjacent to rivers purportedly affect the river course or pattern in multiple ways (Muhs et al. 2000; Liu and Coulthard 2015).

Recent drying of river channels is creating a contemporary sand source for dune construction (Fig. 4.11). With the advent of irrigated agriculture in the western Great Plains, surface water and groundwater sources were exploited causing the water table to drop, a phenomenon evident, for example, throughout much of western Kansas. Irrigation canal diversions and groundwater mining for the widespread use of center pivot irrigation since the 1960s in the Arkansas River valley have dropped the water table substantially, creating a dry river bed and changing channel geometry (Andrzejewski 2015). Other rivers of the region have undergone similar changes, including the Cimarron (Schumm and Lichty 1963; VanLooy and Martin 2005) and North and South Platte rivers (Williams 1978; Eschner et al. 1983; Horn et al. 2012). Though the geologically young, unconsolidated alluvial sediments appear to have contributed sand to dune fields and sand sheets, outcrops of the Ogallala Formation, as an ancient river deposit, are also potential sources due to their wide distribution within the CSGP.

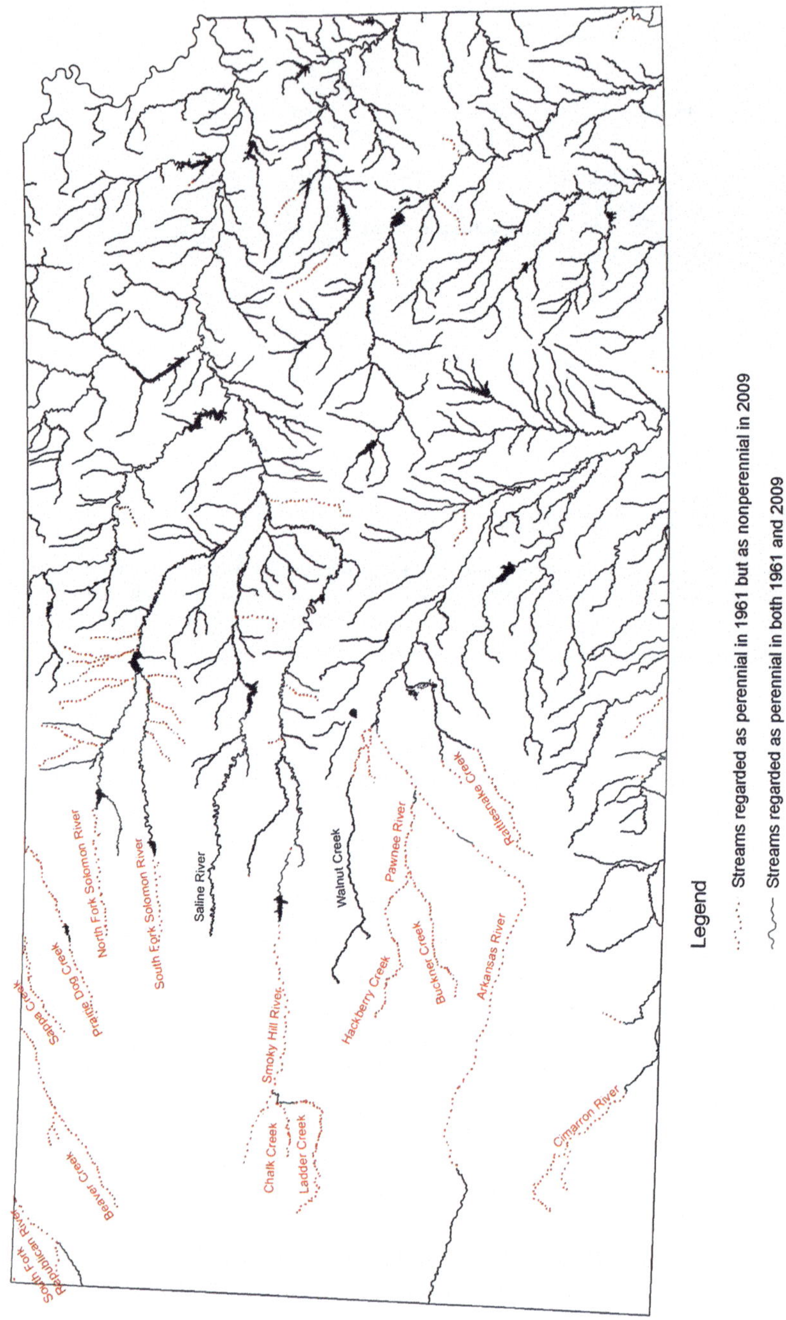


**Fig. 4.11** Streams of western Kansas that were perennial when observed in 1961 but were non-perennial (ephemeral) by 2009 (red). Drawdown of the High Plains Aquifer for irrigation has rendered these channel dry, except on rare occasions during heavy downpours or extended rainy periods. (Kansas Department of Agriculture 2010)

### 4.4.2 Documentation of Sand Provenance

Sand sources first proposed for the CGP and SGP dune fields were a product of observations and theories rather than quantitative geologic data. Regarding the Ferris dune field of Wyoming, Gaylord (1982) proposed that the Tertiary Battle Spring Formation was the primary source, and that the Killpecker dune field 'tail' sands and some Cretaceous through Paleocene sandstone outcrops were secondary sources. Swinehart and Loope (1992) and Loope et al. (1995) speculated that the Lincoln County dune field and possibly the Imperial dune field of southwestern Nebraska were once part of a larger NSH, created by a last glacial-age dune damming of the Platte River system. Failure of this dam would have resulted in the isolation of the two dune fields south of the South Platte River. This theory was based on lacustrine sediments (Diffendal and Leite 1989) terminally dated to about 12.5 ka (Muhs et al. 2000), though such a dune dam would require a massive volume of sand given the sheer size of the Platte River valley and the high valley wall siting of the lacustrine sediments.

Lugn (1968) proposed that the dune fields proximal to the South Platte River in northeastern Colorado (Fort Morgan dune field), as well as others in the region, were derived from the Ogallala Formation; Hunt (1986) agreed but suspected that the South Platte had some input, and Muhs (1985) considered the South Platte River as the primary source. In contrast to the dune fields south of the South Platte River, the Greeley dune field, being on the north side of the river, had to have another source given the northwest paleowinds, and consequently bedrock outcrops to the north were proposed as sources by Madole (1995). Dunes of the GBSP, central Kansas, were first reported to have the Arkansas River as their likely source by Smith (1940) and subsequently Simonett (1960), whereas Stramel et al. (1958) proposed older sand deposits to the south provided the sediment. Arbogast (1996b) and Arbogast and Johnson (1998) documented and radiocarbon dated sandy late Pleistocene alluvium underlying the modern dune field, which suggests that this sediment may have been the dune sand source. As for the dune fields of Texas and New Mexico, Lugn (1968) considered the Ogallala Formation to be the sole source, but Carlisle and Marrs (1982), using a remote sensing approach, proposed topographically-defined, south-to-north wind-transport of sand, which implied the Pecos River was the source.

With the rapid increase in dune studies providing new information on ages of deposition determined by luminescence dating, proximity to sand sources, wind direction indicated by dune orientations, stratigraphy and other observations generated interest in applying more sophisticated approaches to tracing aeolian sand deposits to their source. Despite a tradition of sedimentary geologists using mineralogical maturity to decipher rock history, aeolian geomorphologists belatedly adopted this approach, which assesses the amount of relatively low-resistance minerals (e.g., feldspars, carbonates) in quartz-dominated rock material (Muhs 2004). Winspear and Pye (1996), in a comparison of the absolute oxide content in samples from several blowouts in the Lincoln County dune field (their 'Dickens' dune field) with those of the NSH and the South Platte River, concluded that the South Platte River was the more likely source. The oxide-to-quartz relationship indicated that the Fort Morgan dune field was sourced from the South Platte River (Muhs et al. 1996; Muhs 2004), whereas the Casper dune field was from the North Platte River (Muhs 2004). To the south, Muleshoe dunes were linked to the BDF (Muhs and Holliday 2001; Muhs 2004), but Monahans dune field was problematic in that its mature mineralogy had occurred by in situ weathering, abrasion, or ballistic impact (Muhs 2004). Within the GBSP, concentrations of Ca and Sr, reflective of carbonate minerals, displayed a systematic decrease with increasing distance from the Arkansas River, suggesting increased maturity is a product of abrasion and size reduction (Arbogast and Muhs 2000).

Given that potential sources typically have the same mineralogy as the resultant dunes, Muhs et al. (1996) employed trace element geochemistry (Rb, Sr, Ti, Zr) to determine sources for the Fort Morgan, Wray and Greeley dune fields of northeastern Colorado. All three dune fields yielded the same signature as the South Platte River, but different from the other possible source, the Ogallala Formation. In a study subsequent to that of Winspear and Pye (1996), Muhs et al. (2000) examined the geochemistry of the Lincoln and Imperial dune fields, with findings indicating that each dune field has contributions from both the Nebraska Sand Hills and the South Platte River. Proposed sources of the GBSP have been the Arkansas River or the underlying alluvial terraces, but Arbogast and Muhs et al. (2000) used mineralogy and trace elements to demonstrate that the Holocene dunes consist exclusively of Arkansas River-derived sand. Using their mineralogical and trace element approach, Muhs and Holliday (2001) conducted an analysis of the Southern High Plains dune fields, including the Mescalero, Muleshoe, Lea-Yoakum, and Monahans-Andrews, with results indicating that the BDF, not Pecos River, was the sand source. From this result, they concluded that, due to the low rate of sand yield from the BDF, the dunes are sediment availability-limited. In addition, Muhs and Holliday (2001) observed that the red color of the iron-rich, red-clay grain coatings in the Muleshoe dune field decreases eastward, which they postulated was due to abrasion of the coatings downwind, a finding that was later confirmed through laboratory experiments by White and Bullard (2009).

In Nebraska, Hanson et al. (2009) derived geochemical data showing that the Duncan dunes on the Loup River are linked to that river and its terrace fills rather than to the adjacent Platte River sediment, which is in agreement with the Muhs et al. (1997b) findings for the NSH. Hanson et al. (2009) surmised that, though dune sand could have originated from the Loup River or from the underlying terrace fill, the source is probably the latter given that sand from the river would have had to move several km, which is inconsistent with dune age, location, and reasonable migration rates. Puta et al. (2013) noted that geochemical data indicated similarity between the Stanton dunes and the Elkhorn alluvium as well as to data from other rivers draining the NSH (Muhs et al. 1997b; Hanson et al. 2009). Geochemical data from the Abilene dune field in east-central Kansas ruled out the adjacent Smoky Hill River, but indicated that the source was the late Pleistocene alluvium underlying the dune field (Hanson et al. 2010).

In a compilation of geochemically-based studies of aeolian sand provenance, Muhs (2017) reviewed issues associated with employing mineralogical indicators (elemental concentrations), and took a new approach using K/Rb and K/Ba ratios, reconsidering dune provenance region by region. This approach, while still focused on K-feldspar provenance, largely avoids effects of chemical weathering, and grain abrasion on K-feldspar content. Ratio data derived from the Lincoln County dune field were intermediate between the South Platte River and the NSH, whereas the Imperial dune field ratios were intermediate with a bias toward the South Platte River sediments, which excludes the South Platte River as the sole source as proposed by Winspear and Pye (1996). As for the dune fields of northeastern Colorado, ratio data from the Fort Morgan dune field indicated that possible sources include the South Platte River and Ogallala Formation, as well as a probable yet unidentified source. Data from the Wray dune field fell primarily within the Ogallala Formation signature, as proposed by Lugn (1968) and Hunt (1986). Similarly, ratio data from the Greeley dune field indicated Ogallala Formation and possible Laramie Formation sources. Ratios derived from late Holocene dunes in the GBSP coincided with those from the modern Arkansas River sands but were slightly different from those of the Pleistocene alluvium identified by Arbogast and Johnson (1998) within and to the south of the dunes, agreeing with the earlier Arbogast and Muhs et al. (2000) conclusion. Ratio data from dunes of the Southern High Plains showed that the Muleshoe and Lea-Yoakum dune fields were similar to those of the Blackwater Draw Formation, rather than those of the Pecos River sediment, thereby substantiating the conclusion of Muhs and Holliday (2001). While this benchmark investigation of ratio data from many of the dune fields conclusively identified dune sand provenance, data from others indicated unidentified sources or a previously unappreciated complexity in the provenance history (Muhs 2017). Further, Muhs (2017) and more recently, Muhs and Budahn (2019) cast doubt on the perceived overall importance of the modern and late Pleistocene river systems as sediment sources for the dune fields of northeastern Colorado and Nebraska. They in fact concluded that the Ogallala was the major source for the Greeley dunes (with possible contribution from the Cretaceous Laramie Formation) and for the NSH, rather than the Platte River system.

## 4.5 Historical Dune Activation

### *4.5.1 The 1800s*

With the widespread nature of aeolian sand deposits in the CSGP, early explorers were destined to encounter them, especially given the propensity of these travelers to follow river courses. To get an impression of the level of dune activity during the 1800s, Muhs and Holliday (1995) compiled historical observations of dune conditions as interpreted from early accounts, including those of Pike, Long, Michler, and Fowler, and, despite all the vagaries associated with interpreting these qualitative observations, they were able to conclude that, while the majority of the CSGP dune fields are now mostly stabilized, aeolian activity was occurring over at least parts of the CSGP during the 1800s. Dendroclimatological data document droughts in the early and middle to late 1800s: short droughts occurred during the Pike and Long expeditions (Cook et al. 2007), but three prolonged droughts occurred later—1856-1865, 1870–1877, and 1890–1896, the first of which was called the 'Civil War' or '1860s' drought (Herweijer et al. 2006). Tree-ring records bordering the Great Plains indicate this was indeed a major drought: Stahle et al. (2000) credited the 1860 drought as being the worst since the late 1600s, and Blasing et al. (1988) ranked it as the most severe in over 200 years for the Kansas, Oklahoma and Texas region. Further, weather records from early observation stations and US Army Forts, such as those along the Oregon and Santa Fe trails indicate drought conditions (Mock 1991). Accordingly, OSL ages reported between ca. 200 and 100 years ago (e.g., Forman et al. 2008; Halfen et al. 2012) appear to reflect the droughts of the 1800s.

### 4.5.2 The Dust Bowl

The 1930s' drought, or Dust Bowl on the Great Plains, with it horrendous wind erosion, was the most severe weather-related historical catastrophe in the US (e.g., Worster 2004; Egan 2006). Notwithstanding the cultural perspective on duration, tree-ring records (Frye et al. 2003; International Tree-Ring Database 2018) encompassing the Great Plains indicate that the Dust Bowl was below the temporal scale of prehistoric droughts. The intensity of the Dust Bowl was, however, considerable: Cook et al. (2014b) noted that 1934 was the driest year of the last millennium.

Though the spatial limits of the Dust Bowl were diffuse, Joel (1937) defined the region based on the severity of soil erosion, which included many of the CSGP dune fields (Fig. 4.12). Morton County, located in the southwest corner of Kansas and within the core of the Dust Bowl, experienced severe soil erosion and massive sand transport, with farm buildings and houses being flanked or inundated by silt or sand dunes (Fig. 4.13). Aerial photography of an area immediately south of two abandoned meanders likely cut off by an avulsion during the 1921 flood on the Arkansas River in southwestern Kansas and located 3.2 km west southwest of Syracuse, Kansas provides an impression of the extensive activation of the dune field south of the river (Fig. 4.14). The August 1936 photograph illuminates extensive dune activity (transverse dunes) reflecting southerly winds, whereas the 2014 image shows less than 5% of the area on the west side of the image yet active, though the far eastern side of the image displays an area of major activity, which corresponds to the Syracuse Sand Dunes Park, a facility dedicated to recreational off-road vehicle activity.

To document the healing of the blowouts activated during the Dust Bowl, sites on the Cimarron National Grass Lands (CNGL) in Morton County were tracked temporally to chronicle the recovery rate of the active sand areas (Johnson and Messinger

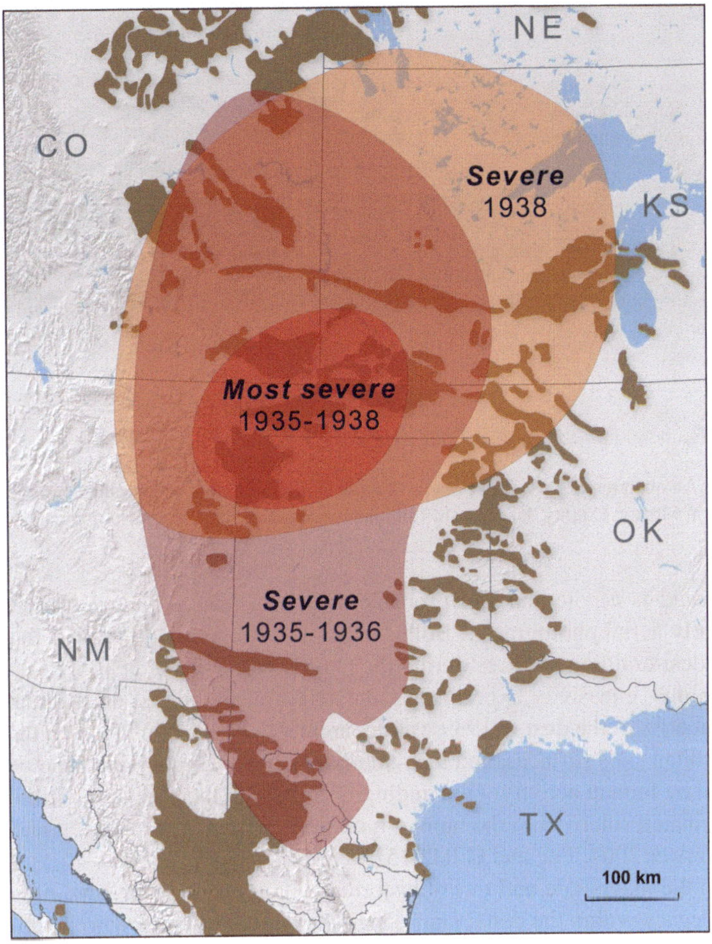


**Fig. 4.12** Geographic extent and variable severity of the Dust Bowl drought of the 1930s. (Adapted from Joel 1937)

2003). Rectified black and white aerial photography (1939–1986), digital orthophotography (1991) and National Agriculture Imagery Program aerial imagery were used to map and temporally track in a GIS environment all blowouts visible on the early photography. An example of the healing from the eastern part of the CNGL illustrates the approximate size of the depicted blowouts for each selected year (Fig. 4.15). For the CNGL, which consists of 437.8 $km^2$, the total area in blowouts in the late 1930s was 48.3 $km^2$ or 11% in 1939, but this does not include the area in active dune crests.

Very few aeolian sand deposits are 'clean,' that is, completely devoid of particles smaller than sand size, which suggests that through winnowing these deposits have the potential to produce airborne fines (silt and clay). Werner et al. (2011) conducted air elutrification experiments on sand samples, findings of which indicated that

**Fig. 4.13** An abandoned Dust Bowl-era farm with aeolian sediment accumulated around the barn and shed, in Morton County, Kansas. (S. Lohman; courtesy of the Kansas Geological Survey)

aeolian sand is a potential and probable source of silt- and clay-sized particles. Using early aerial photography, Bolles et al. (2017) classified land use in order to identify dust-emitting surfaces during the Dust Bowl, with the goal of determining, based on dust emissivity experiments, the origin of mobilized dust. Results indicated that 60% of the dust in 1939 was apparently derived from the uncultivated but active aeolian sand surfaces, and the remainder from areas cultivated and otherwise disturbed by human activity. This finding is counter to the accepted notion that the dust originated solely from the human-modified surfaces (e.g., Bennett and Fowler 1936; Worster 2004; Lee and Gill 2015).

Given the magnitude and recent historical nature of the Dust Bowl, it has left a cultural legacy within the dunal landscape. The buried 'cultural debris' comprises a record within the post-Dust Bowl landscape that serves as a testament to manner in which people lived prior to and during this environmental and societal catastrophe. Guided by available sources, Cordova and Porter (2015) approached the stratigraphic and geomorphic record of the Dust Bowl as archaeologists, in this instance not the prehistoric past, but the contemporary past (cf. Harrison and Schofield 2010).

### *4.5.3 Present Aeolian Activity*

Present CSGP dune and sand sheet activity tends to be limited in scope and scale under prevailing climatic conditions (an exception being the Monahans dunes), with activity commonly expressed as isolated blowouts. For example, Wharton Ranch

**Fig. 4.14** A dune landscape south of the Arkansas River and two abandoned river meanders west of Syracuse, Kansas. An August 1936 aerial photograph of transverse dunes, which are migrating northward under prevailing south winds; a 2014 aerial image showing that most of the area active in the 1930s has healed, except for the southeast part of the image, which is part of the Syracuse Sandhills Park, an off-road vehicle recreational area. (1936 photograph from the US Soil Conservation Service; 2014 image from the USDA Farm Service Agency National Agriculture Imagery Program)

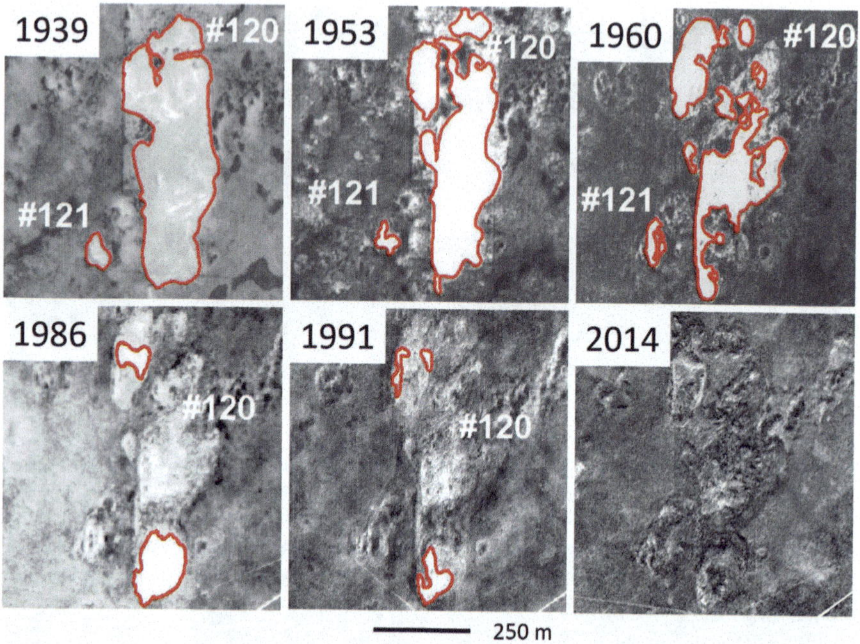


**Fig. 4.15** Temporal sequence of post-Dust Bowl blowout healing in the Cimarron National Grasslands of Morton County, Kansas. Though this area exhibits only a few of the more than 200 blowouts tracked through time, the temporal change was typical for the sample population. The approximate size of the blowout for each selected year in $m^2$: 1939: 260,600; 1953: 221,400; 1960: 170,800; 1986: 56,000; 1991: 14,000, and 2014: negligible. (Modified from Johnson and Messinger 2003)

blowout (Halfen 2012), the largest active blowout in Kansas and located 6 km south of the Arkansas River and 5 km from the Colorado state line, was according to local lore activated during the Dust Bowl. In the 1939 aerial photography, the blowout was ca. 820 m long and 490 m wide, but by summer 2018 the northern boundary had migrated ca. 670 m north, and it had increased in length to ca. 1375 m and width to 590 m. Multiple generations of fencing, erosional remnants of earlier blowout floor levels, and recent exhumation of a late Pleistocene paleosol containing Paleoindian artifacts indicate present blowout degradation (Fig. 4.16).

Recent human activity has initiated or is maintaining aeolian activity throughout much of the western CSGP. Given a general lack of fertility, the lands have been historically used for cattle grazing, which results in destabilization not only by overgrazing but also by cattle movement, especially congregation around cattle tanks, cow paths along fence lines, and where cattle linger on the downwind side of dunes (Fig. 4.17). Since the 1960s, center pivot irrigation has become increasingly common on aeolian sand landscapes in the region, but where these characteristic circular fields have been abandoned, sand often reactivates. Cessation of irrigation in areas in the Arkansas River dune field of southwestern Kansas has resulted in reactivation, as illustrated by an area south of Garden City (Fig. 4.18). Emplacement of oil

**Fig. 4.16** Wharton blowout, southwestern Kansas. (**a**) Aerial view northeast of the Wharton blowout (ca. 1375 m long and 590 m wide); (**b**) An erosional remnant of a former floor of the blowout; (**c**) Remains of the last generation of section line fencing traversing the blowout east-west (for location, see the arrow in (a)) (view east); (**d**) A Paleoindian projectile point found in the surface of an exposed once buried soil radiocarbon-dated to about 13 ka (Halfen 2012). (W. Johnson)

and gas field infrastructure (e.g., well pads, access roads, pipeline corridors) has been disruptive as well, particularly in Texas and adjacent areas of New Mexico, as exemplified by the small Andrews dune field (Fig. 4.19). One of the more recent phenomena impacting dune areas in the region is the development of off-road vehicle parks, which include, for example, Syracuse Sand Dune Park (526 ha; Fig. 4.14) operated by the city of Syracuse, Kansas; Little Sahara State Park (587 ha; Fig. 4.20) in Oklahoma; Mescalero Sands (247 ha) operated by the Bureau of Land Management in Texas; and Kermit Sandhills (private) adjacent to Monahans Sandhills State Park in Texas. Inevitable impacts on these sand areas include the potential destruction of heretofore unrecognized archaeological sites, increased ballistic and especially inter-grain abrasion, obliteration of surface soils, deep rutting, compaction, destruction of macro and micro flora and fauna, increased wind and water erosion, destabilization of adjoining land through unauthorized use and incursion of active sand sheets or dunes, dust production, and of course noise (e.g., Webb and Wilshire 1983; Kutiel et al. 2000).

**Fig. 4.17** Aerial view of cattle-induced activation along a fence line typical of many grazed dune fields. (W. Johnson)

## 4.6 Future Trajectory

As most dune fields of the region have become active multiple times during the late Holocene, there appears to be a high probability for reactivation in the future, especially because the last 3000–2000 years appear more indicative of present climate than do the earlier periods (Muhs et al. 1997a; Schmeisser et al. 2010). Using data available at the time, Muhs and Maat (1993) computed the Lancaster (1988) index of dune mobility (M) values for 40 localities in the Great Plains and found that most yielded M values of 50–100 (inactive: <50; active crests only: 50–100; fully active except for interdunal areas: 100–200, fully active: M > 200) and were consistent with the degree of dune activity detected on aerial photography. To assess the potential increase in dune activity due to greenhouse warming, they conducted sensitivity tests that suggested some dune fields could completely activate, though the level would vary within the region, and many dune fields have areas void of vegetation cover and appear poised for reactivation (e.g., Madole 1994; Muhs and Holliday 1995; Arbogast and Johnson 1998).

Future activation of dune fields also appears inevitable from a recent national climate assessment for the Great Plains (Shafer et al. 2014), given the projected changes in climate (ca. 2050), such as an increase in the frequency of high-temperature days (e.g., quadrupling of 100 ° F days in the SGP), number of warm nights (greatest for Texas), and number of consecutive dry days (greatest in Texas and Oklahoma). More recently, Vose et al. (2017) concluded that paleo-temperature

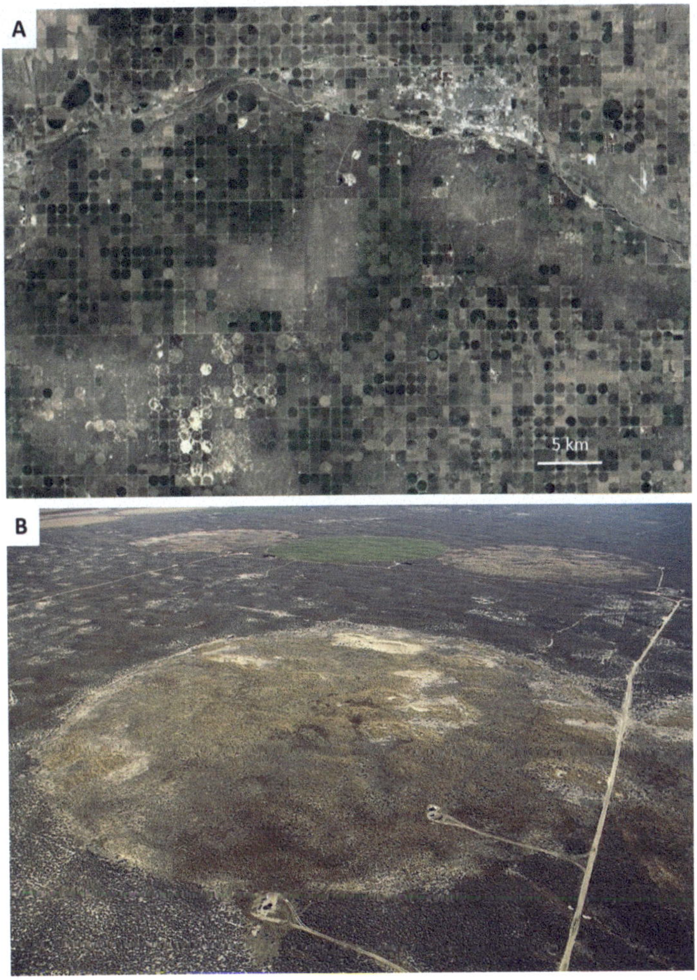


**Fig. 4.18** Abandoned center pivot irrigation fields as loci of sand activation. (**a**) Aerial imagery of congested center pivot irrigation systems in the Garden City, Kansas area. These fields are occasionally abandoned and subsequently often become activated, such as the grouping of fields in the southwestern part of the image (US Geological Survey Landsat); (**b**) Aerial view of blowouts developing in an abandoned center pivot field (ca. 800 m diameter) (W. Johnson)

data indicate that recent decades are the warmest within the last 1500 years and that global climate models project annual average US temperatures to rise, especially in winter. Moreover, Easterling et al. (2017) reported climate model projections indicating that seasonal precipitation will increase slightly in the CGP but decrease in the SGP during winter and spring, and decrease in the CGP and part of the SGP during summer. Similarly, Wehner et al. (2017) projected a decrease in soil moisture over much of the contiguous US, including both the CGP and SGP, but most severe in the CGP.

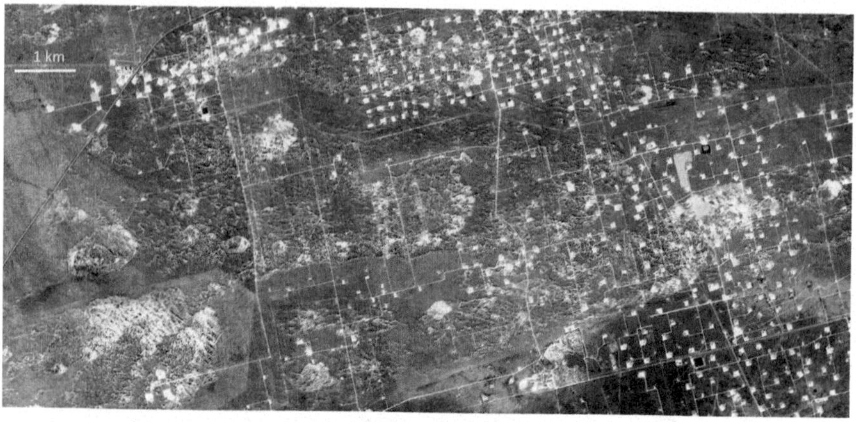


**Fig. 4.19** Active dunes in the Andrews dune field, Texas, some of which appear activated by construction of oil and gas field roads and well pads. (US Geological Survey Landsat)

Remarkably, the 2012 drought of the CGP was an unanticipated singular event (a 'flash drought') and may be an indication of what is to come: precipitation deficits occurring from May through August 2012 were the lowest since record keeping began in 1895, making for a summer drier than any during the Dust Bowl, plus temperatures were above normal (NOAA Drought Task Force Narrative Team 2013). Exceptional drought conditions settled into central and western Nebraska and adjacent Wyoming, western Kansas, panhandle Oklahoma, and the New Mexico-Texas state line area. Proximate causes of the drought were meteorological (reduced Gulf of Mexico moisture and increased subsidence), but underlying causes were elusive (NOAA Drought Task Force Narrative Team 2013; Hoerling et al. 2014).

Given that some of the region's dunes currently exhibit activity (e.g., Monahans) and many others are at or very near the threshold of reactivation, multiyear droughts such as the Dust Bowl will very likely initiate large areas of activation, whereas a megadrought would produce widespread reactivation as evidenced by the MCA event extending to east-central Kansas (Hanson et al. 2010; Halfen et al. 2012) and eastern Nebraska (Hanson et al. 2009; Puta et al. 2013). All this increased dune activity would then yield fine sediment thereby increasing the atmospheric dust load (Bolles et al. 2017), which could magnify drought severity (Cook et al. 2008, 2009, 2013, 2014b).

Unfortunately, the climatological future of the CSGP, according to several models, will most likely include megadroughts, perhaps even more severe than those reconstructed for the MCA (Cook et al. 2014a, 2015, 2016; Dai 2013), which would result in massive reactivations of dune fields and sand sheets. Further, the increasing stress on surface and groundwater resources (Sophocleous 2010; Scanlon et al. 2012; Long et al. 2013) may diminish the extent or even ability of ecosystems and society to adapt.

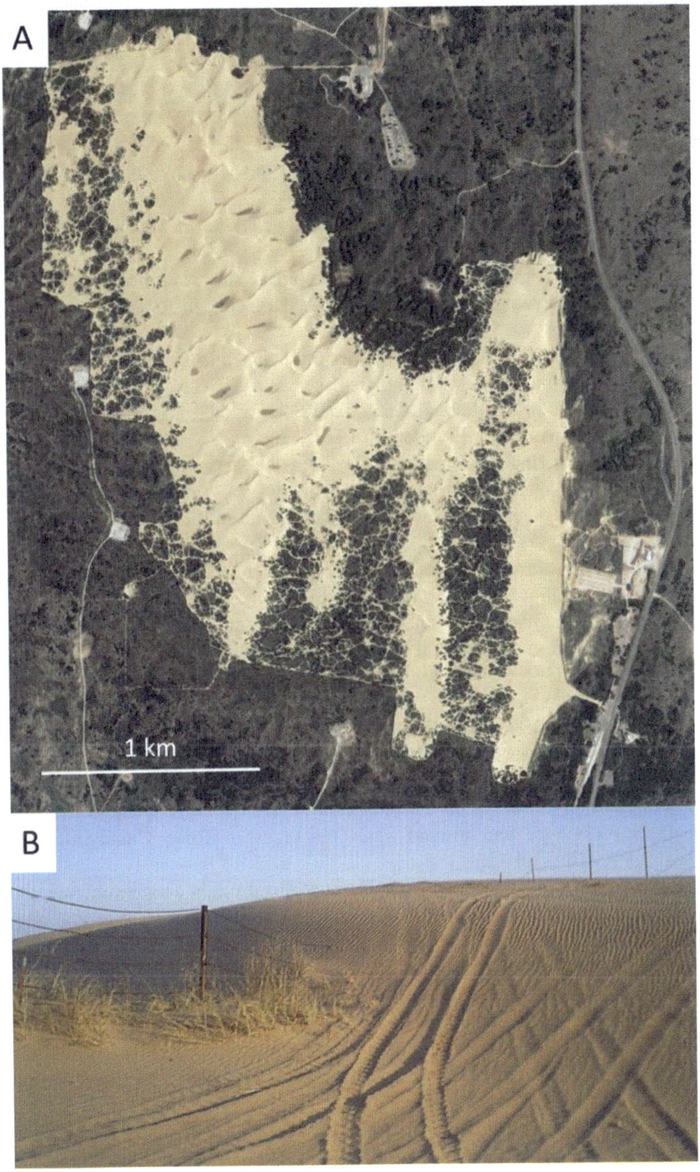


**Fig. 4.20** Little Sahara State Park, an off-road vehicle (ORV) recreational area in northwestern Oklahoma. (**a**) While the extensive maze of trails within the active sand areas is not visible at this scale, trails connecting active areas are distinguishable (US Department of Agriculture Farm Service Agency National Agriculture Imagery Program); (**b**) Fence burial by dune sand activated by off-road vehicles (C. Cordova, Oklahoma State University)

## References

Ahlbrandt TS (1974) Dune stratigraphy, archaeology, and the chronology of the Killpecker Dune Field. Wyoming Geological Survey Report of Investigations 10

Ahlbrandt TS, Swinehart JB, Maroney DG (1983) The dynamic Holocene dune fields of the Great Plains and Rocky Mountain basins, U.S.A. In: Brookfield ME, Ahlbrandt TS (eds) Eolian sediments and processes. Elsevier, New York, pp 379–406

Albanese JP (1974a) Geology of the Casper archaeological site. In: Frison GC (ed) The Casper site. Academic, New York, pp 173–190

Albanese JP (1974b) Geology of the Casper archeological site, Natrona County, Wyoming. In: Wilson M (ed) Applied geology and archeology: the Holocene history of wyoming. Geological Survey of Wyoming, pp 46–50

Albanese JP, Frison GC (1995) Cultural and landscape change during the middle Holocene, Rocky Mountain area, Wyoming and Montana. In: Bettis EA III (ed) Archaeological geology of the archaic period in North America, Geological Society of America special paper, vol 297. Geological Society of America, Boulder, pp 1–19

Andrzejewski KD (2015) Historical metamorphosis of the Arkansas River on the Kansas High Plains. University of Kansas MS thesis, Lawrence

Arbogast AF (1996a) Late Quaternary evolution of a lunette in the central Great Plains: Wilson Ridge. Phys Geogr 17:354–370

Arbogast AF (1996b) Stratigraphic evidence for late-Holocene Aeolian sand mobilization and soil formation in south-central Kansas, U.S.A. J Arid Environ 34:403–414

Arbogast AF (1998) Late Quaternary paleoenvironments and Landscape evolution on the Great Bend Sand Prairie. Kansas Geological Survey Bulleting 242

Arbogast AF, Johnson WC (1996) Geologic map of Stafford County, Kansas. Kansas Geological Survey, Lawrence, Map M-46

Arbogast AF, Johnson WC (1998) Late-Quaternary landscape response to environmental change in south-central Kansas. Ann Assoc Am Geogr 88:126–145.

Arbogast AF, Muhs DR (2000) Geochemical and mineralogical evidence from eolian sediments for northwesterly mid-Holocene paleowinds, central Kansas, USA. Quat Int 67:107–118

Bayne CK (1956) Geology and ground-water resources of Reno County, Kansas. Kansas Geological Survey Bulletin 120

Bennett HH, Fowler FH (1936) Report of the Great plains drought area committee. Government Printing Office, Washington, DC

Berger A, Loutre MF (1991) Insolation values for the climate of the last 10 million years. Quat Sci Rev 10:297–317

Blasing TJ, Stahle DW, Duvick DN (1988) Tree-ring reconstruction of annual precipitation in the south-central United States from 1750–1980. Water Resour Res 24:163–171

Bolles K, Forman SL, Sweeney M (2017) Eolian processes and heterogeneous dust emissivity during the 1930's Dust Bowl Drought and implications for projected 21st-century megadroughts. The Holocene 27:1578–1588

Booth RK, Jackson ST, Forman SL, Kutzbach JE, Bettis EA III, Kreig J, Wright DK (2005) A severe centennial-scale drought in mid-continental North America 4200 years ago and apparent global linkages. The Holocene 15:321–328

Bowen MW, Johnson WC (2012) Late Quaternary environmental reconstructions of playa–lunette system evolution on the central High Plains of Kansas, United States. Geol Soc Am Bull 124:146–161

Bowen MW, Johnson WC, Egbert SL, Klopfenstein ST (2010) A GIS-based approach to identify and map playa wetlands on the High Plains, Kansas, USA. Wetlands 30:675–684

Bowen MW, Johnson WC, King DA (2018) Spatial distribution and geomorphology of lunette dunes on the High Plains of Kansas: implications for geoarchaeological and paleoenvironmental research. Phys Geogr 39:21–37

Brady RG (1989) Geology of the Quaternary dune sands in eastern Major and southern Alfalfa Counties, Oklahoma. Oklahoma State University, PhEd Thesis, Stillwater

Briner JP (2009) Moraine pebbles and boulders yield indistinguishable $^{10}$Be ages: a case study from Colorado. Quat Geochronol 4:299–305

Butler BE, Hutton JT (1956) Parna in the riverine plain of southeastern Australia and the soils thereon. Aust J Agric Res 7:536–553

Carlisle WJ, Marrs RW (1982) Eolian features of the Southern High Plains and their relationship to windflow patterns. Geol Soc Am Spec Pap 192:89–105

Carter VA, Shinker JJ, Preece J (2018) Drought and vegetation change in the central Rocky Mountains and western Great Plains: potential climatic mechanisms associated with mega-drought conditions at 4200 cal yr BP. Clim Past 14:1195–1212

Clarke ML, Rendell HM (2003) Late Holocene dune accretion and episodes of persistent drought in the Great Plains of northeast Colorado. Quat Sci Rev 22:1051–1058

Cook ER, Woodhouse CA, Eakin CM, Meko DM, Stahle DW (2004) Long-term aridity changes in the western United States. Science 306:1015–1018

Cook ER, Seager R, Cane MA, Stahle DW (2007) North American drought: reconstructions, causes and consequences. Earth Sci Rev 81:93–134

Cook BI, Miller RL, Seager R (2008) Dust and sea surface temperature forcing of the 1930s "Dust Bowl" drought. Geophys Res Lett 35:L08710

Cook BI, Miller RL, Seager R (2009) Amplification of the North American "Dust Bowl" drought through human-induced land degradation. Proc Natl Acad Sci 106:4997–5001

Cook BI, Seager R, Miller RL (2011) The impact of devegetated dune fields on North American climate during the late Medieval Climate Anomaly. Geophys Res Lett 38:L14704

Cook BI, Seager R, Miller RL, Mason JA (2013) Intensification of North American megadroughts through surface and dust aerosol forcing. J Climatol 26:4414–4430

Cook BI, Smerdon JE, Seager R, Coats S (2014a) Global warming and 21st century drying. Clim Dyn 43:2607–2627

Cook BI, Seager R, Smerdon JE (2014b) The worst North American drought year of the last millennium: 1934. Geophys Res Lett 41:7298–7305

Cook BI, Ault TR, Smerdon JE (2015) Unprecedented 21st century drought risk in the American Southwest and Central Plains. Sci Adv 1:e1400082

Cook BI, Cook ER, Smerdon JE, Seager R, Williams AP, Coats S, Stahle DW, Diaz JV (2016) North American megadroughts in the Common Era: reconstructions and simulations. WIREs Clim Chang 7:411–432

Cordova C, Porter JC (2015) The 1930s Dust Bowl: geoarchaeological lessons for a 20th century environmental crisis. The Holocene 25:1707–1720

Cordova CE, Porter JC, Lepper K, Kalchgruber R, Scott GF (2005) Preliminary assessment of sand dune stability along a bioclimatic gradient, north-central and northwestern Oklahoma. Great Plains Res 15:227–249

Dai A (2013) Increasing drought under global warming in observations and models. Nat Clim Chang 3:52–58

Daniels JM, Knox JC (2005) Alluvial stratigraphic evidence for channel incision during the Mediaeval Warm Period on the central Great Plains, USA. The Holocene 15:736–747

David PP (1971) The Brookdale Road section and its significance in the chronological studies of dune activities in the Brandon Sand Hills of Manitoba. In: Turnock AC (ed) Geoscience studies in Manitoba. Geological Association of Canada, pp 293–299. (Special Paper 9)

Dean WE, Ahlbrandt TS, Anderson RY, Bradbury JP (1996) Regional aridity in North America during the middle Holocene. The Holocene 6:145–155

Denniston RF, Gonzalez LA, Asmerom Y, Baker RG, Reagan MK, Bettis EA III (1999) Evidence for increased cool season moisture during the middle Holocene. Geology 27:815–818

Denniston RF, DuPree M, Dorale JA, Asmerom Y, Polyak VJ, Carpenter SJ (2007) Episodes of late Holocene aridity recorded by stalagmites from Devil's Icebox Cave, central Missouri, USA. Quat Res 68:45–52

Diffendal RF Jr, Leite MB (1989) Late Quaternary alluvial and lacustrine fills, Keith and Garden counties, Nebraska-The conundrum reconsidered: TER-QUA 1989 Abstracts. Colorado State University

Duller GAT (2004) Luminescence dating of Quaternary sediments: recent advances. J Quat Sci 19:183–192

Easterling DR, Kunkel KE, Arnold JR, Knutson T, LeGrande AN, Leung LR, Vose RS, Waliser DE, Wehner MF (2017) Precipitation change in the United States. In: Wuebbles DJ, Fahey DW, Hibbard KA, Dokken DJ, Stewart BC, Maycock TK (eds) Climate science special report: fourth national climate assessment, volume I: U.S. Global Change Research Program, Washington, DC, pp 207–230

Egan T (2006) The worst hard time: the untold story of those who survived the great American dust bowl. Houghton Mifflin Company, Boston

Enfield DB, Mestas-Nuñez AM, Trimble PJ (2001) The Atlantic multidecadal oscillation and its relation to rainfall and river flows in the continental U.S. Geophys Res Lett 28:2077–2080

Eschner TR, Hadley RF, Crowley KD (1983) Hydrologic and morphologic changes in channels of the Platte River basin; a historical perspective. US Geological Survey Professional Paper 1277-A

Esper J, Cook ER, Schweingruber FH (2002) Low-frequency signals in long tree-ring chronologies for reconstructing past temperature variability. Science 295:2250–2253

Fader SW (1974) Ground water in the Kansas River valley, Junction City to Kansas City, Kansas. Geol Surv Kansas Bull 206:1–12

Fargione J, Kiesecker J, Slaats MJ, Olimb S (2012) Wind and wildlife in the northern Great Plains: identifying low-impact areas for wind development. PLoS ONE 7:e41468

Feathers JK, Holliday VT, Meltzer DJ (2006) Optically stimulated luminescence dating of Southern High Plains archaeological sites. J Archaeol Sci 33:1651–1665

Feng Z-D, Johnson WC, Lu Y-C, Ward PA III (1994) Climatic signals from loess-soil sequences in the Central Great Plains, USA. Palaeogeogr Palaeoclimatol Palaeoecol 110:345–358

Feng S, Oglesby RJ, Rowe CM, Loope DB, Hu Q (2008) Atlantic and Pacific SST influences on medieval drought in North America simulated by the Community Atmospheric Model. J Geophys Res 113:1–14

Fine ST, McGowen SL, Carter BJ, Bement LC, Johnson WC, Simms AR, Halfen AF (2011) Investigation of parna (silt) dune formation in the panhandle of Oklahoma. Soil Science Society of America Annual Meeting Program and Abstracts (abst)

Fine ST, Carter BJ, McGowen SL, Bement LC, Johnson WC, Halfen AF, DeWitt R, Simms AR (2012) Late Pleistocene parna dune formation in the panhandle of Oklahoma. American Quaternary Association Biennial Meeting Program and Abstracts (abst.)

Forman SL, Maat P (1990) Stratigraphic evidence for the late quaternary dune activity near Hudson on the Piedmont of Colorado. Geology 18:745–748

Forman SL, Goetz AFH, Yuhas RH (1992) Large scale stabilized dunes on the High Plains of Colorado: understanding the landscape response to Holocene climates with the aid of images from space. Geology 20:145–148

Forman SL, Oglesby R, Markgraf V, Stafford T (1995) Paleoclimatic significance of Late Quaternary eolian deposits on the Piedmont and High Plains, Central United States. Glob Planet Chang 11:35–55

Forman SL, Oglesby R, Webb RS (2001) Temporal and spatial patterns of Holocene dune activity on the Great Plains of North America: megadroughts and climate links. Glob Planet Chang 29:1–29

Forman SL, Marín L, Pierson J, Gómez J, Miller GH, Webb RS (2005) Eolian sand depositional records from western Nebraska: landscape response to droughts in the past 1500 years. The Holocene 15:973–981

Forman SL, Marín L, Gómez J, Pierson J (2008) Late Quaternary eolian sand depositional record for southwestern Kansas: landscape sensitivity to droughts. Palaeogeogr Palaeoclimatol Palaeoecol 265:107–120

Forman SL, Nordt L, Gomez J, Pierson J (2009) Late Holocene dune migration on the south Texas sand sheet. Geomorphology 108:159–170

Frederick CD (1993) Geomorphic investigations. In: Quigg JM et al (eds) Archaeological and geomorphological investigations at Red Lake dam axis, borrow area, and spillway, Martin County, Texas, Mariah Technical Report 873. Austin, Mariah Associates, Inc, pp 36–49

Frederick CD (1998) Late Quaternary clay dune sedimentation on the Llano Estacado, Texas. Plains Anthropol 43:137–155

Fredlund GG, Jaumann PJ (1987) Late Quaternary palynological and paleobotanical records from the central Great Plains. In: Johnson WC (ed) Quaternary environments of Kansas, Guidebook 5. Kansas Geological Survey, Lawrence, pp 167–178

Fritz SC, Metcalfe SE, Dean WE (2001) Holocene climate of the Americas inferred from paleolimnological records. In: Markgraf V (ed) Interhemispheric climate linkages. Academic, San Diego, pp 241–263

Fryberger SG, Dean G (1979) Dune forms and wind regime. Chapter F. In: McKee (ed) A study of global sand seas: US Geological Survey Professional Paper 1052, pp 137–169

Frye JC, Leonard AB (1964) Relation of Ogallala formation to the southern High Plains in Texas. University of Texas at Austin, Bureau of Economic Geology, Austin, pp 1–25

Frye FK, Stahle DW, Cook ER (2003) Paleoclimatic analogs to twentieth-century moisture regimes across the United States. Bull Am Meteorol Soc 84:901–909

Fullerton DS, Bush CA, Pennell JN (2003) Map of surficial deposits and materials in the eastern and central United States (east of 102 degrees west longitude). U.S. Geological Survey Miscellaneous Investigations Series Map I-2789, scale 1:2,500,000, 1 sheet

Fullerton DS, Colton RB, Bush CA, Straub AW (2004) Map showing spatial and temporal relations of mountain and continental glaciations on the northern plains, primarily in northern Montana and northwestern North Dakota. U.S. Geological Survey Scientific Investigations Map 2843, scale 1:1,000,000, 1 sheet

Gaylord DR (1982) Geologic history of the Ferris dune field, south-central Wyoming. In: Marrs RW, Kolm KE (eds) Interpretation of windflow characteristics from eolian landforms, Geological Society of America special paper 192. Geological Society of America, Boulder, pp 65–82

Gaylord DR (1990) Holocene palaeoclimatic fluctuations revealed from dune and interdune strata in Wyoming. J Arid Environ 18:123–138

Goudie AS, Wells GL (1995) The nature, distribution and formation of pans in arid zones. Earth Sci Rev 38:1–69

Grimm EC, Donovan JJ, Brown KJ (2011) A high resolution record of climate variability and landscape response from Kettle Lake, northern Great Plains, North America. Quat Sci Rev 19–20:2626–2650

Halfen AF (2012) Aeolian dune fields of Kansas and their response to Late Quaternary drought. University of Kansas, PhD dissertation, Lawrence

Halfen AF, Johnson WC (2013) A review of Great Plains dune field chronologies. Aeolian Res 10:135–160

Halfen AF, Fredlund GG, Mahan SA (2010) Holocene stratigraphy and chronology of the Casper dune field, Casper, Wyoming, USA. The Holocene 20:773–785

Halfen AF, Johnson WC, Hanson PR, Woodburn TL, Young AR, Ludvigson GA (2012) Activation history of the Hutchinson dunes in east-central Kansas, USA during the past 2200 years. Aeolian Res 5:9–20

Halfen AF, Lancaster N, Wolfe S (2016) Interpretations and common challenges of aeolian records from North American dune fields. Quat Int 410:75–95

Hall SA, Goble RJ (2006) Geomorphology, stratigraphy, and luminescence age of the Mescalero sands, southeastern New Mexico. In: Land et al (eds) Caves and karst of southeastern New Mexico, New Mexico Geological society, Guidebook 57. New Mexico Geological Society, Socorro, pp 297–310

Hall SA, Goble RJ (2011) New optical age of the Mescalero sand sheet, southeastern New Mexico. N M Geol 33:9–16

Hall SA, Penner WI (2012) Stable carbon isotopes, C3–C4 vegetation, and 12,800 years of climate change in central New Mexico, USA. Palaeogeogr Palaeoclimatol Palaeoecol 369:272–281
Hanson PR, Joeckel RM, Young AR, Horn J (2009) Late Holocene dune activity in the Eastern Platte River Valley, Nebraska. Geomorphology 103:555–561
Hanson PR, Arbogast AF, Johnson WC, Joeckel RM, Young AR (2010) Megadroughts and late Holocene dune activation at the eastern margin of the Great Plains, north-central Kansas, USA. Aeolian Res 1:101–110
Harrison R, Schofield J (2010) After modernity: archaeological approaches to the contemporary past. Oxford University Press, Oxford
Haynes CV Jr (1995) Geochronology of paleoenvironmental change, Clovis type site, Blackwater Draw, New Mexico. Geoarchaeology 10:317–388
Herweijer C, Seager R, Cook ER (2006) North American Droughts of the mid-to late nineteenth century: a history, simulation and implication for Mediaeval drought. The Holocene 16:159–171
Herweijer C, Seager R, Cook ER, Emile-Geay J (2007) North American droughts of the last millennium from a gridded network of tree-ring data. J Clim 20:1353–1376
High Plains Regional Climate Center (HPRCC) (2012). http://www.hprcc.unl.edu/. Accessed 10 July 2018
Hill MG, Holliday VT, Stanford DJ (1995) A further evaluation of the San Jon site. New Mexico. Plains Anthropol 40:369–390
Hills ES (1939) The physiography of north-western Victoria. Proc R Soc Victoria 51:297–323
Hills ES (1940) The lunette, a new land form of aeolian origin. Aust Geogr 3:15–21
Hobbs WO, Fritz SC, Stone JR, Donovan JJ, Grimm EC, Almendinger JE (2011) Environmental history of a closed-basin in the US Great Plains: diatom response to variations in groundwater regimes over the last 8500 cal. yr BP. The Holocene 21:1203–1216
Hoerling M, Eischeid J, Kumar A, Leung R, Mariotti KM, Schubert S, Seager S (2014) Causes and predictability of the 2012 Great Plains drought. Bull Am Meteorol Soc 95:269–282
Holliday VT (1985) Holocene soil-geomorphological relations in a semi-arid environment: The Southern High Plains of Texas. In: Boardman J (ed) Soils and Quaternary landscape evolution. Wiley, New York, pp 325–357
Holliday VT (1987) Eolian processes and sediments of the Great Plains. In: Graf W (ed) Geomorphic systems of North America, Geological Society of America, Geology of North America, vol 2, Boulder, pp 195–202
Holliday VT (1989) The Blackwater Draw Formation (Quaternary): A 1.4-plus m.y. record of eolian sedimentation and soil formation on the Southern High Plains. Geol Soc Am Bull 101:1598–1607
Holliday VT (1990) Soils and landscape evolution of eolian plains: The Southern High Plains of Texas and New Mexico. In: Knuepfer PLK, McFadden LD (eds) Soils and landscape evolution, Geomorphology 3. Elsevier, Amsterdam, pp 489–515
Holliday VT (1995) Stratigraphy and paleoenvironments of late Quaternary valley fills on the Southern High Plains, Geological Society of America Memoir 186. Geological Society of America, Boulder
Holliday VT (1997a) Origin and evolution of lunettes on the high plains of Texas and New Mexico. Quat Res 47:54–89
Holliday VT (1997b) Paleoindian geoarchaeology of the Southern High Plains. University of Texas Press, Austin
Holliday VT (2000) Folsom drought and episodic drying on the Southern High Plains from 10,900–10,200 $^{14}C$ yr B.P. Quat Res 53:1–12
Holliday VT (2001) Stratigraphy and geochronology of upper quaternary eolian sand on the Southern High Plains of Texas and New Mexico, United States. Geol Soc Am Bull 113:88–108
Holliday VT, Meltzer DJ (1995) Geoarchaeology of the Midland (Palaeoindian) site, Texas. Am Antiq 61:755–771

Holliday VT, Rawling JE (2006) Soil-geomorphic relations of lamellae in Eolian sand on the High Plains of Texas and New Mexico. Geoderma 131:154–180

Holliday VT, Gustavson TC, Hovorka SD (1996) Lithostratigraphy and geochronology of fills in small playa basins on the Southern High Plains, United States. Geol Soc Am Bull 108:953–965

Horn JD, Joeckel RM, Fielding CR (2012) Progressive abandonment and planform changes of the central Platte River in Nebraska, central USA, over historical timeframes. Geomorphology 139–140:372–383

Hunt CB (1986) Surficial deposits of the United States. Van Nostrand Reinhold Company, New York

Huntley DJ, Godfrey-Smith DI, Thewalk MLW (1985) Optical dating of sediments. Nature 313:105–107

International Tree-Ring Database (2018). https://catalog.data.gov/dataset/international-tree-ring-data-bank-itrdb. Accessed 15 Feb 2018

Joel AH (1937) Soil conservation reconnaissance survey of the southern Great Plains wind erosion area, USDA Technical Bulletin 556. U.S. Department of Agriculture, Washington

Johnson WC (1991) Buried soil surfaces beneath the Great Bend Prairie of Central Kansas and archaeological implications. Curr Res Pleistocene 8:108–110

Johnson WC, Messinger LG (2003) Historic patterns of dune instability in the Cimarron National Grasslands, southwestern Kansas. Geological Society of America Annual Meeting Program and Abstracts (abst.)

Johnson WC, Willey KL (2000) Isotopic and rock magnetic expression of environmental change at the Pleistocene-Holocene transition in the central Great Plains. Quat Int 67:89–106

Johnson WC, Woodburn TL (2011) Surficial geology of Kearny County, Kansas, Kansas Geological Survey Map Series 62. Kansas Geological Survey, Lawrence

Johnson WC, Willey KL, Mason JA, May DW (2007) Stratigraphy and environmental reconstruction at the Middle Wisconsinan Gilman Canyon Formation type locality, Buzzard's Roost, southwestern Nebraska, U.S.A. Quat Res 67:474–486

Johnson WC, Woodburn TL, Messinger LG (2009) Geologic Map of Morton County, Kansas. Kansas Geological Survey Map Series 116

Johnson WC, Halfen AF, McGowen S, Carter BJ, Bement LC (2010) Silt dunes of Panhandle Oklahoma. Great Plains-Rocky Mountains Section of the Association of American Geographers, Program and Abstracts (abst.)

Johnson WC, Halfen AF, McGowen S, Carter B, Fine S, Bement LC, Simms AR (2012) Last Glacial Maximum development of parna dunes in Panhandle Oklahoma, USA. American Geophysical Union Fall Meeting Program and Abstracts (abst.)

Johnson WC, Halfen AF, Spencer JQG, Hanson PR, Young AR, Mason JA (2019) Late MIS 3 stabilization of dunes in the eastern Central Great Plains, USA. Aeolian Res 36:68–81

Judson S (1950) Depressions of the northern portion of the Southern High Plains of eastern New Mexico. Geol Soc Am Bull 61:253–274

Knight JR, Folland CK, Scaife AA (2006) Climate impacts of the Atlantic multi-decadal oscillation. Geophys Res Lett 33:LI7706

Kocurek G, Lancaster N (1999) Aeolian system sediment state: theory and Mojave Desert Kelso dune field example. Sedimentology 46:505–515

Koop AN, Halfen AF, Johnson WC (2012) Eolian sands of Kansas: a new high-resolution database aiding in research throughout the state. Abstracts of the Association of American Geographers Annual Conference (abst.)

Krueger JP (1986) Development of oriented lakes in the eastern Rainwater Basin region of south central Nebraska. University of Nebraska-Lincoln, MS thesis, Lincoln

Küchler AW (1967) Map: potential natural vegetation of the United States. National Atlas of the United State of America, Department of the Interior, U.S. Geological Survey

Kutiel P, Eden E, Zhevelev Y (2000) Effect of experimental trampling and off-road motorcycle traffic on soil and vegetation of stabilized dunes, Israel. Environ Conserv 27:14–23

Kuzila MS (1994) Inherited morphologies of two large basins in Clay County, Nebraska. Great Plains Res 4:51–63

Kuzila MS, Lewis DT (1993) Properties and genesis of loessial soils across a south-central Nebraska basin. Soil Sci Soc Am J 57:155–161

Labelle JM, Holliday VT, Meltzer DJ (2003) Early Holocene Paleoindian deposits at Nall Playa, Oklahoma panhandle, U.S.A. Geoarchaeology 18:5–34

Laird KR, Fritz SC, Cumming BF, Grimm EC (1998) Early-Holocene limnological and climatic variability in the Northern Great Plains. The Holocene 8:275–285

Lancaster N (1985) Variations in wind velocity and sand transport on the windward flanks of desert sand dunes. Sedimentology 32:581–593

Lancaster N (1988) Development of linear dunes in the southwestern Kalahari, Southern Africa. J Arid Environ 14:233–244

Lancaster N, Wolfe S, Thomas D, Bristow C, Bubenzer O, Burrough S, Duller G, Halfen A, Hesse P, Singhvi A, Tssoar H, Tripaldi A, Yang X, Zarate M (2016) The INQUA dunes Atlas chronologic database. Quat Int 410:3–10

Lee JA, Gill TE (2015) Multiple causes of wind erosion in the Dust Bowl. Aeolian Res 19:15–36

Lepper K, Scott GF (2005) Late Holocene aeolian activity in the Cimarron River valley of west-central Oklahoma. Geomorphology 70:42–52

Litwinionek L, Johnson E, Holliday V (2003) The playas of the Southern High Plains: an archipelago of human occupation for 12,000 years on the North American grasslands. In: Kornfeld M, Osborn AJ (eds) Islands on the plains: ecological, social, and ritual use of landscapes. University of Utah Press, Salt Lake City, pp 21–43

Liu B, Coulthard TJ (2015) Mapping the interactions between rivers and sand dunes: implications for fluvial and aeolian geomorphology. Geomorphology 231:246–257

Long D, Scanlon BR, Longuevergne L, Sun AY, Fernando DN, Save H (2013) Grace satellite monitoring of large depletion in water storage in response to the 2011 drought in Texas. Geophys Res Lett 40:3395–3401

Loope DB (1986) Recognizing and utilizing vertebrate tracks in cross section: Cenozoic hoofprints from Nebraska. Palaios 1:141–151

Loope DB, Swinehart JB, Mason JP (1995) Dune-dammed paleovalleys of the Nebraska Sand Hills: intrinsic versus climatic controls on the accumulation of lake and marsh sediments. Geol Soc Am Bull 107:396–406

Ludvigson GA, Sawin RS, Franseen EK, Watney WL, West RR, Smith JJ (2009) A review of the stratigraphy of the Ogallala Formation and revision of Neogene ("Tertiary") nomenclature in Kansas, Current research in earth sciences, bulletin 256, part 2. Kansas Geological Survey, Topeka

Lugn AL (1968) The origin of loesses and their relation to the Great Plains in North America. In: Schultz CB, Frye JC (eds) Loess and related eolian deposits of the world. University of Nebraska Press, Lincoln, pp 139–182

Madole RF (1994) Stratigraphic evidence of desertification in the west-central Great Plains within the past 1000 years. Geology 22:483–486

Madole RF (1995) Spatial and temporal patterns of late Quaternary eolian deposition, eastern Colorado, U.S.A. Quat Sci Rev 14:155–177

Madole RF, Van Sistine DP, Michael JA (2005) Distribution of Late Quaternary wind-deposited sand in eastern Colorado. USGS Scientific Investigations Map 2875. (map and accompanying pamphlet)

Mandel RD, Hofman JL (2003) Geoarchaeological investigations at the Winger site: A late Paleoindian bison bonebed in southwestern Kansas, USA. Geoarchaeology 18:129–144

Mann ME, Zhang Z, Rutherford S, Bradley RS, Hughes MK, Shindell D, Ammann C, Faluvegi G, Ni F (2009) Global signatures and dynamical origins of the little ice age and medieval climatic anomaly. Science 329:1256–1260

Martin JE, Sawyer JF, Fahrenbach MD, Tomhave DW, Schulz LD (2004) Geologic map of South Dakota. South Dakota Department of Environment and Natural Resources, Geological Survey General Map 10, scale 1:500,000, 1 sheet

Mason JA, Swinehart JB, Goble RJ, Loope DB (2004) Late-Holocene dune activity linked to hydrological drought, Nebraska Sand Hills, USA. The Holocene 14:209–217

Mason JA, Miao XD, Hanson PR, Johnson WC, Jacobs PM, Goble RJ (2008) Loess record of the Pleistocene-Holocene transition on the northern and central Great Plains, USA. Quat Sci Rev 27:1772–1783

Mason, J.A., Swinehart, J.B., Hanson, P.R., Loope, D.B., Goble, R.J., Miao, X., Schmeisser, R.L., 2011. Late Pleistocene dune activity in the central Great Plains, USA. Quat Sci Rev 30, 3858–3870.

Mason JA, Swinehart JB, Loope DB (this volume) The Nebraska Sand Hills. In: Inland dunes of North America. Springer, Cham

Mayer JH, Mahan SA (2004) Late Quaternary stratigraphy and geochronology of the western Killpecker Dunes, Wyoming, USA. Quat Res 61:72–84

McCabe GJ, Palecki MA, Betancourt JL (2004) Pacific and Atlantic Ocean influences on multidecadal drought frequency in the United States. Proc Natl Acad Sci 101:4136–4141

Melton FA (1940) A tentative classification of sand dunes: its application to dune history in the Southern High Plains. J Geol 48:113–174

Miao XD, Mason JA, Swinehart JB, Loope DB, Hanson PR, Goble RJ, Liu XD (2007) A 10,000 year record of dune activity, dust storms, and severe drought in the central Great Plains. Geology 35:119–122

Miller GH, Geirsdóttir Á, Zhong Y, Larsen DJ, Otto-Bliesner BL, Holland MM, Bailey DA, Refsnider KA, Lehman SJ, Southon JR, Anderson C, Björnsson H, Thordarson T (2012) Abrupt onset of the Little Ice Age triggered by volcanism and sustained by sea-ice/ocean feedbacks. Geophys Res Lett 39:L02708

Mock CJ (1991) Drought and precipitation fluctuations in the Great Plains during the late nineteenth century. Great Plains Res 1:26–57

Monnin E, Indermuhle A, Dallenbach A, Fluckiger J, Stauffer B, Stocker TF, Raynaud D, Barnola J-M (2001) Atmospheric CO2 concentrations over the last glacial termination. Science 291:112–114

Muhs DR (1985) Age and paleoclimatic significance of Holocene sand dunes in northeastern Colorado. Ann Assoc Am Geogr 75:566–582

Muhs DR (2004) Mineralogical maturity in dune fields of North America, Africa and Australia. Geomorphology 59:247–269

Muhs DR (2017) Evaluation of simple geochemical indicators of aeolian sand provenance: Late Quaternary dune fields of North America revisited. Quat Sci Rev 171:260–296

Muhs DR, Budahn JR (2019) New geochemical evidence for the origin of North America's largest dune field, the Nebraska Sand Hill, central Great Plains, USA. Geomorphology 332:188–212

Muhs DR, Holliday VT (1995) Evidence of active dune sand on the Great Plains in the 19th century from accounts of early explorers. Quat Res 43:232–237

Muhs DR, Holliday VT (2001) Origin of Late Quaternary dune fields on the Southern High Plains of Texas and New Mexico. Geol Soc Am Bull 113:75–87

Muhs DR, Maat PB (1993) The potential response of eolian sands to greenhouse warming and precipitation reduction on the Great Plains of the U.S.A. J Arid Environ 25:351–361

Muhs DR, Wolfe SA (1999) Sand dunes of the northern Great Plains of Canada and the United States. In: Lemmen DS, Vance RE (eds) Holocene climate and environmental change in the Palliser Triangle: a geoscientific context for evaluating the impacts of climate change on the Southern Canadian Prairies, Geological Survey of Canada Bulletin 534. Natural Resources Canada, Ottawa, pp 183–197

Muhs DR, Stafford TW Jr, Cowherd SD, Mahan SA, Kihl R, Maat PB, Bush CA, Nehring J (1996) Origin of the Late Quaternary dune fields of northeastern Colorado. Geomorphology 17:129–149

Muhs DR, Stafford TW Jr, Swinehart JB, Cowherd SD, Mahan SA, Bush CA (1997a) Late Holocene eolian activity in the mineralogically mature Nebraska Sand Hills. Quat Res 48:162–176

Muhs DR, Stafford TW Jr, Been J, Mahan SA, Burdett J, Skipp G, Rowland ZM (1997b) Holocene eolian activity in the Minot dune field, North Dakota. Can J Earth Sci 34:1442–1459

Muhs DR, Swinehart JB, Loope DB, Been J, Mahan SA, Bush CA (2000) Geochemical evidence for an eolian sand dam across the North and South Platte rivers in Nebraska. Quat Res 53:214–222

Muhs DR, Bettis EA III, Aleinikoff JN, McGeehin JP, Beann J, Skipp G, Marshall BD, Roberts HM, Johnson WC, Benton R (2008) Origin and paleoclimatic significance of late quaternary loess in Nebraska: evidence from stratigraphy, chronology, sedimentology, and geochemistry. Geol Soc Am Bull 120:1378–1407

National Oceanographic and Atmospheric Administration (NOAA) Drought Task Force Narrative Team (2013) An interpretation of the origins of the 2012 central Great Plains drought. NIDIS Assessment Report

Nordt L, Von Fischer J, Tieszen L, Tubbs J (2008) Coherent changes in relative $C_4$ plant productivity and climate during the Late Quaternary in the North American Great Plains. Quat Sci Rev 27:1600–1611

Olson CG, Porter DA (2002) Isotopic and geomorphic evidence for Holocene Climatic, Southwestern Kansas. Quat Int 87:29–44

Olson CG, Nettleton WD, Porter DA, Brasher BR (1997) Middle Holocene aeolian activity on the high plains of west-central Kansas. The Holocene 7:255–261

Prentice IC, Cleator SF, Huang YH, Harrison SP, Roulstone I (2017) Reconstructing ice-age palaeoclimates: quantifying low-$CO_2$ effects on plants. Glob Planet Chang 149:166–176

Puta RA, Hanson PR, Young AR (2013) Late Holocene activation history of the stanton dunes, northeastern Nebraska. Great Plains Res 23:11–23

Reeves CC (1965) Chronology of west Texas pluvial lake dunes. J Geol 73:504–508

Reeves CC (1976) Quaternary stratigraphy and geologic history of Southern High Plains, Texas and New Mexico. In: Mahaney WC (ed) Quaternary stratigraphy of North America. Dowden, Hutchinson and Ross, Stroudsburg, pp 213–234

Rhodes EJ (2011) Optically stimulated luminescence dating of sediments over the past 200,000 years. Annu Rev Earth Planet Sci 39:461–488

Rich J (2013) A 250,000-year record of lunette dune accumulation on the Southern High Plains USA and implications for past climates. Quat Sci Rev 62:1–20

Rich J, Stokes S (2001) Optical dating of geoarchaeologically significant sites from the Southern High Plains and South Texas, USA. Quat Sci Rev 20:949–959

Rich J, Stokes S (2011) A 200,000-year record of Late Quaternary aeolian sedimentation on the Southern High Plains and nearby Pecos River Valley, USA. Aeolian Res 2:221–240

Rich J, Stokes S, Wood WW (1999) Holocene chronology for lunette dune deposition on the Southern High Plains, USA. Z Geomorphol Suppl 116:165–180

Sabin TJ, Holliday VT (1995) Playas and lunettes on the Southern High Plains: morphometric and spatial relationships. Ann Assoc Am Geogr 85:286–305

Scanlon BR, Faunt CC, Longuevergne L, Reedy RC, Alley WM, McGuire VL, McMahon PB (2012) Groundwater depletion and sustainability of irrigation in the US high plains and Central Valley. Proc Natl Acad Sci 109:9320–9325

Schlesinger WH, Reynolds JF, Cunningham GL, Huenneke LF, Jarrell WM, Virginia RA, Whitford WG (1990) Biological feedbacks in global desertification. Science 247:1043–1048

Schmeisser RL, Loope DB, Mason JA (2010) Modern and late Holocene wind regimes over the Great Plains (central U.S.A.). Quat Sci Rev 29:554–566

Schumm SA, Lichty RW (1963) Channel widening and flood-plain construction along the Cimarron River in Southwestern Kansas. US. Government Printing Office, US. Geological Survey Professional Paper 352-D, Washington, DC

Schwalb A, Dean WE (2002) Reconstruction of hydrological changes and response to effective moisture variations from North-Central USA lake sediments. Quat Sci Rev 21:1541–1554

Schwalb A, Dean WE, Fritz SC, Geiss CE, Kromer B (2010) Centennial eolian cyclicity in the Great Plains, USA: a dominant climatic pattern of wind transport over the past 4000 years? Quat Sci Rev 29:2325–2339
Scott GF (1999) Aeolian modification of Pleistocene terraces along the Cimarron River in Major County, Oklahoma. MS Thesis, Oklahoma State University, Stillwater
Seager R, Tzanova A, Nakamura J (2009) Drought in the Southeastern United States: causes, variability over the last millennium, and the potential for future hydroclimate change. J Clim 22:5021–5045
Shafer M, Ojima D, Antle JM, Kluck D, McPherson RA, Petersen S, Scanlon B, Sherman K (2014) Chapter 19: Great Plains. In: Melillo JM, Richmond TC, Yohe GW (eds) Climate change impacts in the United States: the third national climate assessment. U.S. Global Change Research Program, Washington, DC, pp 441–461
Shin S, Sardeshmukh PD, Pegion K (2010) Optimal tropical sea surface temperature forcing of North American drought. J Clim 23:3907–3917
Shuman BN, Serravezza M (2017) Patterns of hydroclimatic change in the Rocky Mountains and surrounding regions since the last glacial maximum. Quat Sci Rev 173:58–77
Simonett DS (1960) Development and grading of dunes in western Kansas. Ann Assoc Am Geogr 50:216–241
Smith HTU (1940) Geologic studies in southwestern Kansas. State Geological Survey of Kansas Bulletin 34
Smith HTU (1965) Dune morphology and chronology in central and western Nebraska. J Geol 73:557–578
Smith LM (2003) Playas of the Great Plains. University of Texas Press, Austin
Smith AJ, Donovan JJ, Ito E, Engstrom DR, Panek VA (2002) Climate-driven hydrologic transients in lake sediment records: multiproxy record of mid-Holocene drought. Quat Sci Rev 21:625–646
Sophocleous M (2010) Review: groundwater management practices, challenges, and innovations in the high plains aquifer, USA—lessons and recommended actions. Hydrogeol J 18:559–575
Soreghan GS, Soreghan MJ, Hamilton MA (2008) Origin and significance of loess in late Paleozoic western Pangaea: a record of tropical cold? Palaeogeogr Palaeoclimatol Palaeoecol 268:234–259
Sorenson CJ, Sallee KH, Mandel RD (1987) Holocene and Pleistocene soils and geomorphic surfaces of the Kansas River valley. In: Johnson WC (ed) Quaternary environments of Kansas, Kansas geological survey guidebook 5. Kansas Geological Survey, Lawrence, pp 93–102
Sridhar V, Loope DB, Swinehart JB, Mason JA, Oglesby RJ, Rowe CM (2006) Large wind shift on the Great Plains during the medieval warm period. Science 313:345–347
Stahle DW, Cook ER, Cleaveland MK, Therrell MD, Meko DM, GrissinoMayer HD, Watson E, Luckman BH (2000) Tree-ring data document 16th century megadrought over North America. EOS Trans Am Geophys Union 81:121–125
Stahle DW, Fye FK, Cook ER, Griffin RD (2007) Tree-ring reconstructed megadroughts over North America since AD 1300. Clim Chang 83:133–149
Stanford D (1979) The Selby and Dutton sites: evidence for a possible pre-clovis occupation of the High Plains. In: Humphrey RL, Stanford DI (eds) Pre-Llano cultures of the Americas: paradoxes and possibilities. The Anthropological Society of Washington, Washington, DC, pp 101–123
Starks PJ (1984) Analysis of the rainwater depressions of clay County, Nebraska. University of Nebraska-Lincoln, M.S. Thesis, Lincoln
Stokes S (1994) Optical dating of selected aeolian sediments from the southwestern United States. Oxford University, DPhil thesis, Oxford
Stokes S, Gaylord DR (1993) Optical dating of Holocene dune sands in the Ferris dune field, Wyoming. Quat Res 39:274–281
Stramel GJ, Lane CW, Hodson WG (1958) Geology and ground-water hydrology of the Ingalls area, Kansas. State Geological Survey of Kansas Bulletin 132

Stuiver M, Reimer PJ, Reimer RW (2019) CALIB 7.1 [WWW program] at http://calib.org. Accessed 25 Jan 2019

Swinehart JB, Diffendal RF Jr (1990) Geology of the pre-dune strata. In: Bleed A, Flowerday C (eds) An atlas of the Sand Hills, Resource atlas no. 5a. University of Nebraska-Lincoln, Lincoln, pp 29–42

Swinehart JB, Loope DB (1992) A giant dune-dammed lake on the North Platte River, Nebraska. Geological Society of America Abstracts with Programs 24(7), A51 (abst.)

Thurmond JP, Wyckoff DG (1998) Late-Pleistocene dunes along the Dempsey Divide, Roger Mills County, Oklahoma. Curr Res Pleistocene 15:139–143

Tian J, Nelson DM, Shen Hu F (2006) Possible linkages of late-Holocene drought in the North American midcontinent to Pacific Decadal Oscillation and solar activity. Geophys Res Lett 33:L23702

United States Environmental Protection Agency (USEPA) (2012) Map: level II ecoregions of the United States. Available online: www.epa.gov/bioweb1/html/usecoregions.html. Accessed 22 July 2018

Van Meerbeeck CJ, Renssen H, Roche DM (2009) How did Marie Isotope Stage 3 and Last Glacial Maximum climates differ?–Perspectives from equilibrium simulations. Clim Past 5:33–51

VanLooy JA, Martin CW (2005) Channel and vegetation change on the Cimarron River, southwestern Kansas, 1953–2001. Ann Assoc Am Geogr 95:727–739

Voelker SL, Stambaugh MC, Guyette RP, Feng X, Grimley DA, Leavitt SW, Panyushkina I, Grimm EC, Marsicek JP, Shuman B, Curry BB (2015) Deglacial hydrolclimate of midcontinental North America. Quat Res 83:336–344

Vose RS, Easterling DR, Kunkel KE, LeGrande AN, Wehner MF (2017) Temperature changes in the United States. In: Wuebbles DJ et al (eds) Climate science special report: Fourth national climate assessment, volume I. U.S. Global Change Research Program, Washington, DC, pp 185–206

Webb RH, Wilshire HG (eds) (1983) Environmental effects of off-road vehicles: impacts and management in arid regions. Springer, New York

Wehner MF, Arnold JR, Knutson T, Kunkel KE, LeGrande AN (2017) Droughts, floods, and wildfires. In: Wuebbles DJ et al (eds) Climate science special report: Fourth national climate assessment, volume I. U.S. Global Change Research Program, Washington, DC, pp 231–256

Wells GL (1992) The aeolian landscape of Central North America from the late Pleistocene. University of Oxford, School of Geography, Oxford

Wells PV, Stewart JD (1987) Cordilleran-boreal Taiga and Fauna on the Central Great Plains of North America, 14,000–18,000 Years Ago. Am Midl Nat 118:94–106

Wendorff F, Krieger AD, Albritton CC Jr, Stewart TD (1955) The Midland discovery. University of Texas Press, Austin

Werner CM, Mason JA, Hanson PR (2011) Non-linear connections between dune activity and climate in the High Plains, Kansas and Oklahoma, USA. Quat Res 75:267–277

White K, Bullard J (2009) Abrasion control on dune colour: Muleshoe Dunes, SW USA. Geomorphology 105:59–66

Williams GP (1978) The case of the shrinking channels—the North Platte and Platte rivers in Nebraska. USGS Geol Circ 781

Williams JW, Shuman B, Bartlein PJ, Diffenbaugh NS, Webb T III (2010) Rapid, time-transgressive, and variable responses to early Holocene midcontinental drying in North America. Geology 38:135–138

Winspear NR, Pye K (1996) Textural, geochemical and mineralogical evidence for the sources of aeolian sand in central and southwestern Nebraska, USA. Sediment Geol 101:85–98

Witty TA (1989) The Ehmke site discussed (14LA311). Unpublished report, Kansas State Historical Society, Topeka

Wolfe SA, Robertson L, Gillis A (2009) Late Quaternary aeolian deposits of Northern North America: age and extent. Geological Survey of Canada, Open File 6006 (CD-ROM)

Wood WW (2002) Role of ground water in geomorphology, geology, and paleoclimate of the Southern High Plains, USA. Groundwater 40:438–447
Woodhouse CA, Overpeck JT (1998) 2000 years of drought variability in the central United States. Am Meteorol Soc Bull 79:2693–2714
Woodhouse CA, Russell JL, Cook ER (2009) Two modes of North American drought from instrumental and paleoclimatic data. J Clim 22:4336–4347
Worster D (2004) Dust Bowl: The Southern Plains in the 1930s. 25th Anniversary edition. Oxford University Press, Oxford
Young NE, Briner JP, Leonard EM, Licciardi JM, Lee K (2011) Assessing climatic and nonclimatic forcing of Pinedale glaciation and deglaciation in the western United States. Geology 39:171–174
Zanner CW, Kuzila MS (2001) Nebraska's Carolina Bays. Geological Society of America Annual Meeting Program and Abstracts (abst.)

# Chapter 5
# The Nebraska Sand Hills

**Joseph A. Mason, James B. Swinehart, and David B. Loope**

**Abstract** The Nebraska Sand Hills form the largest dune field in North America (~50,000 km$^2$), now largely stabilized by native grasses but preserving a rich geologic record of past dune activity and environmental change. This dune field developed on a particularly large undissected portion of the High Plains, with a variety of sand sources exposed nearby. The Nebraska Sand Hills are underlain by the High Plains Aquifer, which sustains a vast number of ecologically important interdune wetlands and lakes. The large barchanoid ridge and barchan dunes of this dune field resemble the active dunes of warm deserts more than the smaller stabilized dunes found elsewhere on the Great Plains. Geologic evidence indicates that the Nebraska Sand Hills alternated between activity and stability over the past 20,000 years, with multiple Holocene episodes of dune activity clearly associated with increased frequency of severe droughts and/or a shift to drier climatic state. Large dunes were also fully active in the cold climate of the Late Pleistocene. Projected climatic change by the late twenty-first century would likely return this major dune field to an active state.

**Keywords** Nebraska Sand Hills · Great Plains · Quaternary paleoclimate · Aeolian sediment provenance · High Plains Aquifer

## 5.1 Introduction

Even visitors who do not recognize the aeolian landforms of the Nebraska Sand Hills are impressed by the unique landscape of this dune field, with grass-covered slopes sweeping up from interdune wetlands and lakes to high dune crests. In fact,

J. A. Mason (✉)
University of Wisconsin, Madison, WI, USA
e-mail: mason@geography.wisc.edu

J. B. Swinehart · D. B. Loope
University of Nebraska-Lincoln, Lincoln, NE, USA



N. Lancaster, P. Hesp (eds.), *Inland Dunes of North America*,
Dunes of the World, https://doi.org/10.1007/978-3-030-40498-7_5

the Nebraska Sand Hills (hereafter, simply referred to as the Sand Hills) stand out as distinctly different, even by comparison with the many smaller stabilized dune fields dotting the Great Plains from Saskatchewan to Texas, because of at least three characteristics. The first is simply the size of the Sand Hills dune field; at about 50,000 $km^2$ (Swinehart 1990) it is the largest dune field in North America and one of the largest in the Western Hemisphere (Muhs and Zárate 2001). The second is the size and morphology of the individual dunes, more typical of large desert dune fields than the episodically active dunes of semiarid regions. No other Great Plains dune field contains many dunes of this scale, or anything near the enormous volume of aeolian sand they represent. The third distinctive characteristic of the Sand Hills involves the diverse and important hydrologic systems that are closely associated with the stratigraphy and geomorphology of this dune field, including a substantial portion of the largest aquifer in North America, the headwaters of several river systems, wetlands, and numerous freshwater and saline lakes (Bentall 1990; Bleed 1990; Bleed and Ginsberg 1990). Like smaller dune fields of the Great Plains, the Sand Hills also preserve native grasslands that have been replaced by agricultural crops across much of the region. Like other Great Plains dune fields, but with richer detail, the Sand Hills preserve a vitally important geologic record of ecosystem and landscape response to climatic change over the past 25,000 years (Loope and Swinehart 2000; Mason et al. 2011; Miao et al. 2007).

## 5.2 Geologic and Geomorphic Setting

The broader geologic context of the Sand Hills is clearly important in understanding why such a large accumulation of aeolian sand developed in this particular setting, although some of the potential connections have received little attention in previous work. The aeolian sand of the Sand Hills dunes rests on strata that accumulated on the western Great Plains from the Eocene through the Pliocene (Fig. 5.1; Swinehart and Diffendal 1990; Swinehart et al. 1985). Most of these Cenozoic continental strata have been well-studied because of their importance for both groundwater hydrology and vertebrate paleontology and several have also been considered as sand sources for the present dunes (Lugn 1935, 1968; Muhs 2017; Muhs et al. 1997b; Muhs and Budahn 2019; Swinehart 1990). The volcaniclastic fluvial and aeolian siltstones and sandstones of the White River Group (Late Eocene-Oligocene) and Arikaree Group (Oligocene-Miocene) are deep in the subsurface within the Sand Hills dune field itself, but crop out to the west and northwest. They are dominated by tephra from large caldera eruptions in the western U.S., mixed with epiclastic material eroded from the Laramide Uplifts of the Rocky Mountains (Hunt 1990; LaGarry 1998; Swinehart et al. 1985). The White River and Arikaree groups are overlain by the Ogallala Group (Miocene), mostly fluvial gravel, sand, and silt eroded from the Rocky Mountains, deposited initially in paleovalleys incised in older rocks of the Great Plains, but ultimately forming an eastward-sloping apron of sediment across much of the region (Diffendal 1982; Swinehart and Diffendal 1990). Cementation by carbonate and silica often makes the Ogallala resistant to

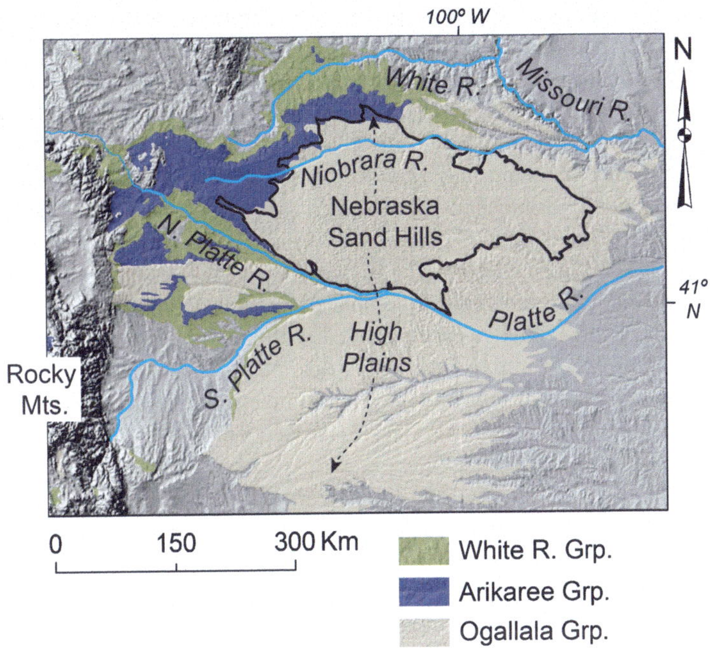


**Fig. 5.1** Geologic and geomorphic setting of the Nebraska Sand Hills on the High Plains. Color shades superimposed on shaded relief map indicate Eocene to Miocene rocks important as potential aeolian sand sources and controls on topography; outcrops of these rocks are common, but they are often overlain by late Cenozoic fluvial and aeolian sediments. Dark line outlines Sand Hills dune field as mapped in Bleed and Flowerday (1990)

erosion, and where it is preserved in the modern landscape it commonly forms a caprock underlying broad tablelands.

Parts of the Great Plains where the Ogallala Group is still largely preserved and the landscape is composed of relatively smooth, eastward-sloping tablelands are referred to as the High Plains (Fig. 5.1). The High Plains tablelands are remnants left by dramatic stream incision after the Miocene, which reshaped the Great Plains landscape (Leonard 2002; Swinehart et al. 1985; Trimble 1980). In Nebraska, an initial phase of incision during the Pliocene was followed by aggradation and widespread deposition of sand and gravel of the Broadwater and Long Pine formations, derived from the Rocky Mountains and deposited by eastward-flowing rivers across much of the present Sand Hills region (Diffendal 1982; Duller et al. 2012; Swinehart and Diffendal 1990). Under the south-central part of the Sand Hills, these coarse fluvial sands and gravels are overlain by Pliocene aeolian sand sheets and associated fine-grained lacustrine and fluvial deposits (Maroney 1978; Myers 1993). Further incision after 2.5 Ma produced the entrenched modern valleys of the North Platte River and its tributaries west and southwest of the Sand Hills (Duller et al. 2012), removing the Ogallala Group and exposing the Arikaree and White River groups over large areas near those rivers (Fig. 5.1). Deep incision and near-complete removal of the Ogallala has also occurred in the White River basin to the north of the Sand Hills. The Sand Hills dunes developed on a particularly large, relatively undissected remnant of the High Plains surface between the areas of deep erosion associated with those two river systems (Fig. 5.1). Within the Sand Hills, only the narrow gorge of the Niobrara River cuts deeply into the High Plains surface today, but the mantle of dune sand may obscure older valleys, since Pleistocene fluvial sediments not clearly associated with modern streams are present under the dune sand, though they are difficult to distinguish from the Pliocene beds in subsurface data (Swinehart and Diffendal 1990).

To summarize, the Sand Hills dune field developed on a particularly extensive remnant of the High Plains surface, covered with and surrounded by a variety of potential sand sources, ranging from the Arikaree and Ogallala groups to Pliocene and Pleistocene sands still present beneath the dunes and fluvial sands transported by the North Platte and Niobrara rivers. Late Cenozoic incision increased fluvial dissection and reworking of the Ogallala and Arikaree groups around the western margins of the dune field, increasing the potential for aeolian mobilization of sand derived from these units, especially the Ogallala Group (Muhs and Budahn 2019). The extent of the relatively undissected surface on which the dune field rests could have allowed long-distance aeolian transport of sand from a variety of sources, and the accumulation of sand to form large dunes may be related to increasing available moisture and vegetation toward the southeast, assuming long-term persistence of a precipitation gradient in that direction. While smaller accumulations of aeolian sand are located on the Great Plains to the north and south of the Sand Hills (Halfen and Johnson 2013; Muhs and Zárate 2001), the much greater extent of the Sand Hills dune field (Fig. 5.2) is likely related to its particular geologic, geomorphic, and (paleo)climatic setting.

## 5.3 Hydrology

Where the Ogallala Group is preserved beneath the High Plains surface, it hosts the High Plains (Ogallala) Aquifer, which extends from South Dakota to Texas. A large fraction of the groundwater in this major aquifer lies beneath the Sand Hills, where it reaches its greatest thickness because of the depositional thickness of the Ogallala Group, limited stream dissection, and high recharge rates through the permeable dune sands (Bleed 1990). Regional groundwater flow in the High Plains Aquifer is generally eastward down the slope of the High Plains, but at the more local scale there is a strong component of flow toward stream valleys (Bleed 1990; Chen and Hu 2004). Even more locally, there are flow systems into wet interdunes (Fig. 5.3a) from areas beneath the surrounding dunes (Chen et al. 2012; Gosselin et al. 1999). Elsewhere in the Great Plains, water from the High Plains Aquifer sustains intensive agriculture, but while there is some irrigated crop production within the Sand Hills, the aquifer is principally important because it sustains perennial streams, lakes, and wetlands (Bentall 1990; Bleed and Ginsberg 1990; Harvey et al. 2007).

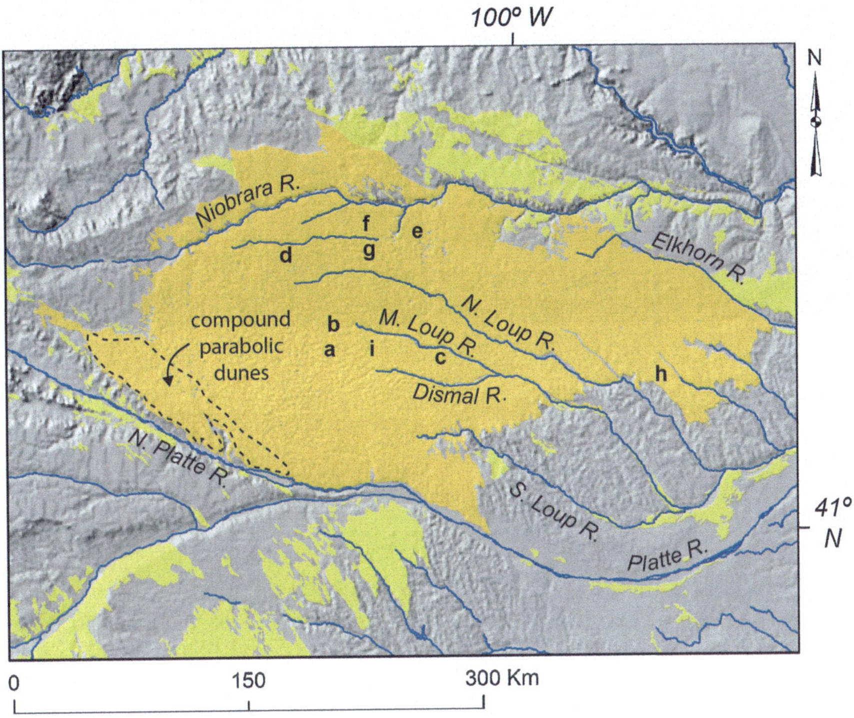


**Fig. 5.2** Nebraska Sand Hills (orange), as mapped in Bleed and Flowerday (1990), with surrounding small dune fields and sand sheets (yellow), major streams, and regional topography. Dashed line delineates area in western Sand Hills where compound parabolic dunes predominate. Letters label locations of Figs. 5.3–5.6: (a) Fig. 5.3a; (b) Fig. 5.3b; (c) Fig. 5.3c–d; (d) Fig. 5.4a, b; (e) Fig. 5.4c; (f) Fig. 5.4d; (g) Fig. 5.5; (h) Fig. 5.6a; (i) Fig. 5.6b–c

Stream networks within the Sand Hills are minimally developed (Fig. 5.2). Main stems of streams with few tributaries flow in incised valleys, some of which appear to be extending headward through interdunes and other low areas. Stream flow is dominated by groundwater discharge, and is much less variable than in adjacent regions where surface runoff is common (Bentall 1990). The north-central and western Sand Hills are dotted with lakes and wetlands, which mark the intersection of the water table with the interdune land surface (Fig. 5.3a). Some western Sand Hills lakes are the result of the blockage of stream valleys by migrating dunes (Loope et al. 1995; Mason et al. 1997). More generally, the migration and changing form of dunes during periods of aeolian activity would have altered topographically influenced local groundwater flow patterns and thus changed the spatial pattern of interdune lakes and wetlands.

In the western Sand Hills, local groundwater flow systems and high evaporation rates combine to concentrate dissolved salts in some lakes (Zlotnik et al. 2010, 2012). The Sand Hills lakes, like the Prairie Potholes of the northern Great Plains, are important stopover points for migratory waterfowl. Interdune wetlands

**Fig. 5.3** Landforms of the Sand Hills (locations in Fig. 5.2). (**a**) Wet interdune with groundwater table at or near surface in foreground to middleground, large barchanoid ridge with superimposed linear dunes in background. (**b**) Oblique aerial photo looking northeast across a large barchanoid ridge, with other barchanoid ridges and interdune lakes visible in the distance; note steep lee (south) face of dune. (**c**) Linear dunes superimposed on a large dome-like dune, view is from crest of linear dune looking south (Nebraska National Forest site of Mason et al. 2011). (**d**) Coring machine collecting subsurface samples on crest of linear dune, same site as (**c**); dune crest has been modified by development of blowouts and small parabolic dunes, creating "choppy" topography. Reddish brown grass that dominates ground cover in (**c**) and (**d**) is the most important native grass of the Sand Hills, little bluestem (*Schizachyrium scoparius*), in fall color

sustained by particularly strong local groundwater flow systems have accumulated up to 7 m of peat over the past 12,500 years (Nicholson and Swinehart 2005).

## 5.4 Modern Climate and Natural Vegetation

The climate of the Sand Hills ranges from sub-humid in the east (average precipitation of ~650 mm year$^{-1}$, 1980–2010), to semiarid in the west (~400 mm year$^{-1}$), with most rainfall in the summer growing season. Mean annual temperature is around 9 °C with a large annual range (data from High Plains Climate Center). Strong winds are common, often associated with strong cyclones that develop in the lee of the Rocky Mountains. Drift potential (DP, Fryberger and Dean 1979), an index of potential aeolian sand transport, is high by global standards (>400 vector units, Ahlbrandt and Fryberger 1980). Strong northwesterly to northerly winds are more common in winter, while in summer southerly to southeasterly winds predominate, although strong winds from either of these two predominant directions can occur in any season. The strong winds that contribute disproportionately to DP are more frequent in winter, so the annual net vector of potential transport (resultant DP) is generally toward the southeast, though it varies somewhat among weather stations in the region, and at some stations is more directly southward or even south-southwestward (Schmeisser et al. 2010; Sridhar et al. 2006).

Much of the Sand Hills dune field remains covered by native grassland (Fig. 5.3), though introduced hay and grain crops are grown in some low-relief areas and pines have been planted on 81 $km^2$ of national forest land in the central Sand Hills. Upland vegetation is dominated by deep-rooted perennial grasses, though forbs and shrubs are also common (Kaul 1990). Some of the most abundant grasses, including the ubiquitous little bluestem (*Schizachyrium scoparius*, Fig. 5.3c, d) are $C_4$ (warm-season) grasses, although there is also an important component of $C_3$ (cool-season) grasses. Stable carbon isotope analyses of bison bones and organic matter in paleosols indicate abundant $C_4$ grasses in the late Holocene as well (Goble et al. 2004; Muhs et al. 1997a). Distinct plant communities occur in wet interdunes and as early colonizers in blowouts and other areas of bare sand (Kaul 1990). Most land in the Sand Hills is privately owned and used for grazing cattle (or in a few cases, bison).

## 5.5 Dune Morphology

The largest bedforms of the Sand Hills are clearly visible on the earliest topographic maps of the region, and in low-resolution satellite images from the 1970s and 1980s. A ground-level view suggests smaller forms superimposed on the large dunes (Fig. 5.3c), which can be delineated to varying degrees on aerial photographs, and now much more clearly defined in LiDAR-based shaded relief maps. On the ground, a choppy surface covered with bare and revegetated blowouts of various sizes is evident on many dunes (Fig. 5.3d). LiDAR data now reveal the extent to which the dunes are partially, or in some cases completely, covered with blowout complexes and small parabolic dunes.

Swinehart (1990) mapped the largest scale of dune forms and described the major morphological types (Figs. 5.4 and 5.5). Large crescentic dunes, including barchanoid ridges and barchans, occupy almost half the area of the dune field and dominate its central core. These are particularly large examples of these common dune forms. The barchanoid ridges have a wavelength on the order of 2–4 km and are from 3 to more than 20 km long (Figs. 5.3b and 5.4a, b). Heights of 50 m are common, and the highest dune in the Sand Hills is a barchanoid ridge rising 135 m above the adjacent interdune. The barchans average about 40 m high, 0.8 km parallel to the migration direction, and 1.2 km wide (relatively large examples shown in Fig. 5.5a, c). Large dome-like dunes occupy about 20% of the dune field; most of these likely originated as large barchans but have now lost an identifiable crescentic

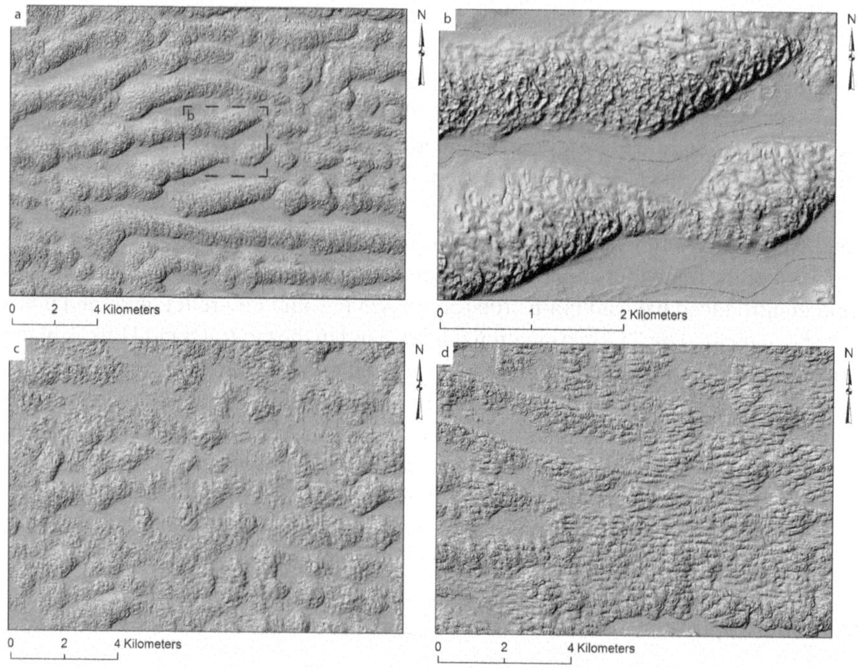


**Fig. 5.4** Dune forms of the Sand Hills, portrayed in shaded relief using LiDAR-based 2-m digital elevation data (locations in Fig. 5.2). (**a**) Large barchanoid ridges, with some dome-like dunes; dashed box is area shown in (*b*). (**b**) Two segments of barchanoid ridges illustrating partial surface modification by development of blowout complexes and small parabolic dunes, with no linear dunes evident; dune to the north has been more extensively (or recently?) modified by blowout development. (**c**) Large area of dome-like dunes, with surface modification by blowouts. (**d**) Linear dunes superimposed on low barachanoid ridges and dome-like dunes, or as primary dune forms in a low-relief landscape in southeast portion of map. East-west orientation of these linear dunes differs from northwest-southeast oriented dunes discussed by Sridhar et al. (2006). (Elevation data from Nebraska Department of Natural Resources)

form and steep lee slopes (Fig. 5.4c). Large parabolic dunes predominate near the southwestern edge of the Sand Hills (Fig. 5.2), while sand sheets form the easternmost extension of the dune field in northeastern Nebraska and also occur locally along the dune field margins. Finally, over a substantial area near the southeastern dune field margin and in many patches elsewhere, linear dunes up to about 15 m high and 150 m wide are either the primary dune form or are superimposed on larger forms (Figs. 5.4d and 5.5a–c). These are relatively straight, steep-sided ridges, in many cases with a shallow trough or series of depressions along the dune crest (Figs. 5.4d and 5.5b). This axial trough, and the generally shorter length and more discontinuous nature of the Sand Hills linear dunes differentiates them from those of major deserts in Australia and southern Africa.

Smaller bedforms are commonly superimposed on the large crescentic, dome-like, and parabolic dunes. Linear dunes are superimposed on larger dunes across large parts of the dune field (Sridhar et al. 2006). The crest and stoss slopes of many

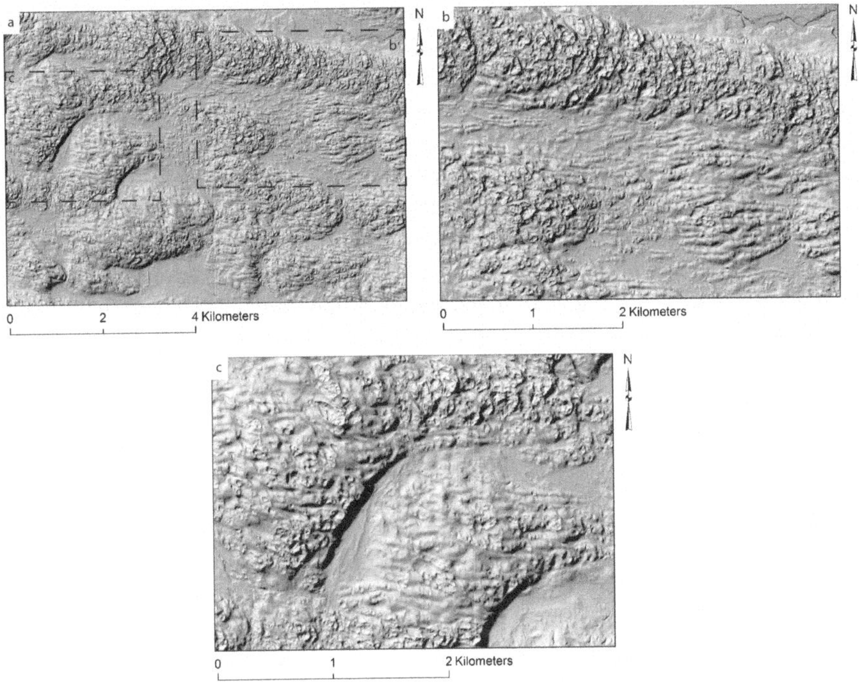


**Fig. 5.5** Dune forms of the Sand Hills, portrayed in shaded relief using LiDAR-based 2-m digital elevation data (locations in Fig. 5.2). (**a**) Barchanoid ridge, dome-like dunes, and two large barchans (marked with upper-case B), with linear dunes superimposed on larger forms and also covering interdune surface in east-central part of map. Dashed lines show areas portrayed in (*b*) and (*c*). (**b**) Linear dunes with well-developed axial troughs, on low-relief interdune surface. Higher dunes to north and west are covered with blowouts and small parabolic dunes, which may have obscured pre-existing superimposed linear dunes. (**c**) Close-up of two barchans in (a), with superimposed linear dunes still evident on the one to the southeast but obscured by blowouts on the one to the northwest

large barchanoid ridges and barchans, and the entire surface of large dome-like dunes, have been reworked into these closely spaced linear dunes (Figs. 5.4d and 5.5). Small parabolic dunes are superimposed on the larger parabolic forms in the southwestern Sand Hills, and are also associated with blowout complexes and other modifications of large crescentic and dome-like dunes across much of the dune field (Fig. 5.4b). Small transverse ridge dunes are superimposed on larger forms in one small area of the southwestern Sand Hills.

Comparison of dune forms with the modern wind regime reveals discrepancies that are potentially important in reconstructing the paleoclimatic conditions that favored past dune activity. As described above, the direction of net potential sand transport (resultant drift direction) in the Sand Hills is generally toward the southeast. Sridhar et al. (2006) noted that in the bimodal modern wind regime, the crests of active dunes would not necessarily be perpendicular to the resultant drift direction. Nonetheless, assuming that dune crest orientations maximize net bedform normal sand transport (Rubin and Ikeda 1990), Sridhar et al. concluded that the modern wind regime at several Sand Hills stations should generally produce dunes with crests oriented southwest-northeast. In many cases these dunes would be oblique, i.e. with some component of dune migration parallel to the crest, rather than transverse. Barchanoid ridges with crests approximating the predicted orientation occur mainly in the westernmost part of the dune field and at its southern margin near North Platte, Nebraska. Eastward across the central core of the Sand Hills, barchanoid ridges shift toward a predominantly east-west orientation. West-northwest to east-southeast orientations are observed toward the eastern limit of the dune forms. The orientations of large barchans do often record southeastward migration, particularly in the westernmost Sand Hills. Further complicating the issue, some large barchans in the central Sand Hills appear to record distinctly different slipface orientations than nearby barchanoid ridges (e.g. Fig. 5.5a).

The linear dunes of the Sand Hills, both primary and superimposed, also have orientations difficult to explain as the work of the modern wind regime. Most of the many linear dunes in the southern Sand Hills are oriented approximately northwest-southeast (Sridhar et al. 2006); however, some significant areas of linear dunes display east-west and southwest-northeast orientations, especially in the northern Sand Hills (e.g. Figs. 5.4d and 5.5a–c). Sedimentary structures exposed by trenching in eight northwest-southeast oriented linear dunes and in larger exposures at two other sites support the interpretation of these dunes as longitudinal, formed by bimodal winds that drove migration approximately parallel to the dune crest (Sridhar et al. 2006; Schmeisser et al. 2010; one of the exposures is shown in Fig. 5b of Mason et al. 2011). If so, the wind regime at the time these northwest-southeast oriented dunes developed probably had a strong southwesterly component that is absent today. That particular paleowind regime does not help explain the differing alignment of linear dunes in some parts of the dune field, however, nor the orientation of barchanoid ridges in the central Sand Hills. In other words, dune forms—in some cases in close proximity to each other—may provide evidence for two or more paleowind regimes substantially different from the modern one.

More data on the varying age, morphology, and sedimentary structures of the linear dunes across the Sand Hills is needed to more confidently interpret the record of wind regimes they preserve. A recent study using ground-penetrating radar (GPR) on lines transverse to WSW-ENE oriented linear dunes in the northern Sand Hills identified 1–2 m thick beds mostly dipping gently toward the south (Larsen 2018). These beds are similar in thickness to the sets of cross-strata observed in a roadcut transverse to a linear dune at the Highway 97 Milepost 81 site (Sridhar et al. 2006, Fig. S4A; Mason et al. 2011, Fig. 5.5b). While it is hard to interpret these results in isolation, they do reveal the potential insight available from additional GPR studies in the Sand Hills.

Interestingly, the compound parabolic dunes (smaller parabolic forms superimposed on larger ones) in the southwesternmost part of the dune field (Fig. 5.2), record southeastward migration (Swinehart 1990) that is fully consistent with the modern wind regime. These dunes occupy the driest part of the Sand Hills, possibly the last to fully stabilize after intervals of dune activity. The orientation of blowouts

**Fig. 5.6** Blowouts in the Sand Hills and stratigraphic record exposed in them. (**a**) Typical blowout, possibly initiated by cattle traffic along fence line visible at right. View is toward the southeast; note sand eroded from blowout and deposited by northerly winds. A similar, but smaller mound of sand at other end of blowout, built by southerly winds is not visible. (**b**) Buried soil exposed in Vinton Blowout (Mason et al. 2011), formed mainly through organic matter accumulation and bioturbation at a time when vegetation cover stabilized a dune surface. (**c**) Stratigraphy in the Vinton Blowout. Buried soil shown in (**b**) is largely preserved toward left, but to the right is increasingly truncated by wind erosion before burial. Sand below the soil was deposited in the late Pleistocene and displays large-scale aeolian cross-strata dipping toward the southeast (see Mason et al. 2011, for details). Sand above soil is late Holocene (J. Swinehart, unpublished data)

and small parabolic dunes superimposed on larger forms across the Sand Hills (Figs. 5.4b, 5.5b and 5.6a) has not been systematically studied using newly available LiDAR data. Qualitatively, at least some blowouts appear consistent with erosion and deposition by both the northwesterly-northerly and the southeasterly components of the modern wind regime, while small parabolic dunes identifiable from LiDAR data suggest a range of westerly to northwesterly wind directions. Muhs and Budahn (2019) inferred northwesterly dune-forming winds from small parabolic dunes observed on aerial photos. All of these observations are broadly consistent with the modern wind regime.

## 5.6 Geologic History

### 5.6.1 *Evolving Geologic Interpretations of the Sand Hills*

The Sand Hills were recognized as aeolian features in the nineteenth century (e.g., Warren 1875, p. 26) but their importance as evidence of Quaternary climate change seems to have drawn little interest until much more recently. Lugn (1935) did not add much insight on that topic, but was the first geologist to seriously consider the origin of the dune field and its relation to the thick loess deposits just to the southeast. He asserted that the dune sand was derived mainly through wind erosion of "older Tertiary and…Pleistocene formations," with the coarser material forming the Sand Hills dunes and the finer particles carried farther downwind to form loess deposits. Lugn's conceptual model was largely speculative, and he promoted some clearly erroneous interpretations based on limited observations, e.g., the idea that the large dunes are cored with Tertiary bedrock (Lugn 1935, p. 161); nonetheless, his work called attention to the development of this major dune field as an important and intriguing geologic problem.

The first more substantive research on the Sand Hills was the work of H.T.U. Smith, much of it carried out in 1949–50, though not published until 1965 (Smith 1965). Smith used aerial photography to map the geomorphology of the dune field, identifying a first series of dunes (the large barchanoid ridges and barchans), a second series (the linear dunes), and a third series (blowouts and other surficial features on the larger dunes). Lacking numerical ages, Smith followed Lugn in assuming that the major phase of dune-building, represented by the first series, coincided with deposition of the Peoria Loess southeast of the Sand Hills during the last glaciation. He concluded that the second series of dunes formed later, but still within the Pleistocene, although he correlated this series with the Bignell Loess, now known to be Holocene (Johnson and Willey 2000; Mason et al. 2003). The third series was assumed to be Recent (Holocene) in age.

Most importantly, Smith (1965) for the first time sought paleoclimatic explanations for the past activity of the dune field. He noted a study of pollen in sediments of an interdune lake (Watts and Wright 1966), which indicated the presence of spruce (*Picea*) in the northern Sand Hills in the late-glacial period. At the same time, Smith pointed out that large dunes like those of his first series, where active, are characteristic of deserts "where average annual rainfall does not exceed a few inches." To reconcile this interpretation with evidence for greater moisture in the Late Pleistocene (e.g. the occurrence of spruce), Smith essentially invoked a changing Late Pleistocene climate, from moister conditions to greater aridity or vice versa. He also recognized the discrepancies between sand transport directions indicated by the Sand Hills dunes and those that would be predicted from the modern winds.

While Smith (1965) left open the possibility of minor ongoing dune activity in the Holocene, associated with his third series, the conventional wisdom of the 1960s and 1970s emphasized a view of the Sand Hills dunes as Pleistocene relics (e.g., Warren 1976; Wells 1983). This view was challenged in the late 1970s by James Swinehart and David Maroney, especially through radiocarbon dating of organics buried by thick aeolian sands, demonstrating substantial middle and late Holocene activity (Ahlbrandt et al. 1983). Ahlbrandt and Fryberger (1980) also applied new developments in aeolian sedimentology to a study of the Sand Hills dunes. Through the 1980s and 1990s, Swinehart, David Loope, and their collaborators focused on reconstructing an increasingly detailed history of Holocene aeolian activity, through radiocarbon dating of the paleosols often observed in upland dune exposures (Fig. 5.6b, c) and of aeolian sand layers in interdune peats, along with work on the morphology and sedimentology of the dunes and the paleoenvironmental record of sediments in Sand Hills Lakes (Loope and Swinehart 2000; Loope et al. 1995; Mason et al. 1997; Swinehart 1990). Daniel Muhs also studied the Holocene chronology of the dunes, along with a major contribution through geochemical and mineralogical studies of aeolian sand provenance and maturity (Muhs 1985; Muhs and Holliday 1995; Muhs et al. 1997b, 1999, 2000).

The most important new development in Sand Hills research, however, was the advent of optically stimulated luminescence (OSL) dating of dune activity, pioneered by Steven Stokes (Stokes and Swinehart 1997). Research since 2000 has produced a large number of new OSL ages from this dune field, allowing reconstruction of the timing of Late Pleistocene and Holocene dune activity with much greater confidence (Forman et al. 2005; Goble et al. 2004; Mason et al. 2004, 2011; McKean et al. 2015; Miao et al. 2007).

The sections that follow describe the current understanding of sand sources, the chronology of dune building and reactivation, and paleoenvironmental controls on dune activity in the Sand Hills, along with an overview of the major research questions still remaining.

### 5.6.2 *Identification of Sand Sources*

The regional scale geologic and geomorphic setting of the Sand Hills on the High Plains surface, reviewed above and portrayed in Fig. 5.1, is significant for understanding both the origin and the present hydrology of the dune field. It is difficult to rule out any of the sand-rich strata beneath or upwind of the dune field on geomorphic grounds alone. Fluvial sediment exposed along the Platte and other rivers systems at low river stages is probably most susceptible to wind-entrainment. The Pliocene aeolian sand and Pleistocene deposits directly underlying the dunes today are also uncemented, however, and presumably were widely exposed before the dune field formed. The fluvial sediments of the Pliocene Broadwater and Long Pine formations are also uncemented but their coarse grain size would likely have limited direct contributions to the aeolian system. While the Ogallala Group is often cemented to some degree, erosion on the dissected margins of the High Plains could have reworked large volumes of sand from this unit into stream deposits that were then available for deflation. The same processes could have contributed aeolian sand from sandstones of the Arikaree Group, widely exposed to erosion to the west and northwest of the dune field.

Most of these potential sources were recognized by Lugn (1935), but the quantitative data useful in more specifically characterizing dune sand provenance were not available until decades later. Muhs et al. (1997b) noted that the Sand Hills dune sands are relatively mature (quartz-rich, feldspar-poor), compared to other Great Plains dune fields, and suggested that this resulted from long-term loss of K-feldspar grains as they were broken down by ballistic impacts when dunes were active. In fact, the Sand Hills dunes have lower K content than most samples of Arikaree or Ogallala sandstones or North Platte River sands. Thus, the Sand Hills dune sand could have been derived from one or more of those sources, and then undergone substantial K-feldspar loss, which Muhs et al. (1997b) interpreted as evidence that the dune field was relatively old. The Pliocene sand sheets, however, have K contents comparable to the present dunes, though they do have higher Rb content.

Muhs (2017) reconsidered the provenance of the Sand Hills dune sand, this time using K/Rb and K/Ba ratios, which should not be affected by K-feldspar depletion through aeolian processes. He found that dune sand samples largely fall within the fields defined by these ratios for Ogallala and Arikaree groups, with only minor overlap with fields defined for Pliocene sands or North or South Platte river sediment. Muhs and Budahn (2019) investigated a wider range of evidence, particularly trace element geochemistry, on provenance of the Sand Hills dunes. These authors concluded that, taken together, K/Rb, K/Ba, Sc-Th-La, Eu/Eu*, $La_N/Yb_N$, As/Sb, and Fe/Sc point to the Ogallala Group as the primary source, with minor contributions from the Arikaree Group, while the Pliocene sands and sediment of the Platte River system were not significant sources.

### *5.6.3 Chronology of Dune-Building and Dune Activity*

The first numerical ages for dune activity in the Sand Hills were produced by radiocarbon dating of organic sediments under dune sand, or organic matter in paleosols within the dunes. These ages inherently pre- or postdate intervals of activity, often with considerable uncertainty, because the slow turnover of soil organic matter means a soil may already have a substantial apparent age before it is buried, and because of the potential for contamination from modern roots in some cases. Nonetheless, radiocarbon dating did demonstrate widespread activation within the past 4000 year, including the accumulation of up to 40 m of aeolian sand at some sites (Ahlbrandt et al. 1983; Swinehart and Diffendal 1990). In addition, Swinehart and Diffendal (1990) reported an age of 13,160 ± 450 $^{14}$C year BP from organic material below a ~ 40 m high dune of the central Sand Hills, demonstrating migration of the dune well after the last glacial maximum (LGM) and possibly in the Holocene. In the 1990s, Swinehart and Loope found aeolian sand beds buried within interdune peats, allowing more precise bracketing of intervals of activity through intensive radiocarbon dating of sedge seeds and other material immediately above and below the sand layers. This research ultimately identified a major interval of drought, water-table drawdown, and aeolian sand sheet advance into desiccated

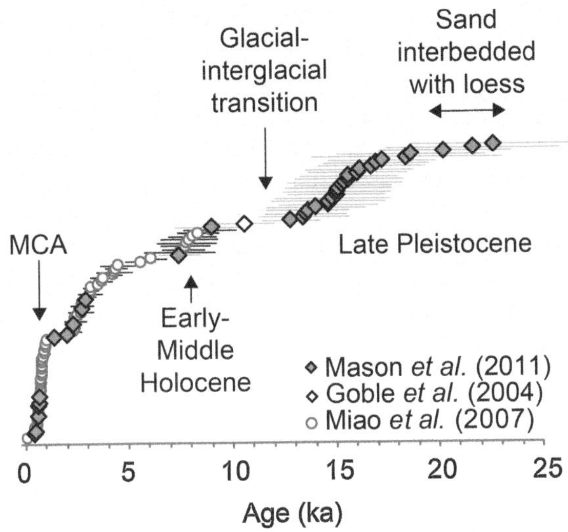


**Fig. 5.7** Summary of OSL ages, each dating deposition of aeolian sand, from Sand Hills and surrounding region of Nebraska and northeastern Colorado (modified from Mason et al. 2011). Ages are plotted by age, in ka (thousands of years ago), and placed vertical from youngest (low) to oldest (high), with symbols indicating original publication. Clusters of similar ages form steep to vertical portions of the staircase-like pattern, while intervals with few ages are marked by gentle slopes or near-horizontal shifts to the right. Major intervals of activity or stability are labeled; see text for details. Three ages in upper right, older than any in the Sand Hills proper, are from sand interbedded with Late Pleistocene loess southeast of the Sand Hills

interdunes in the interval 950–650 cal year BP, along with multiple similar episodes earlier in the Holocene (Mason et al. 2004; Nicholson and Swinehart 2005; Ponte 1995).

The advent of OSL dating allowed direct estimation of the time since sand was exposed to light during aeolian transport. The first OSL ages from the Sand Hills added to the evidence for widespread late Holocene activity (Stokes and Swinehart 1997). More extensive work with this method demonstrated a coherent pattern of activation during Holocene episodes that can be correlated across the Sand Hills and in smaller dune fields nearby. These episodes of widespread activity are evident as clusters of OSL ages, separated by intervals with few ages (Fig. 5.7; Goble et al. 2004; Mason et al. 2004; Miao et al. 2007). The correct interpretation of OSL age clusters—whether they define an entire interval of dune activity or only the final phases of it before stabilization—and even the possibility that clusters are sampling artifacts, have been discussed at length in the literature (e.g., Chase 2009; Leighton et al. 2014; Telfer and Hesse 2013; Xu et al. 2015). In the Sand Hills, however, the ages that define these clusters come in part from sections where clearly identifiable paleosols independently record dune stabilization at times falling between the age clusters (Goble et al. 2004). The loess stratigraphy in the region southeast of the Sand Hills also supports identification of the same episodes of increased aeolian activity, marked by units of rapidly deposited loess separated by paleosols, traceable across numerous loess sections (Mason et al. 2011; Miao et al. 2007).

The best-defined episode of dune activity, recorded by a large number of ages, occurred between 1 ka and 0.6 ka (ka = thousands of years ago), coinciding with the radiocarbon-dated aeolian sand sheet advance across desiccated interdune wetlands (Fig. 5.7). The vast majority of OSL ages from linear dunes fall within this interval, which approximately coincides with the Medieval Climatic Anomaly (MCA). The MCA has been variously defined but is generally considered to fall between 1.2–1.0 ka and 0.7–0.6 ka (e.g., Cook et al. 2004; Feng et al. 2008; Herweijer et al. 2007). A continuous cluster of OSL ages between 4.5 ka and 2 ka, with two peaks around 3.8 ka and 2.3 ka, is also represented at multiple sites in the Sand Hills and nearby dune fields (Fig. 5.7), and in the loess record downwind. In contrast, only two OSL ages fall within the interval 6.5–4.5 ka, also a time of slow to minimal loess accumulation in the region (Miao et al. 2007). An early to middle Holocene age cluster (10–7 ka) is relatively sparse compared to those in the late Holocene, and is largely made up of ages from core samples (Fig. 5.7). These dune ages fall within the most important interval of loess accumulation during the Holocene. The relatively limited representation of this early to middle Holocene dune activity in shallow outcrops of the Sand Hills can be explained by the pervasive reworking of older sands during late Holocene activity, especially the construction of the linear dunes. No OSL ages from shallow exposures or deep cores in the Sand Hills fall within the interval 13–10 ka, around the last glacial-interglacial transition (Fig. 5.7). This observation suggests that 13–10 ka was a time of minimal dune activity, coinciding with minimal loess accumulation and formation of the prominent Brady Soil in the loess-mantled region (Mason et al. 2011).

The oldest cluster of Sand Hills OSL ages is in the Late Pleistocene, between 19 and 13 ka, with most ages in the interval 17–14 ka (Fig. 5.7). These ages are mostly from drill cores, extending as deep as 35 m below the upwind (north or northwest) slopes of large dunes (Mason et al. 2011). Two deep cores, in particular, yielded multiple ages of 17–14 ka, statistically indistinguishable given their estimated errors and with no apparent depth trend, over depth intervals of 20 and 35 m. The simplest explanation of these observations is that the sampled dunes that were fully active and migrating at 17–14 ka, leading to rapid deposition of large volumes of sand, some which is still preserved within upwind portions of the modern dunes. Many of these Late Pleistocene ages are likely to record deposition on the lee slope of large dunes (some are from high-angle cross strata exposed at one site or visible in cores, Fig. 5.6c), yet they were obtained 300 m to >1 km north or northwest of the present lee faces of the large dunes they were sampled from. Thus, substantial Late Pleistocene migration after 14 ka, and/or renewed migration in the early-middle Holocene is likely (Mason et al. 2011), a conclusion also supported by the radiocarbon age reported by Swinehart and Diffendal (1990). Substantial dune building and migration probably also occurred before the oldest OSL ages from the Sand Hills dunes, that is, before 19 ka. Dune sand dating to 20 ka or earlier, based on OSL dating or stratigraphic position, is known from west-central Nebraska south of the Sand Hills (Fig. 5.7; Mason et al. 2011; J. Mason, unpublished field observations) but not from within the dune field itself.

The OSL- and radiocarbon-dated geologic record of the Sand Hills dunes that has been revealed by research since the 1970s is clear and compelling in some respects, but is still incomplete. Given the volume of aeolian sand that accumulated to form this dune field, it seems likely that it first developed before the LGM, and possibly before the last glaciation as a whole. By the time the aeolian sand deposition is first documented, after 19 ka, the large dunes that still characterize the Sand Hills today had already formed. At least some of those large dunes were fully active between 17 and 14 ka, followed by an interval from about 13 to 10 ka when regional aeolian activity may have been minimal, though the best evidence for that is from the regional loess record rather than the Sand Hills. Early-middle Holocene (10–7 ka) aeolian activity is now directly documented within the dune field by OSL dating, but the record is still sparse for that time period.

The Late Holocene is, predictably, the best documented part of the dune field's geologic history. Dune activity after about 4.5 ka was episodic, with the most well-documented episode falling during the MCA. The spatial extent of Late Holocene activity prior to the MCA is uncertain, but it is recorded at multiple sites across the dune field. Since dating consistently indicates the linear dunes were active during the MCA, more can be said about the extent of activity in that interval. Given the extensive area covered by linear dunes, and assuming that the morphodynamic interpretation of these landforms as linear dunes (Sridhar et al. 2006) is valid, their distribution implies fully active dunes over a large fraction of the dune field. The interpretation of Sridhar et al. (2006) also implies that many of the linear dunes are products of a wind regime during the MCA that was different from the present one, as well as the one the large barchans and barchanoid ridges formed in. Some aspects of linear dune morphology remain unexplained, particularly the trough or "crease" along the crest of many of these dunes. In some cases, this feature has the appearance of a series of blowouts along the dune crest, but in other places that explanation seems tenuous. More complete explanation of the morphology of these distinctive linear dunes could potentially provide additional information on the dune field environment during the MCA.

The blowout complexes that now cover large parts of the dunes (Figs. 5.4 and 5.5) suggest frequent but highly localized mobilization of sand, because of local disturbance (e.g. cattle trails along a fence), patchy vegetation cover in dry years, and/or greater wind exposure at high points on the large dunes. Blowout activity continues at present, even in relatively wet years, but its relative importance has almost certainly waxed and waned over time. Aerial photos indicate greater abundance of active blowouts during droughts of 2012, 2001–2002, the mid-1950s, and the 1930s. It appears to be a common view in the Sand Hills that bare sand was more common in the early twentieth and late nineteenth centuries than it is today, though this is difficult to document. Descriptions from the Warren military expedition (1856) also suggest that bare sand was common, possibly reflecting drought at that time (Muhs and Holliday 1995). More information is clearly needed on the dynamics of blowout development, to explore questions such as whether the topography in and around revegetated blowouts gradually becomes smoother over time, allowing an interpretation of relative age (e.g. Figs. 5.4b and 5.5c). Another hypothesis worth testing is that blowout activation preferentially occurs in certain topographic settings but spreads from there across the rest of a dune during droughts. Recently completed LiDAR data acquisition, together with an aerial photo sequence that is now almost 80 years long should be useful in addressing these issues.

### *5.6.4 Environmental Controls on Past Dune Activity*

In summers with adequate moisture, the environmental change required to convert the green, grass-covered Sand Hills into an active sand sea seems enormous. Yet it is now clear that dunes were active across much, if not all, of the dune field even during parts of the late Holocene, when global climate change was quite modest compared to the full range experienced through glacial cycles. The state of the Sand Hills at various times during the late Quaternary can be related to three factors, following Kocurek and Lancaster (1999): sediment supply, sediment availability, and transport capacity of the wind. At this point, most evidence supports identification of sediment availability—the susceptibility of sediment to transport, influenced by conditions such as vegetation density and moisture—as the factor limiting dune activity, for the Holocene at least. The rate and timing of sediment supply, from the potential sand sources discussed earlier, must have been important controls in the initial development of the dune field and the morphological evolution of the dunes. Unfortunately, the well-dated geologic record begins with the interval of Late Pleistocene dune activity centered on 17–14 ka, when dunes as large as those observed today were already present at sites across the dune field. While there certainly could have been ongoing influx of sediment to the dune field, much of the recorded activity of the Holocene involved migration and/or reshaping of the large bedforms present in the late Pleistocene. Thus, sand supply is not likely to have been an important control on Holocene activation and stabilization of the dune field.

It is relatively easy to make a case that transport capacity is not limiting, either, given the predominance of stability today despite a high frequency of strong winds and resulting high DP. Muhs and Holliday (1995) applied a dune mobility index (Lancaster 1988) based on frequency of winds above the threshold for sand mobility and the ratio of potential evapotranspiration to precipitation (PE/P), to the Sand Hills and other Great Plains dune fields. The results suggested that at present, mobility of these dune fields is largely limited by high precipitation rather than low wind speeds, not a surprising result to anyone with field experience in the region. These arguments do not rule out the possibility that there were times in the past when lower-than-present DP favored stability despite other conditions favoring activation, such as low effective moisture.

In the index used by Muhs and Holliday (1995), PE/P is an approximation of moisture available to plants, and it is clearly the grassland vegetation covering the Sand Hills and other Great Plains dunes that limits sediment availability under the present conditions of high PE/P. Since 1900, several severe but short-lived droughts substantially reduced primary productivity of native Great Plains grasslands and led to full activation of dunes in some areas, though not the Sand Hills (Bolles et al. 2017; Melton 1940; Muhs and Maat 1993). A compelling case can be made for either a shift toward more frequent severe droughts, or a longer-term shift of the climate to a drier mean state, as the key drivers of Holocene dune activity in the Sand Hills. This is particularly true for the last major episode of activation during the MCA. Compilations of climatic reconstructions from tree rings indicate that the MCA was characterized by a high frequency of droughts, some severe and extended, across much of the western and central U.S. including the Great Plains (Cook et al. 2004, 2014; Herweijer et al. 2007). The lower average moisture availability produced by these frequent droughts during the MCA is recorded by a variety of other paleoclimatic evidence within and around the Great Plains (Daniels and Knox 2005; Denniston et al. 2007). Within the Sand Hills themselves, the MCA stands out as a time of low lake levels in a composite of reconstructions based on diatom assemblages; sand influx to lake sediment was high in the MCA as well (Schmieder et al. 2011). A particularly direct link between Sand Hills aeolian activity and hydrological drought during the MCA involves the extension of aeolian sand sheets across interdune peatlands at this time, which would have required water table drawdown (Mason et al. 2004). Neither the OSL dating of upland dune sand nor the peatland record have sufficient resolution to indicate whether the extent of dune activity fluctuated as the climate of the MCA fluctuated between severe droughts and wetter periods recorded in tree-ring data, and whether there were lags or hysteresis in this fine-scale response to climate.

There is also evidence, though not as abundant or as clearcut, for dry conditions approximately coinciding with OSL-dated activity between 4.5 and 2 ka (with peaks at 3.8 and 2.3 ka), both in the Sand Hills (Schmieder et al. 2011) and from adjacent regions of the midcontinent (Denniston et al. 2007; Shuman and Marsicek 2016). The early to mid-Holocene (10–6 ka) stands out as a time of persistent dry climate, low lake levels, dust deposition in lakes, and grassland expansion at the expense of forest, in compilations of paleoenvironmental evidence from the Midwest, northern Great Plains and Rocky Mountains (Shuman and Marsicek 2016; Shuman and Serravezza 2017; Williams et al. 2009, 2010). Thus, the aeolian activity at that time, recorded by OSL ages in the Sand Hills and rapid loess accumulation just downwind of the dune field, most likely reflects reduction of vegetation cover by persistent aridity. The subsequent interval of minimal dune activity (~6.5–4.5 ka) is consistent with an abrupt shift to cooler and wetter climate across the midcontinent and eastern U.S. at 5.5 ka (Shuman and Marsicek 2016).

Importantly, while dated sediments in other Great Plains dune fields support the association of Holocene episodes of activity with dry climate, the timing of those episodes varies somewhat across the region (Forman et al. 2001; Halfen and Johnson 2013). For example, while evidence for activation during the MCA is common across the U.S. Great Plains, some other dune fields have produced many more ages indicating activity during the last several hundred years, as compared to the Sand Hills. While this variation may in part reflect differences in sampling strategy, dating methods, or sediment preservation, it is likely also due to spatial variation of Holocene climate change, as recorded by other evidence (Shuman and Marsicek

2016; Williams et al. 2010). In the northern Great Plains of Canada, Late Pleistocene and Holocene dune activity has often been associated with cold, dry intervals, as opposed to warm dry episodes like the MCA (Wolfe et al. 2017).

Perhaps the greatest remaining challenge in interpreting activity of the Sand Hills dunes as a response to paleoenvironmental conditions involves the clear evidence for full activity of the largest Sand Hills dunes from 17 to 14 ka. As reviewed by Mason et al. (2011), paleoecological data and relict permafrost features from areas south of the ice sheet margin demonstrate temperatures at that time far lower than modern, which are also simulated by climate models (though by 14 ka there could have been substantial warming relative to the LGM). Neither climate models nor much of the paleocological data indicate particularly dry conditions, however. In fact, spruce (*Picea*) macrofossils indicate the presence of that tree well south of the Sand Hills around the LGM (Wells and Stewart 1987), which does not suggest an arid environment conducive to full dune activity, although the spruce may have been confined to unusually moist sites. Low atmospheric $CO_2$ levels from the LGM until after 14 ka (Monnin et al. 2001) would have increased plant moisture stress at a given level of effective moisture, possibly reducing vegetation cover on dune sands with low water-holding capacity (Prentice et al. 2017). Lagged vegetation response to warming temperature after the LGM could explain a brief interval of dune activity, but Late Pleistocene OSL ages from the Sand Hills span thousands of years. None of these explanations seems adequate, suggesting that we are still missing some key mechanism or some aspect of the glacial environment that can explain the development and full activity of Sand Hills dunes comparable to those of the world's great warm deserts. This was noted more than 50 years ago by H.T.U. Smith.

We may also be missing some of the more subtle connections and feedbacks between climate, hydrological and ecosystem processes, and dune activity that characterized the Sand Hills during the late Quaternary, and that may provide important clues on how this distinctive landscape will respond to current and future environmental change. For example, particularly strong local groundwater flow systems must have maintained some wet interdunes and lakes even in the dry periods of the Holocene when much of the Sand Hills landscape was barren sand. In fact, dune migration is responsible for the initial development of some lakes, as noted above. McKean et al. (2015) suggest that these local areas of greater soil moisture acted as refugia for plants that could then rapidly colonize adjacent dunes as the climate became more favorable; this may explain a spatial trend of later stabilization after the MCA toward the southeastern part of the dune field where lakes and wetlands are less common. Persistent lakes and wetlands may also have been sites of prehistoric human settlement during dry periods with widespread dune activity (Napier et al. 2017). Given the effects of intensive grazing and trails worn by domestic cattle that can be observed in the Sand Hills today, it is worth considering the effects of bison in the Holocene or of now-extinct megafauna in the Late Pleistocene on dune-stabilizing vegetation, especially in times of climatically driven moisture stress. Wildfire was also a common form of disturbance across much of the Great Plains prior to fire suppression in the twentieth century. Interestingly, fires occurring in recent severe drought years have not triggered dune activity (Arterburn et al.

2018), although these observations clearly do not represent the full range of drought severity even in the late Holocene.

## 5.7 Future of the Sand Hills

Some environmental conditions of the Sand Hills today differ considerably from those of the Holocene or Late Pleistocene, including limited wildfire and managed grazing by cattle that have replaced bison. Nevertheless, the stability of the dunes still depends on the growth of the stabilizing grasses (mostly the same species as in prehistoric times) and the groundwater system has not been greatly altered. Thus, the geologic record of past dune activity is still highly relevant in predicting effects of anthropogenic climate change over the next century on the Sand Hills landscape.

Recent research has increasingly emphasized the likelihood of dry conditions in the central Great Plains by the end of the twenty-first century that will be more severe than those of late Holocene droughts, including those of the MCA (Cook et al. 2015). These predictions are based on a large set of state-of-the-art global climate models (the CMIP5 ensemble) which indicate increased winter and spring precipitation across the central Great Plains, but a decrease in summer, the critical time period for growth of the warm-season grasses that presently stabilize the dunes (Easterling et al. 2017). As a result of higher evapotranspiration rates in a warmer climate, soil moisture, along with decreased summer rainfall, a substantial decrease in summer, fall, and winter soil moisture is predicted for the central Great Plains (Cook et al. 2015; Wehner et al. 2017). These results appear with both high and moderate future carbon emission scenarios, although they are more extreme in the case of high emissions.

Given those predictions, and the Holocene geologic record of Sand Hills dune activity, it is likely that much of the presently stable Sand Hills landscape could be dominated by fully active dunes late in the present century. The geologic record does not have the resolution needed to predict rates or spatial patterns of dune activation, but it does suggest the persistence of some lakes and wet, vegetated interdunes in the newly active dune field. These could be an important component of the future dune field ecosystem and its potential for stabilization if current climate trends are eventually reversed.

## 5.8 Conclusions and Future Research Needs

The Nebraska Sand Hills, the largest dune field in North America, is now largely stabilized; however, a complex history of Late Pleistocene and Holocene dune activity is documented by extensive dating of stratigraphic sections within the dunes themselves and in interdune peatlands. Holocene episodes of dune activity are clearly linked to periods of dry climate (or more frequent severe droughts), while

Late Pleistocene activity may have been at least partly related to low temperatures and low atmospheric $CO_2$, which increased plant moisture stress. At this point there is little positive evidence that changing wind strength played a role in past dune activity; however, dune forms do record past change in the direction of strong winds, so past changes in wind regime need more study. More research is also needed on climatically driven changes in groundwater levels during dry periods, and the possible role of persistent wet interdunes as plant refugia that influenced the spatial pattern of subsequent dune stabilization. Projections of climate change over the twenty-first century increasingly emphasize the likelihood of drought more severe than the dry extremes of the Late Holocene, making widespread reactivation of the Sand Hills dunes a likely prospect. Ongoing research on how climatic, ecological, hydrological, and aeolian processes interact in coming decades is essential for adapting land management to changing conditions, and is also likely to provide important insight in interpreting the geologic record of the Sand Hills.

## References

Ahlbrandt TS, Fryberger SG (1980) Eolian deposits in the Nebraska Sand Hills. U.S.G.S. Professional Paper 1120-A

Ahlbrandt TS, Swinehart JB, Maroney DG (1983) The dynamic Holocene dune fields of the Great Plains and Rocky Mountain Basins, U.S.A. In: Brookfield ME, Ahlbrandt TS (eds) Eolian Sediments and Processes, Developments in Sedimentology 38. Elsevier, Amsterdam, pp 379–406

Arterburn JR, Twidwell D, Schacht WH, Wonkka CL, Wedin DA (2018) Resilience of Sandhills grassland to wildfire during drought. Rangel Ecol Manag 71:53–57

Bentall R (1990) Streams. In: Bleed A, Flowerday C (eds) An Atlas of the Sand Hills. Conservation and Survey Division, University of Nebraska-Lincoln, Lincoln, pp 93–114

Bleed A (1990) Groundwater. In: Bleed A, Flowerday C (eds) An Atlas of the Sand Hills. Conservation and Survey Division, University of Nebraska-Lincoln, Lincoln, pp 67–92

Bleed A, Flowerday C (eds) (1990) An atlas of the sand hills. Conservation and Survey Division, University of Nebraska-Lincoln, Lincoln, Nebraska, 265 pp

Bleed A, Ginsberg M (1990) Lakes and wetlands. In: Bleed A, Flowerday C (eds) An Atlas of the Sand Hills. Conservation and Survey Division, University of Nebraska-Lincoln, Lincoln, pp 115–122

Bolles K, Forman SL, Sweeney M (2017) Eolian processes and heterogeneous dust emissivity during the 1930s Dust Bowl Drought and implications for projected 21st-century megadroughts. The Holocene 27:1578–1588

Chase B (2009) Evaluating the use of dune sediments as a proxy for palaeo-aridity: a southern African case study. Earth Sci Rev 93:31–45

Chen X, Hu Q (2004) Groundwater influences on soil moisture and surface evaporation. J Hydrol 297:285–300

Chen X, Huang YY, Ling MH, Hu Q, Liu B (2012) Numerical modeling groundwater recharge and its implication in water cycles of two interdunal valleys in the Sand Hills of Nebraska. Phys Chem Earth 53–54:10–18

Cook ER, Woodhouse CA, Eakin CM, Meko DM, Stahle DW (2004) Long-term aridity changes in the western United States. Science 306:1015–1018

Cook BI, Smerdon JE, Seager R, Cook ER (2014) Pan-continental droughts in North America over the last millennium. J Clim 27:383–397

Cook BI, Ault TR, Smerdon JE (2015) Unprecedented 21st century drought risk in the American Southwest and Central Plains. Sci Adv 1:e1400082
Daniels JM, Knox JC (2005) Alluvial stratigraphic evidence for channel incision during the Mediaeval Warm Period on the central Great Plains, USA. The Holocene 15:736–747
Denniston RF, DuPree M, Dorale JA, Asmerom Y, Polyak VJ, Carpenter SJ (2007) Episodes of late Holocene aridity recorded by stalagmites from Devil's Icebox Cave, Central Missouri, USA. Quat Res 68:45–52
Diffendal RF Jr (1982) Regional implications of the geology of the Ogallala Group (upper Tertiary) of southwestern Morrill County, Nebraska, and adjacent areas. Geol Soc Am Bull 93:964–976
Duller RA, Whittaker AC, Swinehart JB, Armitage JJ, Sinclair HD, Bair A, Allen PA (2012) Abrupt landscape change post-6 Ma on the central Great Plains, USA. Geology 40:871–874
Easterling DR, Kunkel KE, Arnold JR, Knutson T, LeGrande AN, Leung LR, Vose RS, Waliser DE, Wehner MF (2017) Precipitation change in the United States. In: Wuebbles DJ, Fahey DW, Hibbard KA, Dokken DJ, Stewart BC, Maycock TK (eds) Climate Science Special Report: Fourth National Climate Assessment. U.S. Global Change Research Program, Washington, DC, pp 207–230
Feng S, Oglesby RJ, Rowe CM, Loope DB, Hu Q (2008) Atlantic and Pacific SST influences on Medieval drought in North America simulated by the Community Atmospheric Model. J Geophys Res 113:D11101
Forman SL, Oglesby R, Webb RS (2001) Temporal and spatial patterns of Holocene dune activity on the Great Plains of North America: megadroughts and climate links. Glob Planet Chang 29:1–29
Forman SL, Marin L, Pierson J, Gomez J, Miller GH, Webb RS (2005) Aeolian sand depositional records from western Nebraska: landscape response to droughts in the past 1500 years. The Holocene 15:973–981
Fryberger SG, Dean G (1979) Dune forms and wind regime. In: Mckee E (ed) A Study of Global Sand Seas, USGS Professional Paper 1052. US Geological Survey, Washington, DC, pp 137–169
Goble RJ, Mason JA, Loope DB, Swinehart JB (2004) Optical and radiocarbon ages of stacked paleosols and dune sands in the Nebraska Sand Hills, USA. Quat Sci Rev 23:1173–1182
Gosselin DC, Drda S, Harvey FE, Goeke J (1999) Hydrologic setting of two interdunal valleys in the central Sand Hills of Nebraska. Ground Water 37:924–933
Halfen AF, Johnson WC (2013) A review of Great Plains dune field chronologies. Aeolian Res 10:135–160
Harvey FE, Swinehart JB, Kurtz TM (2007) Ground water sustenance of Nebraska's unique sand hills peatland fen ecosystems. Ground Water 45:218–234
Herweijer C, Seager R, Cook ER, Emile-Geay J (2007) North American droughts of the last millennium from a gridded network of tree ring data. J Clim 20:1353–1376
Hunt RM (1990) Taphonomy and sedimentology of Arikaree (lower Miocene) fluvial, eolian, and lacustrince paleoenvironments, Nebraska and Wyoming: a paleobiota entombed in fine-grained volcaniclastic rocks. In: Lockley MG, Rice A (eds) Volcanism and Fossil Biotas, Geological Society of America Special Paper 244. Geological Society of America, Boulder, pp 69–112
Johnson WC, Willey KL (2000) Isotopic and rock magnetic expression of environmental change at the Pleistocene-Holocene transition in the central Great Plains. Quat Int 67:89–106
Kaul R (1990) Plants. In: Bleed A, Flowerday C (eds) An Atlas of the Sand Hills. Conservation and Survey Division, University of Nebraska-Lincoln, Lincoln, pp 127–142
Kocurek G, Lancaster N (1999) Aeolian system sediment state; theory and Mojave Desert Kelso dune field example. Sedimentology 46:505–515
LaGarry HE (1998) Lithostratigraphic revision and redescription of the Brule Formation (White River Group) of northwestern Nebraska. In: Terry DO Jr, LaGarry HE, Hunt RM Jr (eds) Depositional Environments, Lithostratigraphy, and Biostratigraphy of the White River and Arikaree Groups (Late Eocene-Early Miocene, North America), Geological Society of America Special Paper 325. Geological Society of America, Boulder, pp 63–91

Lancaster N (1988) Development of linear dunes in the southwestern Kalahari, Southern Africa. J Arid Environ 14:233–244
Larsen AK (2018) Timing and formation of linear dunes south of the Niobrara River Valley, North-Central Nebraska Sand Hills. M.S. thesis, University of Nebraska-Lincoln, 95 p
Leighton CL, Thomas DSG, Bailey RM (2014) Reproducibility and utility of dune luminescence chronologies. Earth Sci Rev 129:24–39
Leonard EM (2002) Geomorphic and tectonic forcing of late Cenozoic warping of the Colorado piedmont. Geology 30:595–598
Loope DB, Swinehart JB (2000) Thinking like a dune field: geologic history in the Nebraska Sand Hills. Great Plains Res 10:5–35
Loope DB, Swinehart JB, Mason JP (1995) Dune-dammed paleovalleys of the Nebraska Sand Hills – intrinsic versus climatic controls on the accumulation of lake and marsh sediments. Geol Soc Am Bull 107:396–406
Lugn AL (1935) The Pleistocene Geology of Nebraska. Nebraska Geological Survey Bulletin 10 (2nd series)
Lugn AL (1968) The origin of loesses and their relation to the Great Plains in North America. In: Schultz CB, Frye JC (eds) Loess and Related Eolian Deposits of the World. University of Nebraska Press, Lincoln, pp 139–182
Maroney DG (1978) A stratigraphic and paleoecologic study of some late Cenozoic sediments in the central Sand Hills province of Nebraska. Unpublished Ph.D. dissertation, University of Nebraska-Lincoln, 379 p
Mason JP, Swinehart JB, Loope DB (1997) Holocene history of lacustrine and marsh sediments in a dune-blocked drainage, Southwestern Nebraska Sand Hills, USA. J Paleolimnol 17:67–83
Mason JA, Jacobs PM, Hanson PR, Miao XD, Goble RJ (2003) Sources and paleoclimatic significance of Holocene Bignell Loess, Central Great Plains, USA. Quat Res 60:330–339
Mason JA, Swinehart JB, Goble RJ, Loope DB (2004) Late-Holocene dune activity linked to hydrological drought, Nebraska Sand Hills, USA. The Holocene 14:209–217
Mason JA, Swinehart JB, Hanson PR, Loope DB, Goble RJ, Miao X, Schmeisser RL (2011) Late Pleistocene dune activity in the central Great Plains, USA. Quat Sci Rev 30:3858–3870
McKean RLS, Goble RJ, Mason JB, Swinehart JB, Loope DB (2015) Temporal and spatial variability in dune reactivation across the Nebraska Sand Hills, USA. The Holocene 25:523–535
Melton FA (1940) A tentative classification of sand dunes its application to dune history in the Southern High Plains. J Geol 48:113–174
Miao XD, Mason JA, Swinehart JB, Loope DB, Hanson PR, Goble RJ, Liu XD (2007) A 10,000 year record of dune activity, dust storms, and severe drought in the central Great Plains. Geology 35:119–122
Monnin E, Indermuhle A, Dallenbach A, Fluckiger J, Stauffer B, Stocker TF, Raynaud D, Barnola J-M (2001) Atmospheric $CO_2$ concentrations over the last glacial termination. Science 291:112–114
Muhs DR (1985) Age and paleoclimatic significance of Holocene sand dunes in northeastern Colorado. Ann Assoc Am Geogr 75:566–582
Muhs DR (2017) Evaluation of simple geochemical indicators of aeolian sand provenance: late quaternary dune fields of North America revisited. Quat Sci Rev 171:260–296
Muhs DR, Budahn JR (2019) New geochemical evidence for the origin of North America's largest dune field, the Nebraska Sand Hills, central Great Plains, USA. Geomorphology 332:188–212
Muhs DR, Holliday VT (1995) Evidence of active dune sand on the Great Plains in the 19th century from accounts of early explorers. Quat Res 43:198–208
Muhs DR, Maat PB (1993) The potential response of eolian sands to greenhouse warming and precipitation reduction of the Great Plains of the U.S.A. J Arid Environ 25:351–361
Muhs D, Zárate M (2001) Late quaternary eolian records of the Americas and their paleoclimatic significance. In: Markgraf V (ed) Interhemispheric Climate Linkages. Academic, San Diego, pp 183–216

Muhs DR, Stafford TW, Been J, Mahan SA, Burdett J, Skipp G, Rowland ZM (1997a) Holocene eolian activity in the Minot dune field, North Dakota. Can J Earth Sci 34:1442–1459
Muhs DR, Stafford TW, Swinehart JB, Cowherd SD, Mahan SA, Bush CA, Madole RF, Maat PB (1997b) Late Holocene eolian activity in the mineralogically mature Nebraska sand hills. Quat Res 48:162–176
Muhs DR, Swinehart JB, Loope DB, Aleinikoff JN, Been J (1999) 200,000 years of climate change recorded in eolian sediments of the High Plains of eastern Colorado and western Nebraska. In: Lageson David R, Lester Alan P, Trudgill Bruce D (eds) Colorado and adjacent areas. Geological Society of America, Boulder, pp 71–91
Muhs DR, Swinehart JB, Loope DB, Been J, Mahan SA, Bush CA (2000) Geochemical evidence for an eolian sand dam across the North and South Platte rivers in Nebraska. Quat Res 53:214–222
Myers MR (1993) A Pliocene sand sheet below the Nebraska Sand Hills. M.S. thesis, University of Nebraska-Lincoln, 122 p
Napier TJ, Douglass M, Wandsnider L, Goble RJ (2017) Investigating the human response to the medieval climatic anomaly in the Nebraska Sand Hills: a preliminary study in building occupation histories with OSL dating. Plains Anthropol:1–21
Nicholson BJ, Swinehart JB (2005) Evidence of Holocene climate change in a Nebraska Sandhills wetland. Great Plains Res 15:45–67
Ponte MR (1995) Eolian origin of sand within interdune peat, Central Nebraska Sand Hills. Unpublished M.S. thesis, University of Nebraska-Lincoln
Prentice IC, Cleator SF, Huang YH, Harrison SP, Roulstone I (2017) Reconstructing ice-age palaeoclimates: quantifying low-$CO_2$ effects on plants. Glob Planet Chang 149:166–176
Rubin DM, Ikeda H (1990) Flume experiments on the alignment of transverse, oblique, and longitudinal dunes in directionally varying flows. Sedimentology 37:673–684
Schmeisser RL, Loope DB, Mason JA (2010) Modern and late Holocene wind regimes over the Great Plains (Central USA). Quat Sci Rev 29:554–566
Schmieder J, Fritz SC, Swinehart JB, Shinneman ALC, Wolfe AP, Miller G, Daniels N, Jacobs KC, Grimm EC (2011) A regional-scale climate reconstruction of the last 4000 years from lakes in the Nebraska Sand Hills, USA. Quat Sci Rev 30:1797–1812
Shuman BN, Marsicek J (2016) The structure of Holocene climate change in mid-latitude North America. Quat Sci Rev 14:38–51
Shuman BN, Serravezza M (2017) Patterns of hydroclimatic change in the Rocky Mountains and surrounding regions since the last glacial maximum. Quat Sci Rev 173:58–77
Smith HTU (1965) Dune morphology and chronology in central and western Nebraska. J Geol 73:557–578
Sridhar V, Loope DB, Swinehart JB, Mason JA, Oglesby RJ, Rowe CM (2006) Large wind shift on the Great Plains during the Medieval Warm Period. Science 313:345–347
Stokes S, Swinehart JB (1997) Middle- and late-Holocene dune reactivation in the Nebraska Sand Hills, USA. The Holocene 7:263–272
Swinehart JB (1990) Wind-blown deposits. In: Bleed A, Flowerday C (eds) An Atlas of the Sand Hills. Conservation and Survey Division, University of Nebraska-Lincoln, Lincoln, pp 42–56
Swinehart JB, Diffendal RF Jr (1990) Geology of the pre-dune strata. In: Bleed A, Flowerday C (eds) An Atlas of the Sand Hills. Conservation and Survey Division, University of Nebraska-Lincoln, Lincoln, pp 29–42
Swinehart JB, Souders VL, De Graw HM, Diffendal RF Jr (1985) Cenozoic paleogeography of western Nebraska. In: Flores RM, Kaplan SS (eds) Cenozoic Paleogeography of West-Central United States. Rocky Mountain Section-SEPM, Denver
Telfer MW, Hesse PP (2013) Paleoenvironmental reconstructions from linear dunefields: recent progress, current challenges and future directions. Quat Sci Rev 78:1–21
Trimble DE (1980) Cenozoic history of the Great Plains contrasted with that of the southern Rocky Mountains: a synthesis. The Mountain Geologist 17:59–69

Warren GK (1875) Preliminary Report of Explorations in Nebraska and Dakota in the Years 1855-'56-'57. Reprint. Engineer Department, U. S. Army. Government Printing Office, Washington, DC

Warren A (1976) Morphology and sediments of the Nebraska Sand Hills in relation to Pleistocene winds and the development of aeolian bedforms. J Geol 84:685–700

Watts W, Wright HE Jr (1966) Late-Wisconsin pollen and seed analysis from the Nebraska Sandhills. Ecology 47:202–210

Wehner MF, Arnold JR, Knutson T, Kunkel KE, LeGrande AN (2017) Droughts, floods, and wildfires. In: Wuebbles DJ, Fahey DW, Hibbard KA, Dokken DJ, Stewart BC, Maycock TK (eds) Climate Science Special Report: Fourth National Climate Assessment. U.S. Global Change Research Program, Washington, DC, pp 231–256

Wells GL (1983) Late-glacial circulation over Central North America revealed by aeolian features. In: Street-Perrott A, Beran M, Ratcliffe R (eds) Variations in the Global Water Budget. Springer, Dordrecht, pp 317–330

Wells P, Stewart J (1987) Spruce charcoal, conifer macrofossils, and landsnail and small-vertebrate faunas in Wisconsinan sediments on the High Plains of Kansas. In: Johnson WC (ed) Quaternary Environments of Kansas, Kansas Geological Survey Guidebook Series, vol 5. Kansas Geological Survey, Lawrence

Williams JW, Shuman B, Bartlein PJ (2009) Rapid responses of the prairie-forest ecotone to early Holocene aridity in mid-continental North America. Glob Planet Chang 66:195–207

Williams JW, Shuman B, Bartlein PJ, Diffenbaugh NS, Webb T III (2010) Rapid, time-transgressive, and variable responses to early Holocene midcontinental drying in North America. Geology 38:135–138

Wolfe SA, Lian OB, Hugenholtz CH, Riches JR (2017) Holocene eolian sand deposition linked to climatic variability, Northern Great Plains, Canada. The Holocene 27:579–593

Xu Z, Lu H, Yi S, Vandenberghe J, Mason JA, Zhou Y, Wang X (2015) Climate-driven changes to dune activity during the Last Glacial Maximum and deglaciation in the Mu Us dune field, North-Central China. Earth Planet Sci Lett 427:149–159

Zlotnik VA, Robinson NI, Simmons CT (2010) Salinity dynamics of discharge lakes in dune environments: conceptual model. Water Resour Res 46:W11548

Zlotnik VA, Ong JB, Lenters JD, Schmieder J, Fritz SC (2012) Quantification of salt dust pathways from a groundwater-fed lake: implications for solute budgets and dust emission rates. J Geophys Res-Earth Surf:117, F02014

# Chapter 6
# White Sands

**Ryan C. Ewing**

**Abstract** Situated within the Tularosa Basin and the Rio Grande Rift Zone of central-southern New Mexico, White Sands Dune Field records a unique interplay of geology, hydrology, and climate. The origins of the gypsum sands that make the dune field white are rooted in ancient carbonate and evaporite rocks that surround the basin. Runoff and groundwater supply from the Sacramento Mountains mobilize calcium and sulfur from these rocks to basin playas. Pleistocene pluvial Lake Otero occupied the basin as an ephemerally saline and freshwater lake and hosted a wider ange of fauna including humans, evidenced by extensive trackways. Increased aridity through the Holocene lowered lake levels resulting in deflation of lake strata and production of foredune ridges that accreted to produce to the modern dune field by ~7 kya. The modern dune field is shaped by the antecedent topography, a near surface water table that stabilizes the dunes and limits sediment availability, and a multi-modal wind regime dominated by southwesterly winds with a resultant toward 065° and subordinate northwesterly and southeasterly winds. Changes in dune types across the dune field from dome dunes to crescentic to barchan to parabolic reflect changes in spatial variation in grain size, wind speed, and groundwater salinity. The dominantly white landscape created by the unique history of this basin has given rise to fascinating endemism of white lizards, moths, and other species. Evidence for extensive gypsum dunes in the north polar regions of Mars makes White Sands an excellent analog for planetary studies.

**Keywords** Aeolian · White Sands · Sand dunes · Mars

## 6.1 Introduction

White Sands Dune Field is one of the world's most treasured dune fields. The white sands, composed of 99% gypsum crystals, invite a world-wide audience for inspiration, recreation, and science. The dune field spans across 500 km$^2$ housed within three major United States government entities, and sits just a few miles south of the

R. C. Ewing (✉)
Department of Geology and Geophysics, Texas A&M University, College Station, TX, USA
e-mail: rce@tamu.edu



N. Lancaster, P. Hesp (eds.), *Inland Dunes of North America*,
Dunes of the World, https://doi.org/10.1007/978-3-030-40498-7_6

first nuclear bomb test at the Trinity Site. The active White Sands Missile Range controlled by the US Army comprises most of the Tularosa Basin, which houses White Sands Dune Field. Visitors to the area are tuned to the activity on the range by regular tests that close US-70, which provides access to White Sands National Park. The national park is an entity controlled by the US National Park Service, which facilitates visitation and science in the dune field. The park was established in 1933 by President Herbert Hoover. The eastern part of the basin hosts the Holloman Air Force Base. The intersection of these entities at White Sands compounds an already fascinating and unique geologic, geomorphic, biologic, and human history. Indeed, the rolling white dunes and expansive gypsum plains (Fig. 6.1) result from the intersection of geology, hydrology, and recent climate change and creates a scientific playground for nearly every facet of science. Geologists explain the origins of the dune field and the dynamics of the dunes themselves, and hydrologists resolve water routing through the basin that provides the life blood to playas that produce gypsum and stabilize the dune field from deflation. Paleontologists and anthropologists study ancient human-megafauna interactions, and biologists study species adapted to an entirely white environment, which have become textbook examples of evolution. This natural laboratory continues to be the focus of scientific research as all stakeholders around White Sands turn their attention to how the dune field will respond to changing climate and increased human pressure on land and water resources in the basin. A comprehensive review of the literature on White Sands would occupy at least one book volume; this chapter provides a 'need to know' overview spun from a deep literature base. This chapter gives the reader a road map to begin their science visit or research at White Sands.

## 6.2 Regional Geology, Climate, and Hydrology

White Sands Dune Field owes its existence to the convergence of unique geological, hydrologic, and climatic factors. Each of these factors contributes a necessary ingredient to generate a dune field composed nearly entirely of gypsum sand. Although White Sands is the largest gypsum dune field in the world, the suite of conditions for gypsum sand production extends across the region to give rise to several gypsum-rich basins including the Estancia, Trans-Pecos, and Cuatro Cienegas Basins that link to the regional interplay of geology, hydrology, and climate (Fig. 6.2).

**Fig. 6.1** (continued) Andres Mountains in the background. (**d**) View W across Lake Lucero with a dust devil. (**e**) View across an interdune area during a dust storm that obscured the midday sun. (**f**) Vegetated interdune area within the parabolic dunes at the eastern margin of the dune field. (**g**) View SW across Alkali Flat with standing water forming an ephemeral, microbial mat-lined pond. (**h**) Sunset view across an interdune area within the core crescentic dune field with ridges of accumulated dune cross stratification forming a corrugated interdune surface

**Fig. 6.1** Regional map and photographs of the variety of landscapes within and around the White Sands Dune Field. (**a**) View SE across the tops of the core crescentic dunes. (**b**) Picnic table with wind shield blocking winds from the SW (right). (**c**) View WNW across Alkali Flat with San

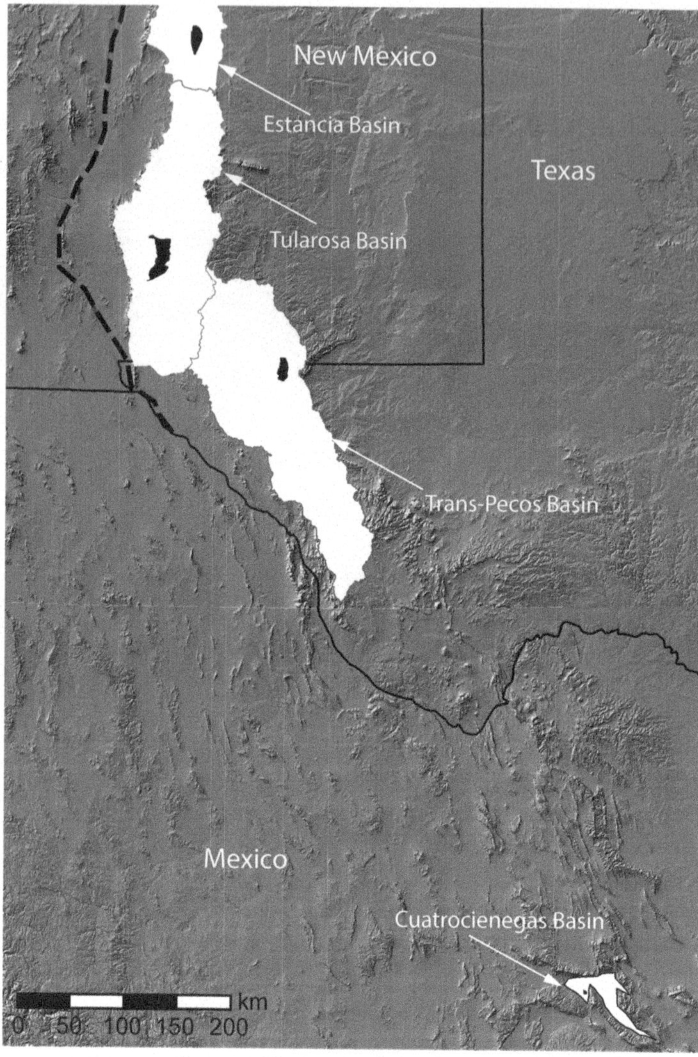


**Fig. 6.2** Topographic hillshade of the southwestern USA and northern Mexico showing basins along the Rio Grande Rift with gypsum dune fields. Dashed black line indicates the principal axis of the Rio Grande Rift. Black areas within the basins (white areas) denote dune fields. Watershed boundaries define basin extent. Base map is a hillshade rendering of 90 m SRTM topography data. (Modified from Szynkiewicz et al. 2010b)

### *6.2.1 Geology*

White Sands Dune Field sits at the intersection of a complex regional geological system that includes Basin and Range tectonics, the Rio Grande Rift, and a thick sequence of Paleozoic carbonate and siliciclastic rocks. The dune field occupies the western part of the Tularosa Basin in central-southern New Mexico, which is a north-south trending fault-bound graben formed at the eastern extent of the Basin and Range Tectonic Province and within the Rio Grande rift zone (Fig. 6.3). The basin is bounded to the east by the Sacramento Mountains and to the west by the San Andres, Organ, and Franklin Mountains. Topographically, the Chupadera Mesa and the elevation of the Hueco Bolson bound the north and south of the basin, respectively. Although the tectonic template for the extensional basin is set by both the basin and range and rift activity, the heat and volcanism associated with the rift is thought to play a larger role in forming White Sands (Szynkiewicz et al. 2010a, b).

The principal axis of the Rio Grande Rift extends over 1000 km north-south from southern Colorado to northern Mexico and occupies the Mesilla Basin just west of the Tularosa (Seager and Morgan 1979; Olsen et al. 1987). Narrow in the north, the rift widens southward into Texas and Mexico. An abrupt widening occurs near Socorro, NM where the rift margin extends eastward to include the Tularosa Basin (Olsen et al. 1987). In this region and southward, the physiographic expression of the rift merges with that of basins formed by Basin and Range extension. The signatures of the rift are thermal and gravitational anomalies associated with rifting and surface features that include lava flows and volcanoes (Seager and Morgan 1979; Olsen et al. 1987; Adams and Keller 1994). The most prominent volcanic feature of the Tularosa Basin is the Carrizozo volcanic fields north of White Sands (Keszthelyi and Pieri 1993) (Fig. 6.3). Heat flow from the rift and to a lesser extent elements such as sulfur and calcium associated with the basaltic mineralogy of the rift-related rocks may contribute to or enhance the composition of the groundwater supply to the Tularosa Basin that gives rise to gypsum sand. However, the carbonate and evaporite suite of rocks associated with the Paleozoic sequences that make up the flanking ranges of the basin and in the basin subsurface dominantly contribute the elements needed to form gypsum in the basin.

The Tularosa Basin is floored by ~2 km of Tertiary basin fill that overlies a ~4 km sequence of Paleozoic rocks, which gives way downward to basement igneous rock (Fig. 6.4). Deposition of the Paleozoic sequence of rocks in the region was accommodated by the ancient Orogrande basin associated with uplift of the ancestral Rocky Mountains. Cambrian-Ordovician dolomites comprise the lowermost unit and give way upward to Silurian cherts, siliciclastics, and dolomites as conglomerates and shales. Limestones, including large bioherms, and mixed clastic-carbonate units make up the much of the Devonian-Mississippian and Pennsylvanian units (Fig. 6.5). Permian units represent a mixed lithology ranging from siliciclastic to carbonate and evaporite, the latter of which is thought to contribute significantly to the formation of White Sands.

The Permian Yeso Formation is widely referenced as the unit that supplies sulfur and calcium ions to the groundwater in the Tularosa Basin that are needed to form gypsum sand (Fryberger 2001). Although the Yeso Formation varies widely across New Mexico and includes siliciclastic sandstones and shales in the north and west of New Mexico, around the Tularosa Basin, it is composed of dominantly of evaporites. The presence of the evaporite in this location and the variation in lithology across the state reflects a general change in paleoenvironments from terrestrial

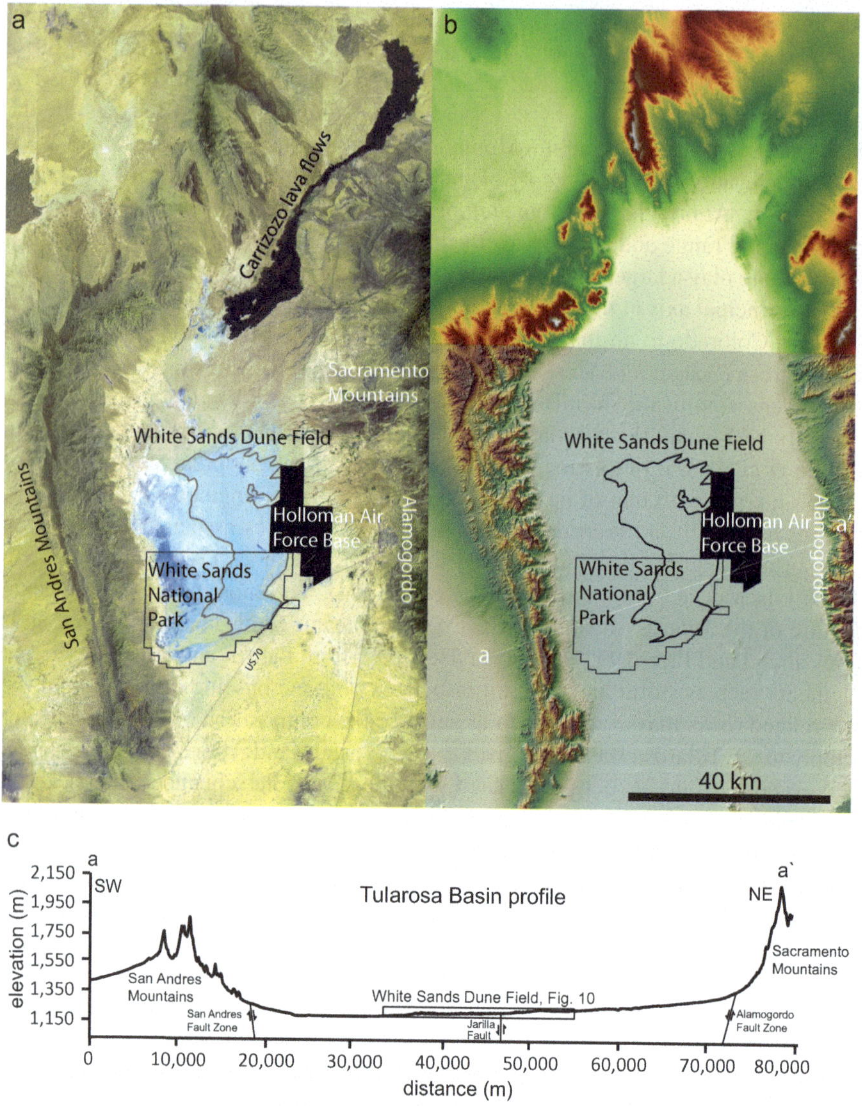


**Fig. 6.3** Location map of White Sands Dune Field within the Tularosa Basin and basin topographic profile. (**a**) Landsat 8 ETM+ false color image highlights the gypsum dune field in blue with boundaries highlighted by thin black line. The black feature at the top of the image is the Carrizozo lava flow. (**b**) Digital elevation model (DEM) showing the NS trending San Andres Mountains to the west and the base of the Sacramento Mountains to the east. DEM is a composite of SRTM 30 m data in the north of the basin and IFSAR 5 m data in the south denoted by a break in the shading. (**c**) Topographic profile across the Tularosa Basin with locations of major fault zones, White Sands Dune field, and Fig. 6.10 noted

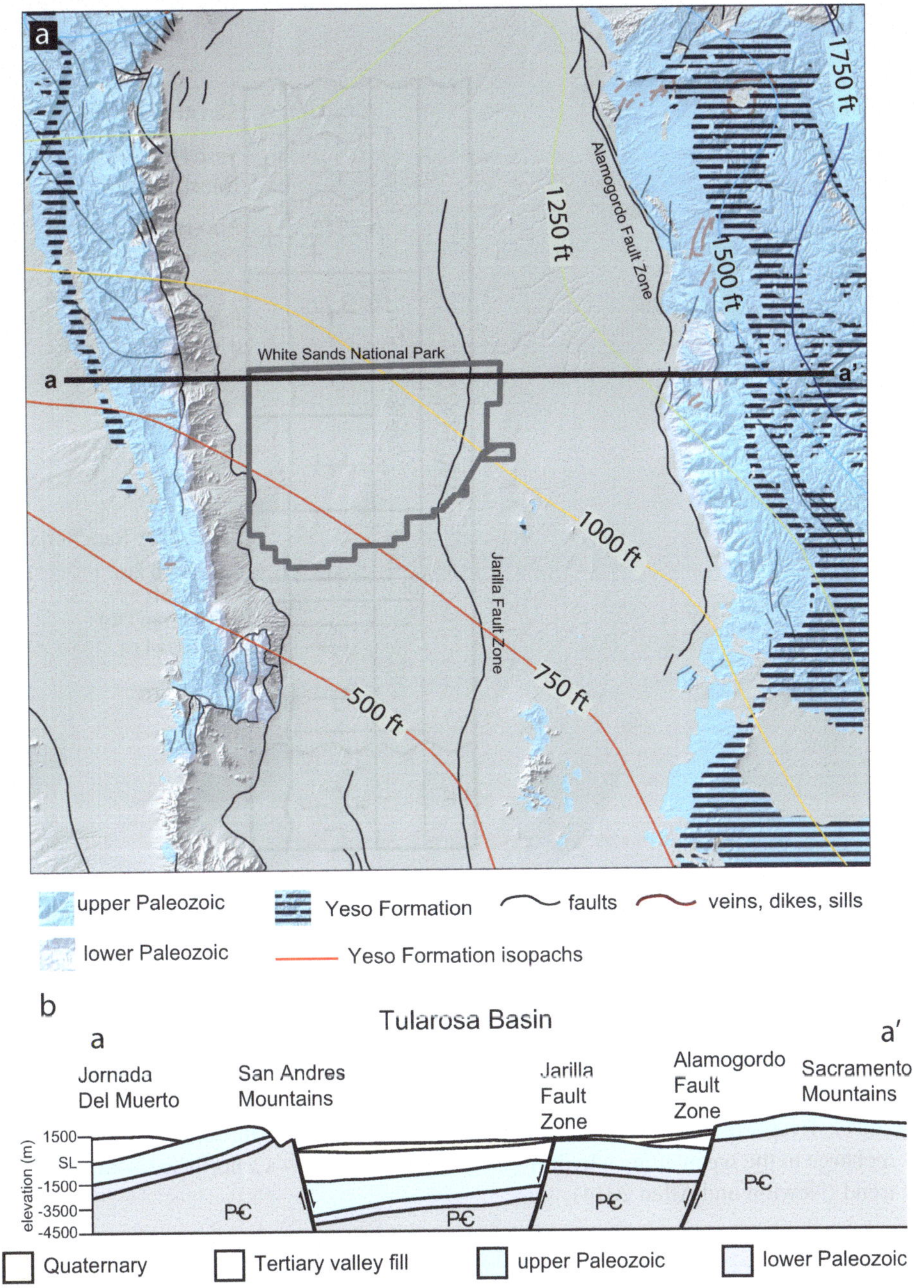


**Fig. 6.4** Paleozoic geological map and schematic cross-section of the Tularosa Basin. (**a**) Early Paleozoic units (purple) outcrop athe base of the Sacramento Mountains and top of the San Andres Mountains. Late Paleozoic units outcrop across the Sacramento Mountains and at the top of the San Andreas Mountains. The line-pattern area denotes the extent of the outcropping Yeso Formation, which is thought to be the contribute evaporites to the basin that form gypsum, which supplies the dune field. Isopach lines estimating the thickness of the Yeso Formation are shown on the map grading from 500 ft in the SW to 1750 ft thick in the NE (modified from Kottlowski 1963; Fryberger 2001). (**b**) Schematic cross-section a-a′ shows the basin structure and the extent of the Paleozoic rocks in the subsurface (modified Fryberger 2001)

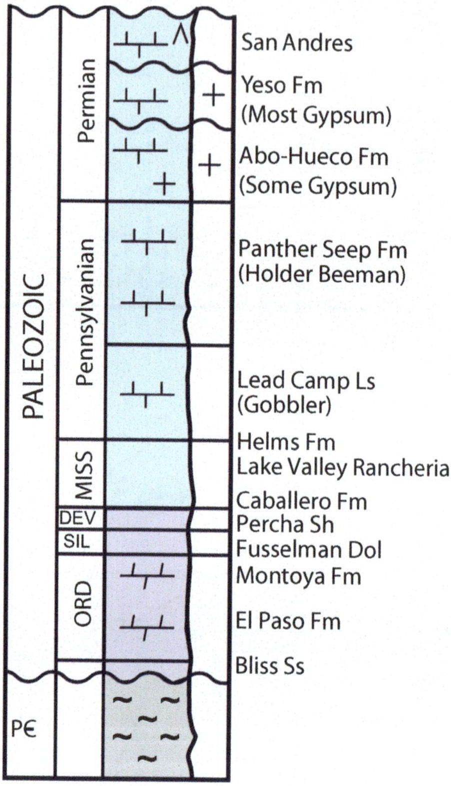


**Fig. 6.5** Geological stratigraphic column showing Paleozoic formations within the Sacramento Mountains

aeolian sand dunes and sand sheets in the north and west to marine carbonates in the south and east. The thickest evaporites in the sequence occur during retreat of the terrestrial facies and incursion of the marine facies where laminated gypsum and bedded halite developed in a shallow, hypersaline seaway and coastal bordering salinas and sabkhas. The thickest Yeso Formation occurs northwest of the Tularosa Basin at Carrizozo, where it reaches 457 m thick and tapers to less than 229 m thick directly beneath White Sands Dune Field (Fig. 6.4) (Kottlowski 1963; Fryberger 2001). Evaporites are mobilized from the Yeso Formation through groundwater recharge to the basin along a hydrologic gradient that follows a northeast-southwest trend (Newton and Allen 2014).

## 6.2.2 *Hydrology*

The hydrology of the Tularosa Basin and White Sands Dune Field affects both the formation and survival of the dune field. Supplied by the Permian Yeso Formation, calcium and sulfur required to create gypsum arrive into the basin largely from groundwater flow from the northeast toward the southwest (Fig. 6.6). Lesser

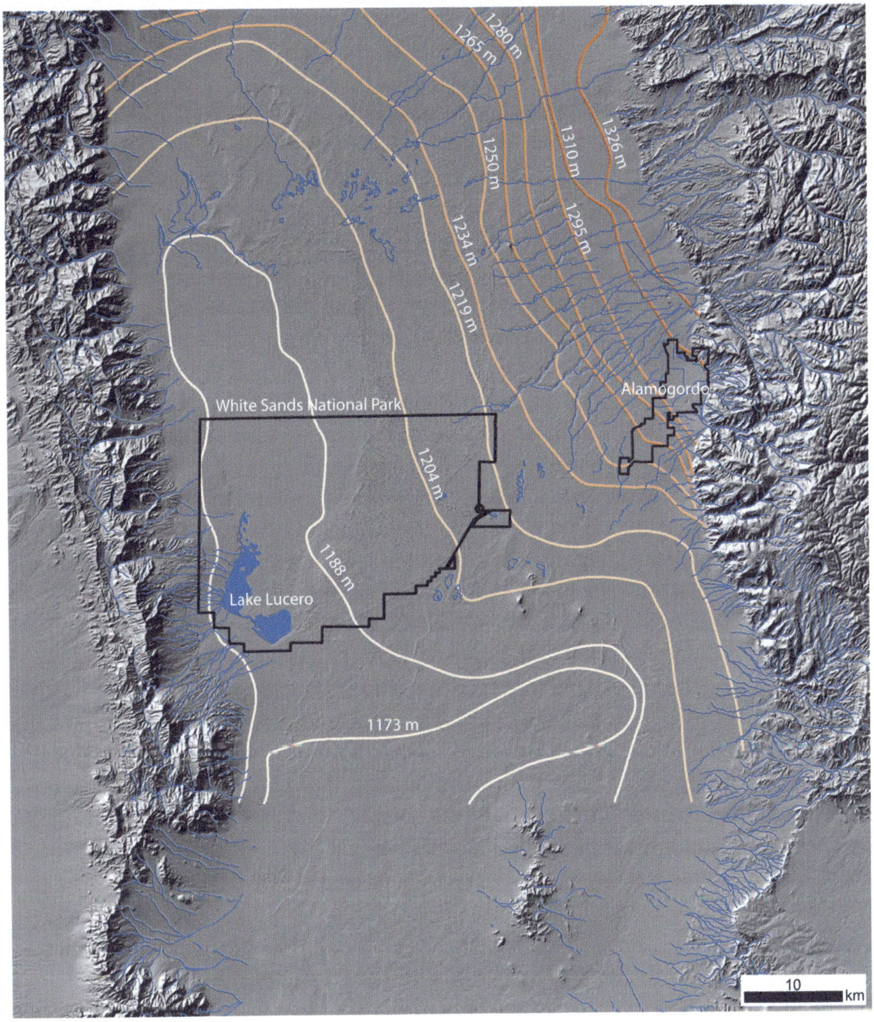


**Fig. 6.6** Hydrologic features of the Tularosa Basin. Streams are shown in blue. Water table contours are shown as an orange gradient rising from 1173 m in the southwestern basin to 1326 m in the northeast. The hydrologic gradient toward the southwest is thought to bring evaporite-rich waters derived from the Yeso Formation. (Modified from Newton and Allen 2014)

amounts derive from groundwater brought up to the surface along faults in the playas that extend down through the subsurface Paleozoic sequences (Szynkiewicz et al. 2010b; Newton and Allen 2014). Minimal surface runoff reaches the center of the basin, rather, most water infiltrates before reaching the lowest points in the basin on the southwestern side. Ephemeral runoff from the largest stream in the basin, Lost Creek, which issues from the Sacramento Mountains, terminates within the dune field, with other smaller creeks dying out before reaching the dune field margin. Notably, basement rocks and lower Paleozoic rocks comprise the east-facing slopes of the San Andres Mountains in the western basin indicating that runoff and groundwater from this range supplies little of the source material for gypsum in the basin.

The proximity of the regional groundwater table to the surface defines the expression of the landscape in the basin. Where the water table intersects the surface in the northern and eastern part of the basin, rare ephemeral springs emerge at the toe of the Carrizozo lava flow and the northern flanks the dune field. More prominently, playas occupy the southwestern part of the basin where the water table sits at and just below the surface. Lake Lucero and Alkali Flat are the most active playas in the basin (Fig. 6.6). The lack of a fluvial system draining the Tularosa Basin facilitates the formation of these extensive playas, however, some groundwater flow exits the basin to the south toward the Rio Grande River indicating that the Tularosa Basin is not an entirely closed system (Fig. 6.6) (Embid and Finch 2011; Newton and Allen 2014). The western flanks of these playas are marked by several meters-high deflationary shorelines that give way westward to largely siliciclastic alluvial fans draining the San Andres Mountains. The abruptness of the western shorelines and the angular geometry of the perimeter of these playas and associated sub-basin playa shorelines imply that the tectonic structure of the basin may be an antecedent condition that controls some aspects of the groundwater supply to the playas (Fig. 6.6) (Olsen et al. 1987; Szynkiewicz et al. 2010b). These evaporitic playas produce much of the gypsum supplied to the dune field in the current climate.

Lake Lucero, the largest and most active of the playas, sits at the lowest elevation in the basin, 1185 m. The groundwater table at Lake Lucero is within 50 cm of the surface throughout the year and regularly rises above the surface during precipitation events, during which time fish, shrimp, and other animals emerge (Fig. 6.7). Because of the proximity to the groundwater table, a suite of minerals forms at the playa surface and within the playa subsurface that provide sand-to-pebble sized crystals that source the dunes, and clay-to-silt sized particles that act as a major North American dust source (Fig. 6.7). The surface is dominated by micron-scale thenardite ($Na_2SO_4$), mm-scale gypsum crystals, and up-to-meter-scale selenite ($CaSO_4 \cdot 2H_2O$) crystals (Fig. 6.7). The subsurface is dominated by gypsum with layers of mirabilite ($Na_2SO_4 \cdot 10H_2O$). Alkali Flat, a deflationary playa that extends across much of the western part of the basin to the upwind margin of the dune field activates during heavy precipitation events that drive surface flooding, but nominally supplies minor amounts of gypsum to the field.

A dual aquifer system within the dune field consists of the deeper, regional groundwater aquifer and a shallow, perched aquifer (Allen et al. 2009; Newton and Allen 2014). The shallow water table within the dune field occurs <1 m below the interdune surface throughout the year. This aquifer is principally recharged through precipitation (Newton and Allen 2014) with interdune areas flooding during severe precipitation events. Water beneath the dunes has the lowest total dissolved solids and is the youngest water in the basin at an average of a few hundred years, supporting the precipitation model for recharge (Newton and Allen 2014). Water beneath the interdune areas, however, has a higher total dissolved solids load and is much older at nearly 10,000 years. Although the groundwater beneath the interdunes

**Fig. 6.7** Photographs of Lake Lucero. (**a**) Lake Lucero after a 2006 heavy precipitation event (Photo courtesy David Bustos). (**b**) Lake Lucero in 2015 dry. Note dust lifting from the playa surface. (**c**) Centimeter-scale gypsum crystals on the playa surface (squares in scale are 1 cm). (**d**) Thenardite 'puffy' crusts formed on the playa surface. This material contributes to the dust load from the basin. (**e**) Erosional shoreline of Lake Lucero exposing strata from Pleistocene Lake Otero

likely represents a regional groundwater supply, pump tests show no connectivity between the shallow aquifer and the deeper aquifer (Newton and Allen 2014).

Moisture and early cementation bind the near-surface gypsum sediments that sit within the unsaturated zone above the water table in the interdunes (Schenk and Fryberger 1988; Newton and Allen 2014). Within the dunes, precipitation drives cohesion at a few cm below the surface. The cohesion of these sediments prevents activation and deflation of the dune field and gives rise to an overall sediment availability-limited dune field (Schenk and Fryberger 1988; Kocurek et al. 2007). Springtime high winds frequently strip most of the sand from the dune stoss slopes, leaving exposed an indurated, crusted dune stoss slope.

The groundwater supply across the basin remains a focus of study. Increased concern about water use from human development along with climate changes toward increased drought, potentially impact the stability of the field. A lowering of the water table would release sediment for transport thereby increasing dune transport rates and, likely, dust production. Such effects could impact the government and civilian infrastructure and population in the basin.

### 6.2.3 Climate

White Sands sits within a typical hot and arid climate regime in the southwestern United States. Mean daily temperatures vary between 15 and 38 ° C with highly variable precipitation of ~270 mm/yr. Much of the precipitation occurs from July through September associated with the North American or Southwestern monsoon, which is a weather pattern consisting of an increase in precipitation largely in the form of thunderstorms across much of northern Mexico and the southwestern US. The monsoon is driven by intense summer heating over the southwestern US and northern Mexico region that reverses wind flows and draws moisture from the southwest from Gulf of California and eastern Pacific and from the southeast from the Gulf of Mexico. This moisture plays a key role at White Sands by recharging the perched aquifer, stabilizing the dune surfaces, and creating a new sediment supply to the dune field from precipitation of gypsum in Lake Lucero and on Alkali Flat.

Winds at White Sands are highly variable with three prominent modes from the southwest, northeast, and south (Fig. 6.8a, b) (Fryberger 2001; Rachal and Dugas 2009; Ewing et al. 2015; Pedersen et al. 2015). The southwesterly mode dominates the overall flow frequency and intensity with a resultant toward 065°. Northerly winds have a resultant direction toward 165°; and southeasterly winds blow toward 345° (Pedersen et al. 2015). The southwesterly winds dominantly blow during the windy season from March to May, but also occur during December to February. The northerly winds arrive as fronts in both the winter months and during the summer monsoonal season. Southerly winds are most prominent during the summer from June through September (Pedersen et al. 2015).

Threshold shear velocity required to move 0.4 mm sand at White Sands calculated from the 10 m meteorological station at Holloman Air Force Base with a

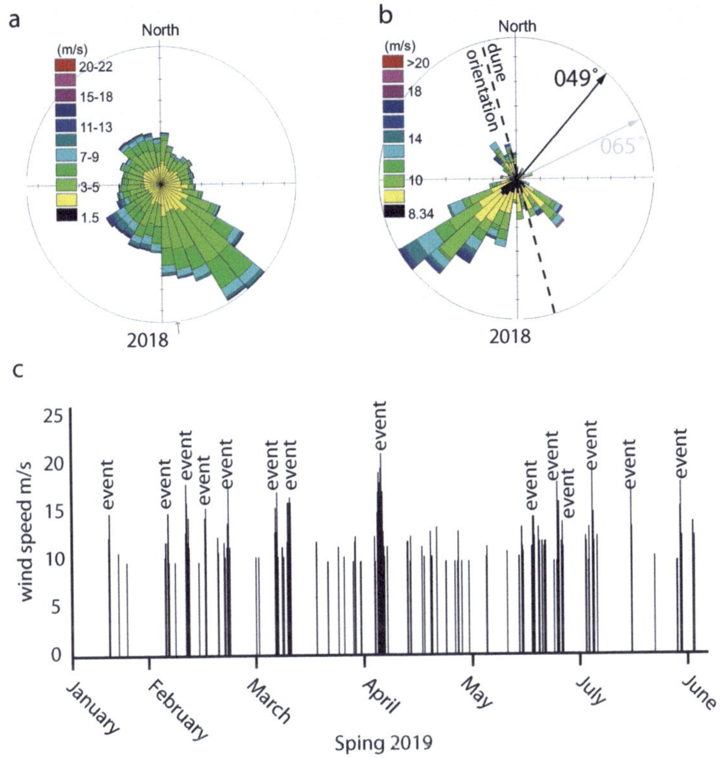


**Fig. 6.8** Wind distribution at White Sands Dune Field. (**a**) Wind speed and direction plotted on a rose diagram of all 2018 winds collected hourly at the KHMN station on Holloman Air Force Base. (**b**) 2018 wind speeds and direction above the 8.34 m/s threshold wind speed to move 0.4 mm grains (Eastwood et al. 2012; Pedersen et al. 2015). Note the dominant southwesterly wind (**c**) Spring 2019 wind events at White Sands as defined by Pedersen et al. (2015)

roughness of 0.2 mm is 0.3139 m/s, which is equivalent to a wind speed of 8.34 m/s at 10 m height (Pedersen et al. 2015). A frequency-magnitude analysis of the winds at White Sands shows that the dominant winds doing work on the field occur at a shear velocity of 0.39 m/s (Jerolmack et al. 2011), which sits between 0.35 and 0.45 m/s shear velocity measured at 4 heights up to 80 cm in the field by Eastwood et al. (2012). The threshold is exceeded somewhat infrequently between 3% and 6% of the time (Pedersen et al. 2015). At White Sands, as measured from hourly averaged data from the KHMN station, calm days with no transporting winds occurred 1.5 times more frequently than days with at least one wind measurement above the threshold (Pedersen et al. 2015). Pedersen et al. (2015) indicate that only 81 wind events, defined as wind speeds measured above threshold separated by at least 12 h below threshold, occur per year with an average duration of 4.7 h for the 3 year period between 2007 and 2010. Although the average wind speed of the wind events was 10 m/s and ranged upward to around 15 m/s for the study, interannual variability exists in the number and intensity of such events (Fig. 6.8c).

The southwesterly winds typically produce the highest wind speeds and dominate the sand flux in the dune field (Fryberger 2001; Pedersen et al. 2015). This wind mode drives the overall NW-SE orientation of the dune pattern (Fig. 6.8b) (Fryberger 2001; Ewing et al. 2006; Rachal and Dugas 2009; Pedersen et al. 2015). The other modes, however, play a role in deforming the shape of the dune as the dunes migrate (Ewing et al. 2015; Swanson et al. 2016) and give rise to superimposed protodunes on the flanks of the dune stoss slopes and southeasterly migration of the dune sinuosity (Ewing et al. 2015; Pedersen et al. 2015).

## 6.3 Late Pleistocene and Holocene History

### *6.3.1 Late Pleistocene*

Like many basins in the southwestern United States, during the Last Glacial Maximum (LGM) the Tularosa Basin was cooler and wetter than today and hosted a large pluvial lake, Lake Otero (Betancourt et al. 2001; Allen 2005; Allen et al. 2009). Based on limited exposures of lake sediments around the basin, Lake Otero is thought to have extended across the western half of the basin up to the 1204 m elevation mark in the basin during highstand (Fig. 6.9) (Allen et al. 2009). Deflationary scarps expose up to 10 m of lake strata, but more typically 5–7 m, along the western flanks of Lake Lucero and Alkali Flat, and 1–2 m exposure within drainages across the paleolake-occupied area of the basin (Allen et al. 2009). Lake strata consist of decimeter- to meter-scale beds of gypsiferous mud, siliciclastic mud, fine-grained gypsum and interlaminated to thickly interbedded gypsum sand and siliciclastic mud (Fig. 6.7e) (Allen et al. 2009). Rarely, beds 5–15 cm thick contain an abundance of calcium carbonate. Ostracodes, mollusks, formanifera, and gastropods and, rarely, fish fossils, exist throughout various intervals within the sequence (Allen et al. 2009). The lake sequence indicates a primarily saline body of water that favored precipitation of evaporite minerals gave way to a somewhat fresher body of water during highstands that hosted a diversity of aquatic organisms, and, as evidenced by footprints in the muds, invited land mammals and potentially humans (Allen et al. 2006; Lucas et al. 2007; Bustos et al. 2018). The freshest waters likely concentrated seasonally, during heavy precipitation and runoff into sub basins formed at the margins of the lake (Holliday et al. 2019).

### *6.3.2 Holocene*

The pluvial environment in the Tularosa Basin characteristic of the late Pleistocene gave way to a dry, deflationary environment during the Holocene driven by increased aridity across the southwestern United States (Allen and Anderson 2000; Anderson

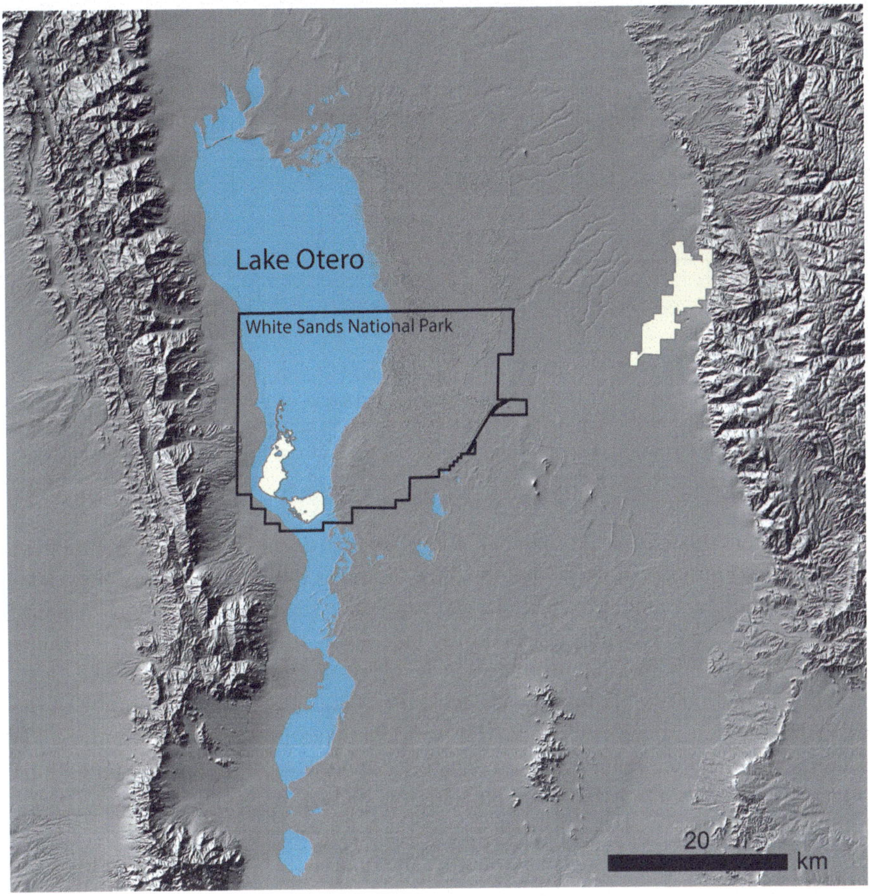


**Fig. 6.9** Extent of Lake Otero (blue) in the Tularosa Basin based on estimates that high stand occurred at the 1204 m shore line (Allen et al. 2009). Langford (2003) and Baitis et al. (2014) suggest that the lake may have extended up to 1216 m. Lake Lucero is highlighted in yellow within the blue Lake Otero region and Alamogordo is shown yellow on the right of the image. Basemap is a hillshade rendering of IFSAR 5 m DEM

et al. 2002). Decreased precipitation and increased evaporation drove falling lake and groundwater levels that exposed Lake Otero sediments and made them available to deflation by the wind (Allmendinger 1972; Langford 2003; Kocurek et al. 2007). A sediment core taken in the center of the modern dune field shows 8.5 m of aeolian strata overlying Lake Otero strata dated at 7.3 ka indicating deflation and the beginning of the dune field by this time. Evidenced by numerous, successively lower in elevation, erosional paleo-shorelines across the basin, the drop in the water table appeared episodic. Episodic deflation of Lake Otero strata is thought to have provided sediment to the dune field and built associated accumulations of dune strata (Langford 2003; Baitis et al. 2014).

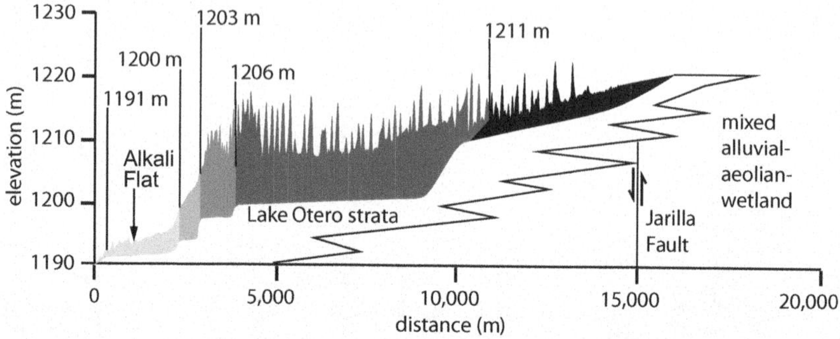


**Fig. 6.10** Topographic profile across White Sands Dune Field show location and elevations of shorelines interpreted by Baitis et al. (2014). The shaded areas of the dune field show the most recent model of dune field accumulation (Baitis et al. 2014) in which dune accumulation accreted onto Lake Otero strata during lake-level still stands. (Modified from Baitis et al. 2014 and Newton and Allen 2014)

The current model to explain the ~8.5 m of aeolian accumulation and the modern dune field topography suggests that the dune field was built by basinward progradation of dune-field segments, which derive from sands shed from lake shorelines into foredune ridges and lee dune fields during lake level still stands (Fig. 6.10) (Baitis et al. 2014). Falling lake and groundwater levels resulted in deflation of Lake Otero strata and creation of deflationary surfaces, like Alkali Flat. The topography of the modern dune field signals evidence of this model by a series of elevated ridges that occur across the dune field, interpreted as relict dune ridge topography. Some of the ridges occur at elevations that correlate with recognized shorelines around the basin. Further supporting evidence of this model derives from sulfur isotope data measured through Lake Otero strata and across the dune field, which show an increase in δ34S values upward through the strata and downwind across the dune field (Szynkiewicz et al. 2010a). Consistent with the Baitis et al. (2014) model, this data suggests that minimal mixing among the dunes has occurred since the formation of the dune segments.

## 6.4 Dune Field

The White Sands Dune Field emerges abruptly from the eastern end of Alkali Flat and terminates abruptly into a vegetated plain composed of loess on the eastern side of the basin (Fig. 6.11). The upwind and downwind margins are oriented NW-SE and are sinuous and parallel, resulting in an average dune-field width of ~13 km that ranges between 11 and 14 km. At its southern border, an amalgamated parabolic dune ridge marks the abrupt termination of the dune field and a southward landscape transition into paleo-playas and sparsely vegetated sand sheets of mixed siliciclastic and gypsum sand. This landscape dominates the southern part of the Tularosa

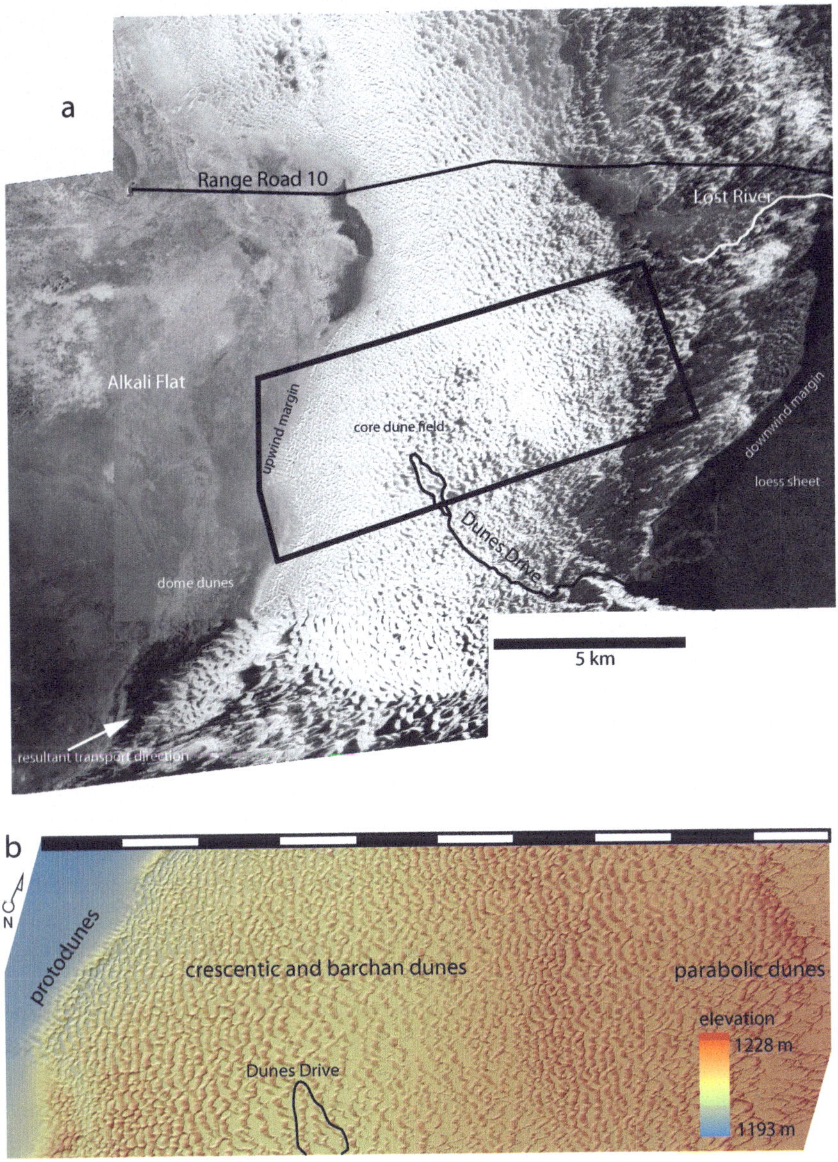


**Fig. 6.11** Landscape features and dune types at White Sands Dune Field. (**a**) Satellite image of White Sands Dune Field within White Sands National Park. The Dunes Drive is the access road in the park. The black rectangle represents the area of lidar collected in 2007–2010, and 2015 and shown in (**b**). (**b**) Color-shaded topography across the core area of White Sands Dune Field showing transitions in dune type from the upwind margin on the western side to the downwind margin on the east

Basin before giving way to and quartz sand nebkha dunes in the north of the Hueco Bolson. The northern border transitions diffusely into isolated gypsum dune fields, sparsely vegetated sand sheets, and ephemeral wetlands. Defined by these boundaries, the total area of the active dune field is ~500 km$^2$ (Ewing et al. 2006).

Dune types and scales vary across the basin (Fig. 6.11). The most notable transition in dune type and scale occurs from southwest to northeast along the resultant wind direction (McKee 1966). Isolated dome dunes composed of medium to coarse, angular gypsum sand (Allmendinger 1972; Fryberger 2001) form on Alkali Flat. Many of these dunes emerge around the 1195 m shoreline, but few migrate across the ~2 km flat between the 1195 m contour and the dune field. Rather, the dome dune sand appears to dissipate and reappear as medium to coarse, subangular sand patches across Alkali Flat before reaching the upwind dune field margin. The sand patches amalgamate and transition to protodunes (Phillips et al. 2019) before generating the first dunes in the field, which are crescentic dunes with complex crestlines (i.e., star and linear components to the pattern exist). Barchanoid and crescentic dunes characterize the core area of the dune field around Dunes Drive, which is 3.5 km downwind from the upwind margin. Around 8 km downwind, a progressive increase in vegetation drives a shift from barchanoid to into vegetated parabolic dunes (McKee 1966; Langford et al. 2009; Reitz et al. 2010; Jerolmack et al. 2012).

The spatial distribution of dune types at White Sands has been recognized since McKee's (1966) seminal observations, but the origin of the transition from dome dunes to parabolic dunes across the basin remains debated. McKee recognized that the dunes formed within a uniform, largely unidirectional wind, and suggested that the variation in dune type arose from an increase in dune maturity downwind and changes in sediment availability; consistent with the origins of the known dune types (e.g., barchan dunes form in sand availability-limited environments, Wasson and Hyde 1983). Fryberger (2001) similarly recognized the variation in dune type across the basin and suggested the transition from unvegetated dunes to vegetated dunes ~8 km into the dune field arose from an increase in freshness of the ground water, which allowed the vegetation to take hold. The origin of the fresh groundwater arose from runoff from the Sacramento Mountains and precipitation (Langford et al. 2009; Newton and Allen 2014). Langford et al. (2009) tested Fryberger's hypothesis through a series of groundwater salinity tests and found that salinity was indeed lower by a factor of three in the parabolic dunes. Jerolmack et al. (2012) challenged the notion that a change in the groundwater salinity is the sole cause of the transition in dune types and rather attribute the transition from crescentic to barchan to parabolic as the result of a spatial decrease in sediment flux. They suggest the decrease in flux occurs because of the development of an internal boundary layer at the upwind margin of the dune field. The boundary layer occurs because of the abrupt change in the landscape-scale topography at the upwind margin of the dune field where the planar, flat Alkali Flat deflation plain transitions into the roughness of the 5–10 m high dunes of the dune field. Jerolmack et al. (2012) suggested the vegetation freshened the water by extracting salts from the groundwater and this accounted for the change in salinity across the dune field. The effect of the landscape roughness on atmospheric flows was later described as a mixing layer, rather than a true internal boundary layer through use of large eddy simulation (LES) (Anderson and Chamecki 2014). Baitis et al. (2014) suggest the transitions are linked to variable Pleistocene lake level stands in which the modern dune field responds to the sediment supply linked to the antecedent shoreline topography. Pelletier (2015) combine the shoreline and boundary layer hypotheses and indicate that preexisting shoreline topography could trigger accelerations in the atmospheric flows across the dune field that drive spatial changes in sediment flux.

### *6.4.1 The Upwind Margin*

The upwind margin of the White Sands Dune Field represents a remarkable landscape transition (Fig. 6.12). Five to ten meter high dunes emerge from an otherwise flat landscape and record the spatial and temporal development of dunes. The transition occurs over a ~700 m wide sand ramp that is variable in width northwest to southeast along the length of the dune field. The width of the margin reduces to less than 100 m where the margin curvature nearly parallels the resultant transport direction of the dune field. The elevation changes by ~13 m across the margin from 1197 m in Alkali Flat to 1210 m at the top of the first dunes.

Sand patches accumulating behind vegetation, deflated salt crusts, hollows, and from no obvious topographic or aerodynamic influence represent the harbingers of the upwind margin of the dune field. A spatial increase in the density and ultimately the amalgamation of sand patches leads to the formation of a continuous sand sheet at the transition from the Alkali Flat hardpan to the upwind sand-alluviated margin. The sand sheet is characterized by medium to coarse grained gypsum sand and very coarse grains of variable mineralogy, including dolomite ($CaMg(CO_3)_2$) and calcite ($CaCO_3$) (Fenton et al. 2017). The wide range of grain sizes results in the formation of discontinuous quasi-regularly spaced coarse-grained patches with ripples separated by better sorted medium grain sand (Fig. 6.12). Individual coarse-grained ripples range in height from a few cm to decimeters (Fenton et al. 2017).

As sand cover becomes more continuous toward the NE, the sand sheet organizes into decameter wavelength and centimeter amplitude hummocky topography interpreted as protodunes (Phillips et al. 2019). The protodunes have a median grain size of 0.67 mm on the surface with interiors characterized by 0.23 mm sand. The juxtaposition of coarse-grains on the surface and medium grains on the interior of the bedforms indicates that they are zibars (e.g., Nielson and Kocurek 1986; Kocurek and Ewing 2016), but from a process perspective these bedforms become dunes and are therefore described as protodunes. The protodunes increase in both wavelength and amplitude across the margin. Phillips et al. (2019) showed a 60% increase in wavelength and a 2000% increase in amplitude in one location where a series of 5 protodunes formed leading to dune initiation. Measurements in other locations reveal this is representative of the upwind margin. The initial protodunes are symmetric to slightly asymmetric in profile and give way to strong asymmetric shape at

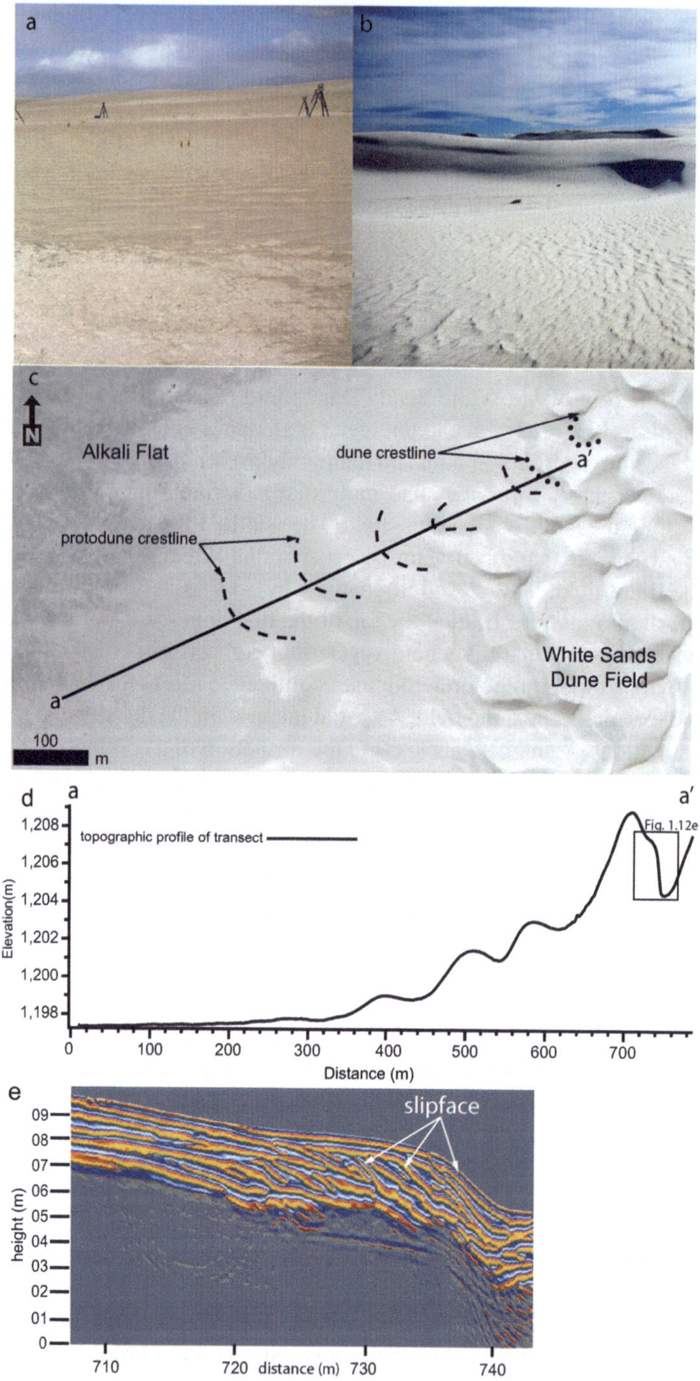


**Fig. 6.12** Photos, topography, and strata of the protodune-to-dune transition at the upwind margin of White Sands Dune Field. (**a**) View NE from Alkali Flat to the dune field across several protodunes. Inter-protodune areas are visible as bright areas between protodunes. (**b**) First dune with

the slipface formation stage. The protodunes migrate at an increasing rate from their emergence to the first dune. The rates range from 3.0 to 6.0 m/year across the margin.

Visual observation, trenching and ground penetrating radar (GPR) reveal widespread dune cross stratification and low angle cross stratification within the subsurface of the upwind margin. Areas on the upwind margin barren of active sand expose dune cross-stratification at the surface. GPR data show high-angle reflections in the subsurface interpreted as dune-cross stratification. Exposure of the cross stratification increases onto Alkali Flat where it dominates the surface expression of the deflation plain (Fig. 6.12) (Fryberger 2001; Szynkiewicz et al. 2010a). The presence of cross stratification beneath the sand ramp and on Alkali Flat signals the former presence of the dune field. Flat-lying to low-angle cross stratification comprise the most of the surficial deposits on Alkali Flat and on the sand ramp. These strata originate from playa processes and the migration of protodunes (Phillips et al. 2019). The superposition of the low-angle deposits over high-angle dune cross stratification across the ramp gives way downwind to dune cross stratification within the modern, active dunes.

GPR reflections of the first dunes formed along the upwind margin reveal slipface forming processes as a protodune transitions to a dune (Fig. 6.12). An initial slipface forms and disappears several times before the slipface becomes a permanent feature of the first dune. The disappearance of the slipface likely occurs through the development of an erosional reactivation surface generated by the northerly and easterly reversing winds or through dune interactions. Once established, the dune height increases rapidly by sand trapping and scour at the base of the lee slope. At White Sands the slipface height grows from a few decimeters at the outset of dune growth to over 20 meters within one dune wavelength generating initial dunes that average 7 m high and 120 m in wavelength (Ewing and Kocurek 2010).

### *6.4.2 Central Dune Field*

Crescentic and barchanoid dunes comprise the unvegetated dunes of White Sands Dune Field. Within 2 km of the upwind margin, sinuous crested dunes oriented at 345° with interdune troughs dominate. The presence of interdune troughs, rather than interdune flats indicates relatively high sediment availability in this area of the

**Fig. 6.12** (continued) slipface formed at the beginning of the dune field. (**c**) Image of upwind margin with crestlines of protodunes (dashed lines) and first dunes (stippled lines) highlighted (Modified from Phillips et al. 2019). (**d**) Topographic profile across the protodunes shown in (**c**) (modified from Phillips et al. 2019). Note the strong increase in amplitude and modest increase in wavelength. (**e**) GPR profile across the protodune-to-dune transition (modified from Phillips et al. 2019). Note the development of multiple slipfaces preserved in the strata before establishment of the modern slipface dune field compared to regions down wind. From the margin, dunes grow in height to an average of 7 m at 1.5 km downwind, beyond which the dunes decay in height toward a dune-field-wide average of 3.5–4.6 m (Ewing and Kocurek 2010; Baitis et al. 2014). The highest relief of a dune in the field is 14 m and occurs within the first 2 km (Baitis et al. 2014). The growth of dune height and wavelength in this region arises from frequent dune-dune interactions, which, within the first 3.5 km of the dune field, are dominated by mergers and lateral linking (Ewing and Kocurek 2010). Although the transverse crescentic dune pattern is consistent with the dominant southwesterly winds, crestlines that extend for two to three dune wavelengths in the resultant transport direction and orthogonal to the primary crest orientation reflect the influence of seasonal winds and local variations in sediment availability. These longitudinal crests may emerge as a 'fingering mode' component of the pattern in response to the variable wind regime and local changes in sediment availability, such as a transition from transport-limited interdune troughs to availability-limited interdune flats (Courrech du Pont et al. 2014). The intersection of these dunes give the appearance of a complex, star-dune patterns near the upwind margin (Pedersen et al. 2015).

Beyond 2 km from the upwind margin and around The Dunes Drive, dunes transition to crescentic and barchan dunes separated by expansive interdune flats (Figs. 6.11 and 6.13). Dunes in this region tend to be shorter in mean and maximum height, more widely spaced, and more continuous along crest (Baitis et al. 2014). The wider spacing between dunes is largely due to the large interdunes; dune length and sinuosity remains similar to other areas of the field. Barchan dunes in this area emerge where dune collisions truncate a downwind crest and a portion of the impacted dune detaches and migrates across the interdune flat as a protodune or barchan dune. Rarely, crescentic dunes calve barchan dunes from their defects. Calving, as in many barchanoid dunes (e.g., Elbelrhiti et al. 2005), arises from protodunes migrating across the stoss slopes (Fig. 6.13). These protodunes are ~20 m in wavelength, consistent with that predicted from linear stability analysis (e.g., Claudin and Andreotti 2006), and are oriented downslope, nearly orthogonal to the curvature of the dune brinkline. These bedforms account for a significant amount of sand moved across the dune stoss slope (Lee et al. 2019).

The expansive, availability-limited interdune areas of this region are typical of a wet aeolian system and a unique feature of White Sand Dune Field. By proximity to the near surface water table, the interdune areas are stabilized by the dampness of the capillary fringe and early diagenesis through gyspum cementation of grains (Schenk and Fryberger 1988). The lowest and wettest areas adjacent to the upwind dune's lee slope host efflorescent crusts and microbial mats (Fig. 6.13) (Schenk and Fryberger 1988; Fryberger 2001; Kocurek et al. 2007). Rising from the low areas toward the downwind dune's stoss slope, accumulation of dune deposits create the so-called corrugated surface, which is an expression of the former lee slope in plan-view (Fig. 6.13d). Mapping of the ridges on the surface signals the former location of a dune; in many cases the dune directly downwind (Kocurek et al. 2007; Szynkiewicz et al. 2010a; Brothers et al. 2017). The largest, ~1–3 m wavelength ridges and swales denote grainflow stratification overlain by thick wind-ripple

**Fig. 6.13** Core crescentic and barchan dunes near Dunes Drive in the middle of White Sands Dune Field. (**a**) Color-shaded topography of dunes around dunes drive. Note the road and picnic tables resolved in the lidar-derived topography (1 m/pixel spatial resolution). White arrow denotes same location as indicated by the white arrow on (**c**). (**b**) View south across the crest of a crescentic dune. (**c**) Aerial view of flooded interdune areas around Dunes Drive (Photo courtesy of David Bustos). Note the picnic tables under water. White arrow refers to same location as in (**a**). (**d**) Photo of corrugated surface of interdune. The ridges and valleys of the surface reflect changes in the induration of the surface that typically correlates with types of aeolian stratification (e.g., grainflow vs wind ripple). (**e**) Photo of laminated microbial mats exposed in a trench with an interdune area. (**f**) Topographic profile across a barchan dune. Location of profile noted in (**a**)

stratification. This wavelength is thought to mark seasonal wind cycles whereby the grainflow stratification signals the dominant, transverse westerly winds and the wind ripple stratification indicates northerly winds from winter fronts (Kocurek et al. 2007; Eastwood et al. 2012).

### 6.4.3 *Parabolic Dunes*

Parabolic dunes line the downwind, eastern, and southern margins of the dune field and emerge as the main dune from deflation off of the southeastern arm of Lake Lucero (Fig. 6.14) (Fryberger 2001; KellerLynn 2012; Kocurek and Ewing 2016). The parabolic dunes have a typical shape with heavily vegetated anchored arms that point into the southwesterly wind and an active, less vegetated snout that points down wind. The parabolic dunes span a wide range of sizes up to 500 m in length and 8 m in height. The parabolic dunes migrate between 0.6 and 2.4 m/yr. (McKee and Douglass 1971); a rate much slower than that of the unvegetated core crescentic dunes. Interestingly, the downwind margin of the dunes remains largely static with minimal migration over the past 60 years despite the apparent slow annual migration of the parabolic dunes (Pelletier and Jerolmack 2014). The southern margin parabolic dunes transition from most active at the downwind margin near the national park headquarters to completely inactive at the southern margin. The dunes terminate abruptly into a sharp vegetated inactive sand ridge that parallels US 70 from southwest to northeast. The parabolic dunes that emerge from Lake Lucero transition over ~3 wavelengths from semi-vegetated barchanoid dunes with vegetated horns and ultimately into fully vegetated parabolic. These parabolic dunes appear to remain active for ~7 km downwind before dissipating into a fully stabilized parabolic dune field. The vegetation free, crescentic and barchan dunes of the central dune field transition downwind into parabolic dunes, which extend across the last ~4 km of the dune field before terminating abruptly into a downwind loess

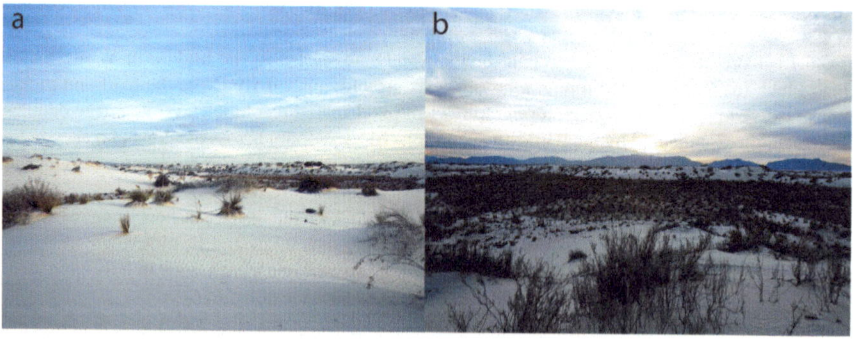


**Fig. 6.14** Photos of parabolic dunes. (**a**) Photo from active snout of parabolic dune. (**b**) Photo across vegetated interdune area toward active snout of parabolic dune

sheet. This transition in patterns comprises one of the most striking landscape patterns at White Sands. The inversion process of the barchan to parabolic dunes is thought to occur due to the decay in sediment transport rate decreases across the dune field (Reitz et al. 2010; Jerolmack et al. 2012). The drop in transport rate allows pioneering species of plants to take hold and begin to grow midway across the field. Remote sensing data show that erosion/deposition rates in sand transport drop below the growth rate of the vegetation at the spatial location in the field when the parabolic patterns begin to emerge (Durán and Herrmann 2006; Reitz et al. 2010).

Despite decades of study, the dunes of White Sands remain the focus of intense research. In 2007, the White Sands National Park supported by the National Park Service and National Science Foundation initiated a highly successful effort to collect data and establish protocols to monitor dune movement. Airborne light detection and ranging (lidar) data was collected over the dune field during June 2007, June 2008, January 2009, September 2009, June 2010, and August 2015 to monitor changes in the dunes. More flights are planned. The publically available data have served a wide population of researchers and continue to be a resource for the park managers and scientists. To monitor weather and wind, 10 m meteorology towers collect data within and around the dune field, and to monitor changes in water table and ground moisture, wells are distributed throughout the basin. This on-going effort provides new opportunities to understand the intricacies of this unique system and its resilience to increased climate and human pressure.

## 6.5 Moths, Lizards, Megafauna, and Humans

The white landscape of the playas and dunes at White Sands stages a unique biologic history. Species adapted to the white landscape exist today (Rosenblum and Harmon 2011; Rosenblum et al. 2010) surrounded by ancient gypsum lake strata that contain one of largest collections of Pleistocene megafuna tracks and trackways in North America (Allen et al. 2006). Recent evidence shows that the megafauna did not roam alone in the region, but interacted with early human occupants of North America (Bustos et al. 2018). Pottery artifacts and hearths built into the gypsum dunes show that humans used White Sands as a base throughout much of its recent history.

Several examples of species that have adapted to the white landscape exist (Fig. 6.15) (Rosenblum 2005; Rosenblum et al. 2010; Metzler 2014). Of those blanched lizards and moth stand out. Varying degrees of blanched, white lizards exist within the white dune field, the surrounding darker albedo environments, and within marginal environments demonstrating adaptation to the variably white environment, likely to avoid predation. These lizards are a prime example of convergent evolution in which the specific mechanisms that lead to the reduction in melanin production to adapt to the white environment differed among several species of white lizards that exist at White Sands (Rosenblum et al. 2010; Rosenblum and Harmon 2011). The over 30 new species of moths that have been discovered at

**Fig. 6.15** Photos of animals with remarkable endemism to the white landscape and animal tracks at White Sands. (**a**). Light-colored lizard on dune stratification. (**b**). White moth (Photo courtesy Eric Metzler). (**c**). Mammoth trackway exposed in Alkali Flat (Photo courtesy David Bustos). (**d**). A trackway composed of sloth and human footprints (Photo courtesy David Bustos)

White Sands since 2006 show a similar endemism to the white landscape (Metzler 2014). Moths that live entirely within the dunes are pale in color or absent color, whereas, those that are found in neighboring darker environments are multicolored or less pale than those in the dunes.

Megafauna tracks and trackways found throughout Alkali Flat in Lake Otero strata signal playa environments during the Pleistocene that invited species including mammoth, sloths, felids, canids, camelids, and ungulates (Fig. 6.15) (Morgan and Lucas 2006; Urban et al. 2018; Bustos et al. 2018). The tracks typically preserve in relief as dolomite casts or as depressions. The tracks have a limited lifetime once they reach the surface because of deflation and erosion of the surface. Thus, new tracks are constantly exposed from deflation into older lake strata as exposed tracks erode. Many of the tracks are exposed only part of the year due to changes in moisture conditions throughout the playa that highlight or obscure the tracks. During periods of high surface salt efflorescence production the tracks remain invisible, but as moisture conditions change, the tracks reappear (Bustos et al. 2018).

In a remarkable discovery, Bustos et al. (2018) report on tracks interpreted as an example of human predation of a sloth. The sloth tracks show evidence of defensive behavior and attempts to evade human stalkers. Barefoot humans apparently followed and attacked the sloth as evidenced by an array of overlapping human-sloth footprints in a so-called flailing circle (Fig. 6.15). The co-location of these tracks also highlights the presence of humans in this area when the now extinct sloths were alive. The age probability of this window points to a time around 10–12 ka.

More recent human activity in the region is highlighted by a series of occupation sites around the dunes (Worman et al. 2019). Principal indicators of human occupation are hearths or ovens built on and around the dunes that date between 3960 and 770 $^{14}C$ yr BP (Worman et al. 2019 and references therein). Many of the remnant hearths appear to be built on the dunes, which hardened to plaster from firing of the gypsum. A rough correlation exists between the age of the encampments and position in the dune field such that the oldest occur in the southwest and youngest in the northeast. This likely reflects the advancing front of the parabolic dunes and access to freshwater, which is and would have been a scarce resource within the basin.

## 6.6 Planetary Analogs

As a well-studied dune field, White Sands is frequently cited as an analog for aeolian processes found on other worlds, such as Mars where sand dune activity is present today and was so in its deep past (Bridges et al. 2012; Ewing et al. 2017; Grotzinger et al. 2005, 2015). Discoveries of gypsum on Mars, and in particular within Olympia Undae in the Martian north polar dune fields, gave rise to new significance of White Sands as an analog for Mars. Olympia Undae Dune Field is the largest on Mars covering some 700,000 km$^2$ and possesses evidence for gypsum (Fig. 6.16) (Langevin et al. 2005; Fishbaugh et al. 2007; Horgan et al. 2009). The specific origins of gypsum in the region remains unknown, but is thought to relate to water derived from the polar ice cap during warm periods (Fishbaugh et al. 2007), volcanic and hydrothermal deposits (Tanaka et al. 2008), dune accumulation of gypsum sediment from elsewhere on Mars or from ancient surrounding deposits (Horgan et al. 2009), or from glaciers (Massé et al. 2012). Though these worlds are far off, White Sands contains extensive gypsum deposits associated with an ancient source material and a unique hydrologic and volcanic history which parallels ideas about the north polar region of Mars (Szynkiewicz et al. 2010a).

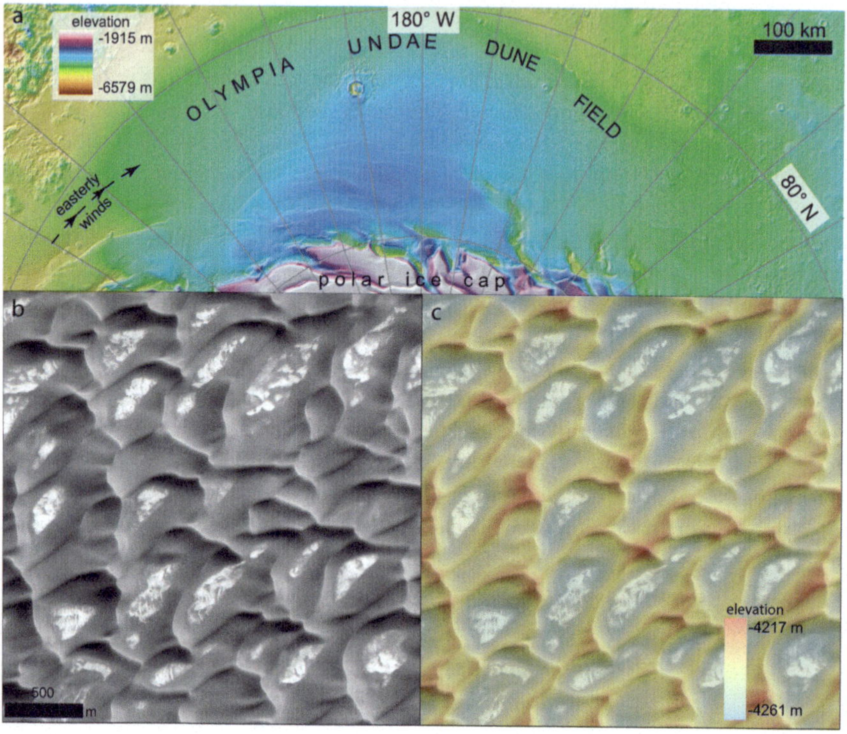


**Fig. 6.16** Sand dunes in the north polar region of Mars. (**a**) Mars Orbiter Laser Altimeter (MOLA) color-shaded topography of Olympia Undae Dune Field, the largest dune field on Mars, and with high abundances of gypsum. (**b**). High Resolution Imaging Science Experiment (HiRISE) satellite image of dunes within Olympia Undae Dune Field. Note preserved stratification in bright inter-dune areas. (**c**) Color-shaded topography of dunes shown in (**b**)

**Acknowledgements** Thanks to David Bustos, White Sands National Park Natural Resources Manager, for long-term and continued support facilitating science. Direct support for research at White Sands included in this chapter was provided by task agreements to the author from the National Park Service through the Chihuahuan Desert Network.

## References

Adams DC, Keller GR (1994) Crustal structure and basin geometry in south-central New Mexico. Special Paper 241. Geological Society of America, Boulder

Allen BD (2005) Ice age lakes in New Mexico. New Mexico's ice ages. N M Mus Nat Hist Sci Bull 28:107–113

Allen BD, Anderson RY (2000) A continuous, high-resolution record of late Pleistocene climate variability from the Estancia basin, New Mexico. Geol Soc Am Bull 112(9):1444–1458

Allen BD, Love DW, Myers RG (2006) Preliminary age of mammal footprints in Pleistocene lake-margin sediments of the Tularosa Basin, south-central New Mexico. N M Geol 28:61–62

Allen BD, Love DW, Myers RG (2009) Evidence for late Pleistocene hydrologic and climatic change from Lake Otero, Tularosa Basin, south-central New Mexico. N M Geol 31(1):9–25
Allmendinger RJ (1972) Hydrologic control over the origin of gypsum at Lake Lucero, White Sands National Monument, New Mexico. MS thesis, New Mexico Institute of Mining and Technology, Socorro, New Mexico, 182pp
Anderson W, Chamecki M (2014) Numerical study of turbulent flow over complex aeolian dune fields: the White Sands National Monument. Phys Rev E 89(1):013005
Anderson RY, Allen BD, Menking KM (2002) Geomorphic expression of abrupt climate change in southwestern North America at the glacial termination. Quat Res 57(3):371–381
Baitis E, Kocurek G, Smith V et al (2014) Definition and origin of the dune-field pattern at White Sands, New Mexico. Aeolian Res 15:269–287
Betancourt JL, Rylander KA, Peñalba C et al (2001) Late Quaternary vegetation history of Rough Canyon, south-central New Mexico, USA. Palaeogeogr Palaeoclimatol Palaeoecol 165(1–2):71–95
Bridges NT, Ayoub F, Avouac JP et al (2012) Earth-like sand fluxes on Mars. Nature 485(7398):339
Brothers SC et al (2017) Stratigraphic architecture resulting from dune interactions: white sands dune field, New Mexico. Sedimentology 64(3):686–713
Bustos D, Jakeway J, Urban TM et al (2018) Footprints preserve terminal Pleistocene hunt? Human-sloth interactions in North America. Sci Adv 4(4):eaar7621
Claudin P, Andreotti B (2006) A scaling law for aeolian dunes on Mars, Venus, Earth, and for subaqueous ripples. Earth Planet Sci Lett 252(1–2):30–44
Courrech du Pont S, Narteau C, Gao X (2014) Two modes for dune orientation. Geology 42(9):743–746
Durán O, Herrmann HJ (2006) Vegetation against dune mobility. Phys Rev Lett 97(18):188001
Eastwood EN, Kocurek G, Mohrig D et al (2012) Methodology for reconstructing wind direction, wind speed and duration of wind events from aeolian cross-strata. J Geophys Res Earth 117(F3)
Elbelrhiti H, Claudin P, Andreotti B (2005) Field evidence for surface-wave-induced instability of sand dunes. Nature 437(7059):720
Embid EH, Finch ST (2011) White Sands National Monument Inventory of Water Rights and groundwater evaluation data. John Shomaker & Associates, Inc. Report prepared for White Sands National Monument
Ewing RC, Kocurek GA (2010) Aeolian dune interactions and dune-field pattern formation: White Sands Dune Field, New Mexico. Sedimentology 57(5):1199–1219
Ewing RC, Kocurek G, Lake LW (2006) Pattern analysis of dune field parameters. Earth Surface Processes and Landforms: The Journal of the British Geomorphological Research Group 31(9):1176–1191
Ewing RC, McDonald GD, Hayes AG (2015) Multi-spatial analysis of aeolian dune-field patterns. Geomorphology 240:44–53
Ewing RC, Lapotre MGA, Lewis KW et al (2017) Sedimentary processes of the Bagnold Dunes: implications for the eolian rock record of Mars. J Geophys Res Planets 122(12):2544–2573
Fenton LK, Bishop JL, King S et al (2017) Sedimentary differentiation of aeolian grains at the White Sands National Monument, New Mexico, USA. Aeolian Res 26:117–136
Fishbaugh KE, Poulet F, Chevrier V (2007) On the origin of gypsum in the Mars north polar region. J Geophys Res Planets 112(E7):1–17
Fryberger S (2001) Geological overview of White Sands National Monument (https://web.archive.org/web/20061006061948fw_/http://www.nps.gov/archive/whsa/Geology%20of%20White%20Sands/GeoHome.html)
Grotzinger JP, Arvidson RE, Bell JF et al (2005) Stratigraphy and sedimentology of a dry to wet eolian depositional system, Burns formation, Meridiani Planum, Mars. Earth Planet Sci Lett 240(1):11–72
Grotzinger JP, Gupta S, Malin MC (2015) Deposition, exhumation, and paleoclimate of an ancient lake deposit, Gale crater. Mar Sci 350(6257):aac7575

Holliday VT, Harvey A, Cuba MT et al (2019) Paleoindians, paleolakes and paleoplayas: landscape geoarchaeology of the Tularosa Basin, New Mexico. Geomorphology 331:92–106

Horgan BH, Bell JF, Noe Dobrea EZ (2009) Distribution of hydrated minerals in the north polar region of Mars. J Geophys Res Planets 114(E1):1–27

Jerolmack DJ, Reitz MD, Martin RL (2011) Sorting out abrasion in a gypsum dune field. J Geophys Res Earth 116(F2):1–15

Jerolmack DJ, Ewing RC, Falcini F et al (2012) Internal boundary layer model for the evolution of desert dune fields. Nat Geosci 5(3):206

KellerLynn K (2012) White Sands National Monument: geologic resources inventory report. Natural Resource Report. NPS/NRSS/GRD/NRR—2012/585. National Park Service, Fort Collins, CO, USA

Keszthelyi LP, Pieri DC (1993) Emplacement of the 75-km-long Carrizozo lava flow field, south-central New Mexico. J Volcanol Geotherm Res 59(1–2):59–75

Kocurek G, Ewing RC (2016) Trickle-down and trickle-up boundary conditions in eolian dune-field pattern formation. In: Autogenic dynamics and self-organization in sedimentary systems, Special publication 106. SEPM, Tulsa, pp 5–17

Kocurek G, Carr M, Ewing R et al (2007) White Sands Dune Field, New Mexico: age, dune dynamics and recent accumulations. Sediment Geol 197(3–4):313–331

Kottlowski FE (1963) Paleozoic and Mesozoic strata of southwestern and southcentral New Mexico. State Bureau of Mines and Mineral Resources, New Mexico Institute of Mining & Technology, Socorro

Langevin Y, Poulet F, Bibring JP et al (2005) Sulfates in the north polar region of Mars detected by OMEGA/Mars Express. Science 307(5715):1584–1586

Langford RP (2003) The Holocene history of the White Sands dune field and influences on eolian deflation and playa lakes. Quat Int 104(1):31–39

Langford RP, Rose JM, White DE (2009) Groundwater salinity as a control on development of eolian landscape: an example from the White Sands of New Mexico. Geomorphology 105(1–2):39–49

Lee DB, Ferdowsi B, Jerolmack DJ (2019) The imprint of vegetation on desert dune dynamics. Geophys Res Lett 46(21):12041–12048

Lucas SG, Allen BD, Morgan GS (2007) Mammoth footprints from the upper Pleistocene of the Tularosa Basin, Doña Ana County, New Mexico. N M Mus Nat Hist Sci Bull 42:149–154

Massé M, Bourgeois O, Le Mouélic S et al (2012) Wide distribution and glacial origin of polar gypsum on Mars. Earth Planet Sci Lett 317:44–55

McKee ED (1966) Structures of dunes at White Sands National Monument, New Mexico (and a comparison with structures of dunes from other selected areas) 1. Sedimentology 7(1):3–69

McKee ED, Douglass JR (1971) Growth and movement of dunes at White Sands National Monument, New Mexico. US Geol Surv Prof Papr 750-D:D108–D114

Metzler EH (2014) The remarkable endemism of moths at White Sands National Monument in New Mexico, USA, with special emphasis on Gelechioidea (Lepidoptera). J Asia-Pac Biodivers 7(1):e1–e5

Morgan GS, Lucas SG (2006) Pleistocene vertebrates from southeastern New Mexico. In Caves and Karst of Southeastern New Mexico. New Mexico Geological Society Guidebook, 57th Annual field conference, pp 317–335

Newton BT, Allen B (2014) Hydrologic investigation at White Sands National Monument. New Mexico Bureau of Geology and Mineral Resources, Aquifer Mapping Program

Nielson J, Kocurek G (1986) Climbing zibars of the Algodones. Sediment Geol 48(1–2):1–15

Olsen KH, Baldridge WS, Callender JF (1987) Rio Grande rift: an overview. Tectonophysics 143(1–3):119–139

Pedersen A, Kocurek G, Mohrig D et al (2015) Dune deformation in a multi-directional wind regime: White Sands Dune Field, New Mexico. Earth Surf Process Landf 40(7):925–941

Pelletier JD (2015) Controls on the large-scale spatial variations of dune field properties in the barchanoid portion of White Sands dune field, New Mexico. J Geophys Res Earth 120(3):453–473

Pelletier JD, Jerolmack DJ (2014) Multiscale bed form interactions and their implications for the abruptness and stability of the downwind dune field margin at White Sands, New Mexico, USA. J Geophys Res Earth 119(11):2396–2411

Phillips JD, Ewing RC, Bowling R et al (2019) Low-angle eolian deposits formed by protodune migration, and insights into slipface development at White Sands Dune Field, New Mexico. Aeolian Res 36:9–26

Rachal DM, Dugas DP (2009) Historical dune pattern dynamics: White Sands Dune Field, New Mexico. Phys Geogr 30(1):64–78

Reitz MD, Jerolmack DJ, Ewing RC et al (2010) Barchan-parabolic dune pattern transition from vegetation stability threshold. Geophys Res Lett 37(19):1–5

Rosenblum EB (2005) Convergent evolution and divergent selection: lizards at the White Sands ecotone. Am Nat 167(1):1–15

Rosenblum EB, Harmon LJ (2011) "Same same but different": replicated ecological speciation at White Sands. Evolution: International Journal of Organic Evolution 65(4):946–960

Rosenblum EB, Römpler H, Schöneberg T et al (2010) Molecular and functional basis of phenotypic convergence in white lizards at White Sands. Proc Natl Acad Sci 107(5):2113–2117

Schenk CJ, Fryberger SG (1988) Early diagenesis of eolian dune and interdune sands at White Sands, New Mexico. Sediment Geol 55(1–2):109–120

Seager WR, Morgan P (1979) Rio Grande rift in southern New Mexico, west Texas, and northern Chihuahua. In: Rio Grande rift: tectonics and magmatism 1487-106. American Geophysical Union, Washington, DC

Swanson T, Mohrig D, Kocurek G (2016) Aeolian dune sediment flux variability over an annual cycle of wind. Sedimentology 63(6):1753–1764

Szynkiewicz A, Ewing RC, Moore CH et al (2010a) Origin of terrestrial gypsum dunes—implications for Martian gypsum-rich dunes of Olympia Undae. Geomorphology 121(1–2):69–83

Szynkiewicz A, Moore CH, Glamoclija M (2010b) Origin of coarsely crystalline gypsum domes in a saline playa environment at the White Sands National Monument, New Mexico. J Geophys Res Earth 115(F2):1–14

Tanaka KL et al (2008) North polar region of Mars: advances in stratigraphy, structure, and erosional modification. Icarus 196(2):318–358

Urban TM, Bustos D, Jakeway J (2018) Use of magnetometry for detecting and documenting multi-species Pleistocene megafauna tracks at White Sands National Monument, New Mexico, USA. Quat Sci Rev 199:206–213

Wasson RJ, Hyde R (1983) Factors determining desert dune type. Nature 304(5924):337

Worman FS, Kurota A, Hogan P (2019) Dunefield geoarchaeology at White Sands National Monument, New Mexico, USA: site formation, resource use, and dunefield dynamics. Geoarchaeology 34(1):42–61

# Chapter 7
# Great Sand Dunes

**Andrew Valdez and James R. Zimbelman**

**Abstract** Great Sand Dunes is located in the Rio Grande Rift where subsidence creates a depositional basin that supplies sediment to an aeolian system. Prevailing winds from the southwest blow across the San Luis Valley transporting sand to the Sangre de Cristo mountain front where the wind regime becomes complex and vertically growing dune forms develop. Fluvial processes also shape Great Sand Dunes. Water transports sand into the system and within the system it moves sand in the direction opposite to the wind. Equilibrium between the two transport mediums causes the deflection of streams flowing around the dunefield where erosion creates large dune forms as the streams truncate the dunefield. Once the streams get past the dunefield, the sand they deposit can be blown back into the dunefield creating other large dune forms. Water also supports the biological diversity in the park and provides recreational value. Water use in the San Luis Valley has historically been greater than water supply, so current management efforts are trying to attain sustainable water use levels. Increased knowledge of the aeolian system has improved the story that the National Park Service passes on to the visiting public. The variety of dune types present in the park, the elevation of the park, the large volume of sand that accumulates there, and the proximity of the dunes to the adjacent mountains all serve as valuable analogs to aeolian deposits on Mars.

**Keywords** Great Sand Dunes · Colorado · Geologic processes · Hydrology · Aeolian · National Park Service · Planetary analogues · Mars

A. Valdez (✉)
Great Sand Dunes National Park and Preserve, National Park Service, Mosca, CO, USA
e-mail: andrew_valdez@nps.gov

J. R. Zimbelman
Center for Earth and Planetary Studies, National Air and Space Museum, Smithsonian Institution, Washington, DC, USA
e-mail: zimbelmanj@si.edu



N. Lancaster, P. Hesp (eds.), *Inland Dunes of North America*,
Dunes of the World, https://doi.org/10.1007/978-3-030-40498-7_7

## 7.1 Introduction

Like all landscapes, Great Sand Dunes is the product of complex and diverse geologic processes. Some, such as rifting, can create a depositional environment capable of supplying sand to build individual dunes and even more extensive sand deposits as well as accommodation space for these deposits. Others, such as sand transport by wind and water, can move and deposit sand while other processes, such as the growth of vegetation and cementation by evaporate minerals, can inhibit sand movement. All of these natural mechanisms are present at Great Sand Dunes (GSD), and they contribute to an aeolian system that has developed along a topographic gradient that is superimposed on a hydrologic system. (Fig. 7.1). Similar sand dune environments can be found on other planetary bodies (Lorenz and Zimbelman 2014); GSD provides a remarkable setting in which to study active aeolian process for comparison to aeolian features observed on other planets.

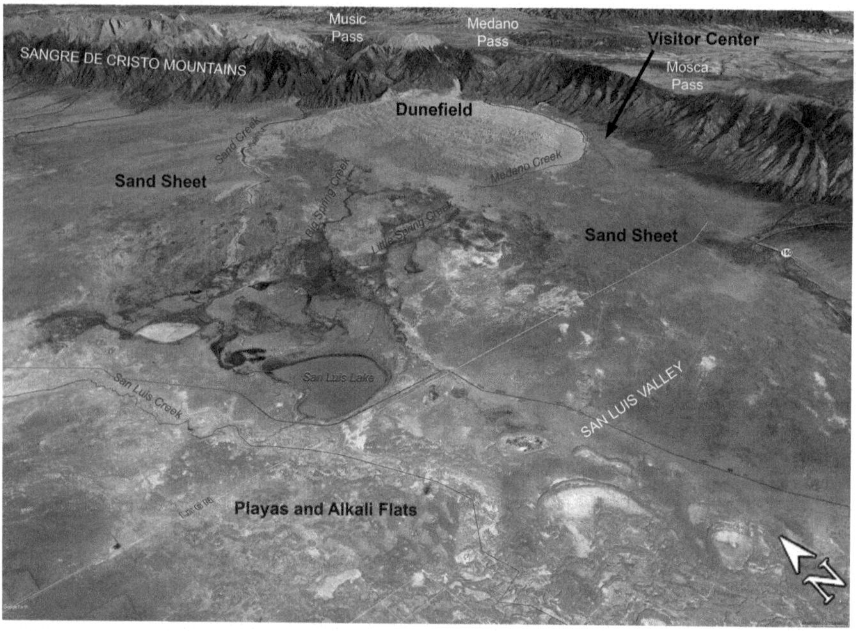


**Fig. 7.1** Oblique view of Great Sand Dunes aeolian and hydrologic systems. Aeolian transport begins in the playas and alkali flats, where lunettes develop on the northeast side of the playas. Sand is transported across the sand sheet as indicated by parabolic dunes. Wind regime is altered by the mountain passes and sand accumulates as reversing and star dunes in the dunefield. Stream flow that originates in the passes flows around the dunefield, modifying the dunefield perimeter. Groundwater recharge occurs along the mountain front at a level the produces groundwater discharge areas downgradient producing gaining streams such as Big and Little Spring Creeks. Except during very wet periods, the playas are the terminus of the hydrologic system and water exits via evapotranspiration. Vertical exaggeration = 1.5

### 7.1.1 Location

Great Sand Dunes is located in southcentral Colorado in the San Luis Valley (SLV). It is a high elevation desert surrounded by the San Juan Mountains on the west and the Sangre de Cristo Mountains on the east (Fig. 7.2). The valley floor ranges in elevation from 2290 to 2440 m and has an average precipitation of 19 to 28 cm. The surrounding mountains rise up to 4372 m and are wetter with precipitation ranging from 28 cm in the foothills to 135 cm on mountain summits. The headwaters of the Rio Grande are in the San Juan Mountains west of the SLV. The valley floor can be subdivided into Alamosa Basin, the northern part of which is very flat. The central portion has the San Luis Hills. The southern part is covered by the flood basalts of the Taos Plateau.

### 7.1.2 A Unique Place

The lowermost portion of the GSD aeolian system is a depression that serves as the terminus for local streams (Fig. 7.3). Here evaporation is a dominant process and as a result, groundwater has become saline, producing alkali flats. As the sand surface rises above the capillary fringe, the salts are no longer present and the sand is loose, but then shrubs and grasses become established, increasing the surface roughness and decreasing sand mobility. The result is a sand sheet that extends from the alkali flats to the Sangre de Cristo Mountains, where it often pinches out as sand ramps. At the mountain front, specifically where there are passes through the mountains, wind patterns become very complex and sand dunes that once migrated across the valley floor collect in a dunefield that contains the tallest dunes in North America. Streams that flow out of these passes are deflected around the dunefield, and in so doing, they erode into the dunes to carry sand upwind where it can be blown back to the dunefield. The stream-supplied sand helps produce the large dune forms seen today along Medano Creek.

The wind's ability to move and sort sand creates a unique and harsh environment that life must adapt to. Vegetation that grows on the dunes must be able to efficiently collect moisture and survive burial or deflation caused by a shifting sand surface. Some plants do this by having an extensive root system while others produce large, hairy seeds that may show early signs of speciation (Andrew et al. 2012). Animals must also cope with the unique nature of the sand; at GRSA, six species of endemic insects have evolved to do precisely that.

Great Sand Dunes is an awe-inspiring landscape. Like many other unusual landscapes, it is protected by the National Park Service. These special places are managed with the intent of preserving the area so that current and future visitors can enjoy them and be inspired by them. Unlike most geologic wonders, GSD is a dynamic resource with measureable changes occurring to the primary resource (sand dunes) on an annual basis. Therefore, the dunes must be actively managed

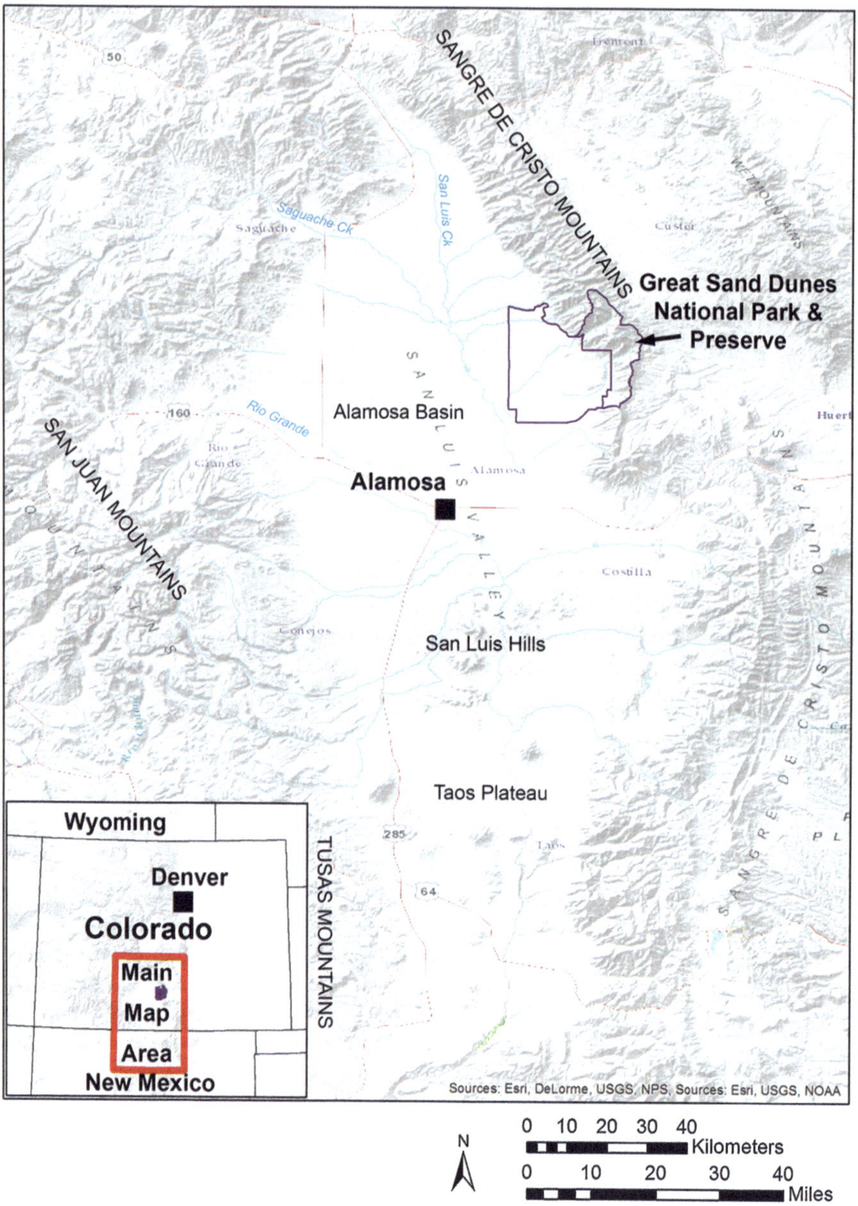


**Fig. 7.2** Location of Great Sand Dunes National Park & Preserve in the northeast portion of the San Luis Valley, Colorado. The San Luis Valley can be subdivided from north to south into the Alamosa Basin, the San Luis Hills, and the Taos Plateau. The Alamosa basin is very planar and covered by fluvial and aeolian deposits. The San Luis Hills are erosional remnants of the San Juan Mountains. The Taos Plateau is comprised of flood basalts. The Sangre de Cristo Mountains border the San Luis Valley on the east and the San Juan and Tusas mountains on the west

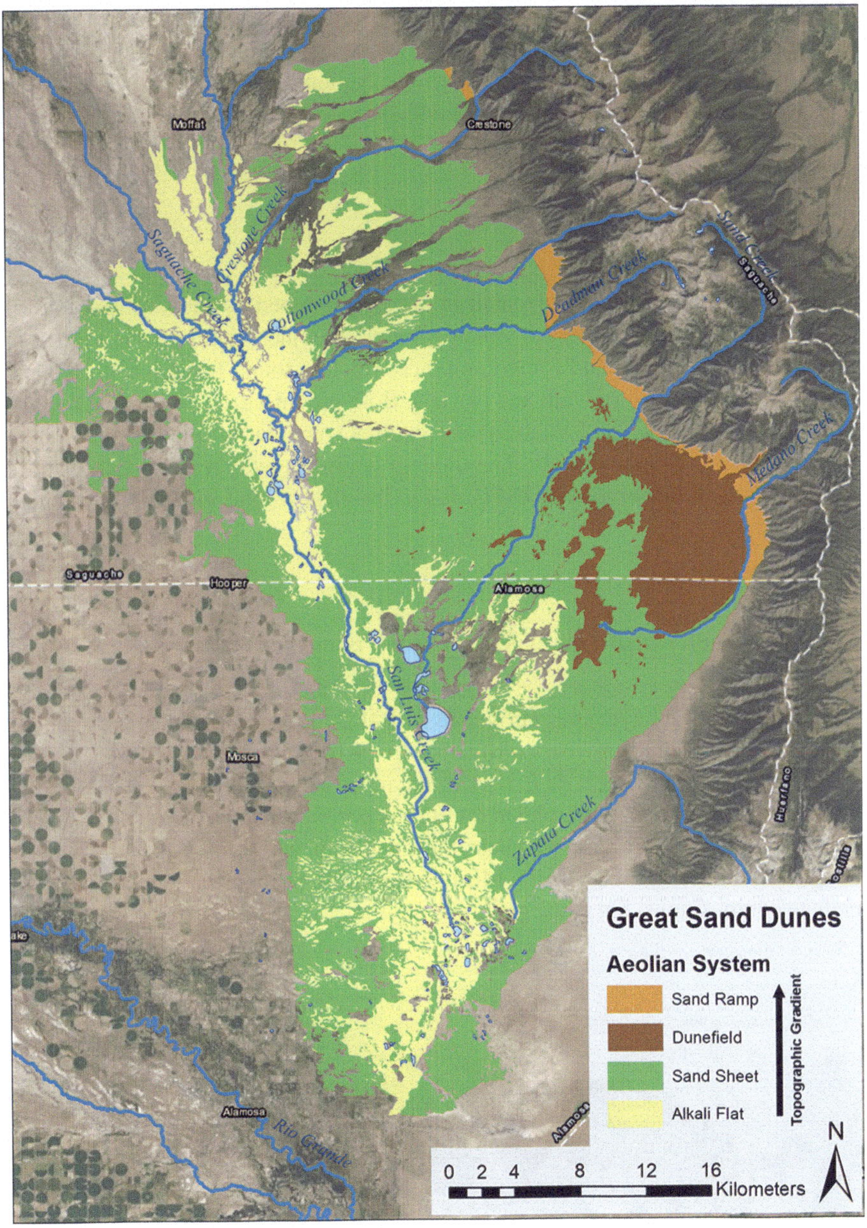


**Fig. 7.3** Great Sand Dunes Aeolian System. The deposits are developed along a topographic gradient. Alkali flats are in the lowest areas that are dominated by evaporation. Where ground levels rise above the influence of evaporation, sand is stabilized by vegetation forming a sand sheet. Along the mountain front is the dunefield and where the sand sheet ramps onto the mountain front are sand ramps

with the same attention that other dynamic resources demand, such as biological and hydrological resources. GSD is an area where an ever-increasing knowledge of the resources has made a difference in how the National Park Service (NPS) is able to protect the natural systems and explain the story behind how the dunes formed.

## 7.2 Regional Geology

The San Luis Valley (SLV) is located in a region mapped by the U.S. Geological Survey (USGS) as the Rocky Mountain geologic province, but it also has much in common with the Basin and Range province which is found further south in New Mexico and to the west in Utah and Nevada. Both provinces are dominated by orogenic processes and to understand them requires some background about what is occurring on a continental scale. The modern landscape within Great Sand Dunes National Park & Preserve (GSDNPP) is the result of a long history of mountain building.

### *7.2.1 Orogeny and Other Mountain Processes*

The oldest rocks are Precambrian gneiss (1.8–1.7 billion years old) and igneous intrusions (1.7–1.4 billion years old) (Lindsey 2010) and they are exposed on the western slope of the Sangre de Cristo Mountains (Sangres) that is the backdrop to GSD (Fig. 7.4). The metamorphism and intrusions were likely the result of continental accretion as the SLV was near the margin of the North American Craton (Whitmeyer and Karlstrom 2007). Most of the intrusive rocks have felsic or intermediate compositions, but some gabbro is present.

**Fig. 7.4** Great Sand Dunes landscape view. On the valley floor, the dunefield is surrounded by a vegetated sandsheet. Stabilized dunes can be seen as the hilly sand sheet surface. Behind the dunefield are the Sangre de Cristo Mountains. Most of the mountain front is composed of Precambrian metamorphic and igneous rocks. Younger sedimentary rocks, such as the Pennsylvanian/Permian Crestone Conglomerate cap peaks north of the dunefield

The Precambrian is capped by much younger rocks, separated by an unconformity representing over 1 billion years. Previously deposited marine rocks were removed by the Ancestrial Rockies Orgeny, an uplift that began about 320 Ma as the ancient continents of Gondwana and Laurentia collided (Gonzales and Karlstrom 2011). Deposition occurred adjacent to the uplift producing the Minturn (Pennsylvanian) and the Sangre de Cristo (Pennsylvanian and Permian) Formations. The Crestone Conglomerate member of the Sangre de Cristo Formation is an attractive rock with large Precambrian clasts in a hematite-cemented, sandy matrix. It is well cemented, but highly fractured, resulting in the jagged appearance of some of the peaks behind GSD, such as Crestone Peak (Fig. 7.4). Peaks formed by Precambrian rocks, such as Cleveland Peak, tend to be smoother. During the Mesozoic, Ancestral Rockies were mostly eroded away and regional elevation lowered to the point where transgression once again covered Colorado with a shallow sea.

The next episode of mountain building was the Laramide Orogeny which occurred 70 to 40 Ma when an ocean plate, called the Farallon Plate, slid under western North America (Dickenson 1981) and compressed crustal rocks. Evidence of the Laramide can be seen in the Sangres as folded rock layers and by older rock units thrust over younger ones (Lindsey 2010). Many of the current drainages within the Sangres are eroded along former thrust faults and secondary tear faults that splay off the main thrust (Webster 2000).

The Larimide was followed by a period of volcanism that produced the San Juan Mountains (San Juans) 35 Ma to 15 Ma. There were 3 episodes of volcanism (Lipman and Mehnert 1975) that created a large volcanic plateau covering much of SW Colorado. The first episode was a series of centralized volcanos that erupted intermediate composition rocks such as andesite interbedded with volcaniclastic deposits. Volcanism evolved to caldera eruptions that produced numerous ignimbrites (volcanic tuffs). The last episode was relatively small and consists of a thin veneer of basalt that caps some of the ridges in the San Juan mountains.

The final orogenic process is rifting. The mantle beneath western North America has uplifted the area, applying extensional forces on the crust creating the Rio Grande Rift. One of many rift systems in the Basin and Range Province. The Rio Grande rift extends from New Mexico into central Colorado and is responsible for the formation of the SLV and Sangre de Cristo Mountains.

Rifting in the SLV is believed to have begun around 25 Ma (Thompson et al. 2015) and is still considered a modern process with an average regional extension rate of 1.2 mm/year (Berglund et al. 2012; Murray 2015). Rates may increase during large magnitude seismic events and at GSD, paleosiesmolgy work identified magnitude 7.2 earthquakes at 10–15 ka and 5 ka (McCalpin 2006). Most of the displacement occurs on the east side of the SLV where the Sangre de Cristo Fault defines the boundary between the San Luis Basin and the Sangre de Cristo Mountains. In a simple sense, as the crust pulls apart breaking it into blocks, the block that is the San Luis Basin rotates downward. The Sangres block rotates upward. This tilting led to the asymmetry of the basin structure, which is hinged on the west and dips toward the east where faults accommodate rift displacement. The orientation of the Sangre de Cristo horst also is tilted eastward resulting in the oldest rocks exposed on their west side, the side seen behind GSD, and younger rocks on some ridge tops on the east side.

Buried under the flat plain of the SLV floor, the subsurface structure of the San Luis Basin graben is complex as it is broken up into several north-south trending blocks. On the eastern margin, adjacent to the Sangre de Cristo fault and not too deep in the subsurface is a half graben (Watkins 1996) that extends several miles west of the Sangres before an another fault drops to the much deeper Crestone Graben. West of the Crestone Graben, but still in the subsurface is the Alamosa Horst. Further south in the SLV, the horst rises above the basin fill forming the San Luis Hills. West of the Alamosa Horst is another structural low, the Monte Vista Graben. Not as deep as the Crestone Graben, it tilts upward toward the west until it surfaces as the foothills of the San Juans with little evidence of being fault bound.

The total vertical displacement between Sangre de Cristo Horst and the Crestone Graben could be as much as 8000 m (Kirkham and Magee 2020), and the basin fill in the Crestone Graben is believed to be more than 6000 m thick (Drenth et al. 2011, 2016). The Monte Vista Graben has 1100 m of fill. The basin fill consists of pre-rift sediments from the Laramide uplift, San Juan volcanic flows and another thick sequence of fluvial sediments, the Santa Fe Formation, which are derived from erosion of the San Juans and Sangres. Above the Santa Fe Formation is the Alamosa Formation, a series of lacustrine clay deposits found throughout most of the north-central portions of the San Luis Basin, but they pinch out near the margins. Above the Alamosa Fm is 15 to 100 m of Quaternary Alluvium and the Great Sand Dunes aeolian deposits.

As rifting progressed, faults began to act as conduits to the mantle leading to another episode of volcanism 5 Ma to 1 Ma. Most of the eruption centers were in northern New Mexico, but they do extend into Colorado. The flows spread across the valley floor, stacking on each other forming the Taos Plateau. The plateau impounded the Rio Grande, flooding the northern SLV and creating Lake Alamosa. This lake existed from 3.5 Ma to 0.4 Ma and its extent expanded and contracted as runoff into the SLV varied. When full, it was one of the largest high elevation lakes in North America (Machette et al. 2013; Ruleman et al. 2016). The variable size of Lake Alamosa produced the interbedded clays and sands of the Alamosa Formation. The clays, known locally as the Blue Clay, are indicative of reducing conditions and the incomplete breakdown of organic materials in lake sediments. Sedimentation eventually filled Lake Alamosa and continued rifting accommodated the burial of the Alamosa Formation with younger fluvial and aeolian sediments.

Glaciation during the past 1.8 Ma has also shaped the mountains surrounding GSD, but most glacier-related landforms seen today are related to the two most recent glacial periods. The Bull Lake glaciation occurred 190 ka to 130 ka and the Pinedale glaciation was 30 ka to 12 ka. They carved cirques and U-shaped valleys. Glacial periods can be important to dune building as the cold temperatures likely reduce vegetation on the sand deposits and outwash enhances sediment supply.

### 7.2.2 *Rifting Sets the Stage*

Rifting is a key process in the development of GSD. The subsidence of the San Luis Basin creates accommodation space for sediment deposition making the SLV capable of supplying sand for extensive aeolian deposits. Similar settings throughout the Basin and Range Province also have active dunes, such as White Sands (NM), Death Valley (CA), Sand Mountain (NV), and Little Sahara (UT) (Zimbelman and Williams 2007; Lancaster, this volume). Many visitors to GSDNPP are surprised to find such large dunes at the base of the Rocky Mountains. Rifting is the reason they are here. Most other Colorado valleys are mostly erosional and lack the sand supply to build extensive aeolian deposits.

Within the San Luis Basin, it is the Crestone Graben that in instrumental in why GSD developed where it has. Subsidence in this graben is greater than deposition, producing an internally drained area that is known as the Closed Basin. Streams draining the San Juans and Sangres at the northern end of the SLV flow into the Closed Basin, intermittently supplying water and sediment to an extensive playa system. These playa areas, also referred to as the sump, are believed to be the immediate source (on the valley floor) of the sand as aeolian deposits have developed along their length, but are thicker and much more extensive east of the sump (Madole et al. 2008) extending to the Sangres.

The playas are effective at sorting sand from other sediments as the sand is deposited along the shoreline; when the playas dry, the sand deposits are exposed to the wind and can blow across the dry lakebed often collecting on the downwind side. Currently most of the closed basin's streams are diverted for agricultural use, so stream flow no longer has the opportunity to reach the playas. Precipitation is cyclical so in prehistoric times, playa filling and drying likely correlated to wet and dry periods. Glacial outwash would have likely provided large influxes of sediment and even directed the Rio Grande into the Closed Basin (Madole et al. 2008; Johnson 1967).

The source of the sand can also be determined using regional geology knowledge. Ultimately the sand is derived from rocks in the San Juan and Sangre de Cristo Mountains. The San Juans are composed of volcanic rocks and the Sangres of metamorphic and intrusive igneous. The bulk composition of the sand in the dunefield is 52% volcanic rock, 29% quartz, 9% feldspar, 10% other (Wiegand 1977). Using mineralogy and crystal size to determine provenance suggests that most of the sand originates in the San Juans (Hutchison 1968). Most quartz sand grains are clear, matching the quartz phenocrysts of the ignimbrite. The ages of the source rocks vary greatly; San Juan rocks are relatively young (Cenozoic), and Sangres rocks are old (Precambrian). Using zircon chronology dating yields a result of 70% San Juans source and 30% Sangres source (Madole et al. 2008). The San Juans are the more distal mountains to GSD with minimum distance of 60 km (37 miles), but they also have a larger drainage basin and greater streamflow (Hearne and Dewey 1988), so it is reasonable that more sand comes from the San Juans.

### 7.2.3 Formation and Age of the Sand Dunes

Lake Alamosa may be the key to understanding how old the dunes are. Many early estimates on the age of GSD assumed they were post Pinedale glaciation (less than 12 ka). Rich Madole, USGS, used geomorphic and stratigraphic data to refine the estimate. North of the dunefield, he identified lag deposits as Bull Lake age (~130 ka), which are underlain by older aeolian deposits. Well logs indicated sand in the subsurface, and Madole bracketed the oldest sands as being younger than the uppermost clays of the Alamosa Formation. The oldest sand deposits were younger than 440,000 years when Lake Alamosa drained (Machette and Marchetti 2006), which matches drill information done by the NPS. At the southern base of the dunefield, a 40 m (130 ft) core was collected and the entire extent of the core was aeolian sand. Further west on the sand sheet, two wells penetrated the sand at depths of 88 m (290 ft), and 100 m (330 ft). Cores were not collected from the two deeper wells, but well cuttings indicate that the overwhelming majority of sediment removed from the well bore was sand similar to what is seen at the surface. The aeolian sand was followed by 3 to 4 m of coarse fluvial sand, followed by the uppermost blue clay, suggesting that the aeolian environment has been in place since shortly after Lake Alamosa times.

Madole used OSL and $^{14}C$ dating to provide more precise ages of surficial deposits in his effort to produce a geologic map of Great Sand Dunes (Madole et al. 2016). Unit Qes1, latest Holocene, consists of unvegetated sand on the sandsheet. Qes2, late Holocene, occurs in parts of the sandsheet with loose sand but is mostly stabilized by vegetation. Qes3, late and mid Holocene, are parts of the sandsheet that are vegetated, but also have calcium carbonate within 1.5 m of the surface. Unit Qgsd is the large, active dunes located within the dunefield, aged latest Holocene to middle Pleistocene (Madole et al. 2016). https://pubs.er.usgs.gov/publication/sim3362

## 7.3 Park History

### 7.3.1 Native Americans and Early Settlers

Great Sand Dunes has a long record of human activity that used the resources of the aeolian system. In 2019, over half a million visitors came to see the dunes and play in Medano Creek. Stepping backward in time, the earliest evidence of human visitors comes from the Folsom Age, 9,000 years ago when hunters used the hummocky, stabilized dune surface to stalk megafauna such as *Bison antiquus* (Jodry and Stanford 1996). They were followed by Archaic groups that hunted smaller game and collected Indian Rice Grass and Blowout Grass seeds that grow abundantly on the sand sheet. The sand sheet is rich with archeological resources because it was a food and water source, and it is a mobile environment subject to deflation, exposing archeological sites. The dunefield is largely depositional, making artifacts harder to find, but one can speculate that these early groups may have played on the dunes like the modern visitors.

Historic Native American tribes knew about these dunes and some consider them sacred. The Apaches called them Seinanyedi (it goes up and down) (White 2005), and the Utes referred to them as Saa waap maa nache (sand that moves) (Naranjo 2016). There are Pueblo tribes who believe their point of emergence (sipapu) is from a lake/pond near GSD. The Utes and Apaches most likely used the SLV during warm months as it can get bitterly cold during the winter. Some evidence of their visits to GSD includes Culturally Modified Trees, which are ponderosa pine with patches of bark peeled off near the tree's base. They consumed the cambium layer which is nutritious and has medicinal qualities (Martorano 1999). The Pueblo tribes had permanent settlements located to the south in New Mexico, thus historically have been more physically separated from GSD then the Utes and Apaches, but they maintain the connection through prayers that mention the dunes.

The first written description of GSD comes from the journal of American explorer Zebulon Pike, who on January 28, 1807 came over the Sangre de Cristo Mountains from the east, discovering these dunes. In doing so, he wandered into New Spain and was later arrested by the Spanish and his journals were confiscated. In his journal reconstruction, he wrote that the dunes appeared "exactly that of a sea in a storm, (except as to color) not the least sign of vegetation" (Carter 1978).

Territorial claim to the SLV changed from Spain to Mexico in 1821, following the Mexican War of Independence, then from Mexico to the USA in 1848 after the Mexican-American War. Settlement of the Alamosa Basin portion of the SLV began in 1850s as Hispanics from the Taos and Espanola (NM) areas began to settle on the margins of the basin establishing towns such as San Luis and Conejos. More settlers came during a land promotion in the 1860s and a gold rush in the 1880s. In 1874 the first known photograph of GSD was taken by noted western photographer William Henry Jackson. He captioned the photo, "A curious and very singular phase of nature's freak." (Rowlands and Geary 1998). In 1878 railroads began serving the SLV and towns began to be established on the valley floor in response to that (Spencer 1925).

### 7.3.2 *National Monument*

By the early 1900s agriculture was firmly established in the SLV and it supported a growing population that began to recreate at GSD. Gold prospecting also migrated to GSD as sluice boxes were set up on Medano Creek to process the sand. Local citizens began to worry about the impacts that mining could have on the dunes. That prompted a woman's group, the P.E.O. Sisterhood, San Luis Valley Chapter, to start a letter writing campaign in 1930 that called for protection of GSD. The letters reached state and national politicians and in 1931, NPS staff evaluated these dunes and concluded that they should be included in the National Park System. Roger Toll reported that he knew of no other sand dunes in the U.S. that were superior to the dunes of the SLV (Toll 1931). On March 17, 1932, President Herbert Hoover signed a Presidential Proclamation, establishing Great Sand Dunes National Monument (GSDNM). This formally established the dunefield's name as Great Sand Dunes. Locals today and mostly likely back then, just refer to them as 'the sand dunes.' The national monument boundary encircled the dunefield as that is the most visually spectacular part of the aeolian system and where visitation was occurring. The goal of the NPS was to preserve and protect the dunes and provide for the enjoyment of the visiting public. The response of the P.E.O. Sisterhood was, "Someday our children's children will thank us for our foresight in saving this magnificent nature wonder" (Trimble 1975).

NPS management at GSDNM started slowly but grew as visitation increased. Mining activity stopped but GSDNM was managed remotely by staff in Santa Fe and Mesa Verde until 1940 when seasonal staff began to work on site. In 1946, the first full-time, permeant employee began work at GSDNM. The monument had close to 9000 visitors that year. By 1956 visitation grew to 52,700 and the NPS began a national effort to improve visitor facilities with the goal of finishing by 1966. It was thus called Mission66, the program that built the Visitor Center, campground, and staff housing. The access road was paved in 1960. In 1966 visitation reached 157,000 and in 1990 visitation was 272,000.

In the early 1990s NPS management of GSDNM fundamentally changed when water development proposals on the adjacent Baca Ranch began to appear (Fig. 7.5). Prior management efforts were focused on developing and providing visitor services, but the water development issue forced the NPS to begin more active management of the nature resources it was charged with protecting.

The first proposed water project was by American Water Development Inc. (AWDI) whose project wanted to pump 200,000 acre-feet of groundwater annually. AWDI's groundwater modeling predicted that after 20 years of pumping, the water table at the Baca Ranch/GSDNM boundary could be lowered by 50 m. At the time, the NPS was not sure how that would affect the natural system, but was alarmed by the potential for such change. In 1991, AWDI applied for well permits from the Colorado Division of Water Resources. The NPS and many other local parties objected to the issuance of the permit, so the application went to Water Court. In Water Court the applicant needs to address 3 things; (1) that the water is available, (2) that it would be put to beneficial use, and (3) that it can be done without injuring existing water rights. It was the third point that AWDI could not demonstrate to the satisfaction of the Water Court and the application was denied. Expert testimony by hydrologist Eric Harmon convinced the Court that lowering the water table beneath GSDNM would decrease flow in streams and that would impact the instream flow water right held by the NPS (HRS 1999). Other water rights holders in the local ranching community were able to show the same. After the failure to secure the well permits, AWDI filed for bankruptcy and sold the Baca Ranch.

The fate of the AWDI permit application was decided in Water Court, so water rights effects were the critical issue, not potential effects on GRSA's natural resources, but the debate on what groundwater withdrawal would do to the dunes

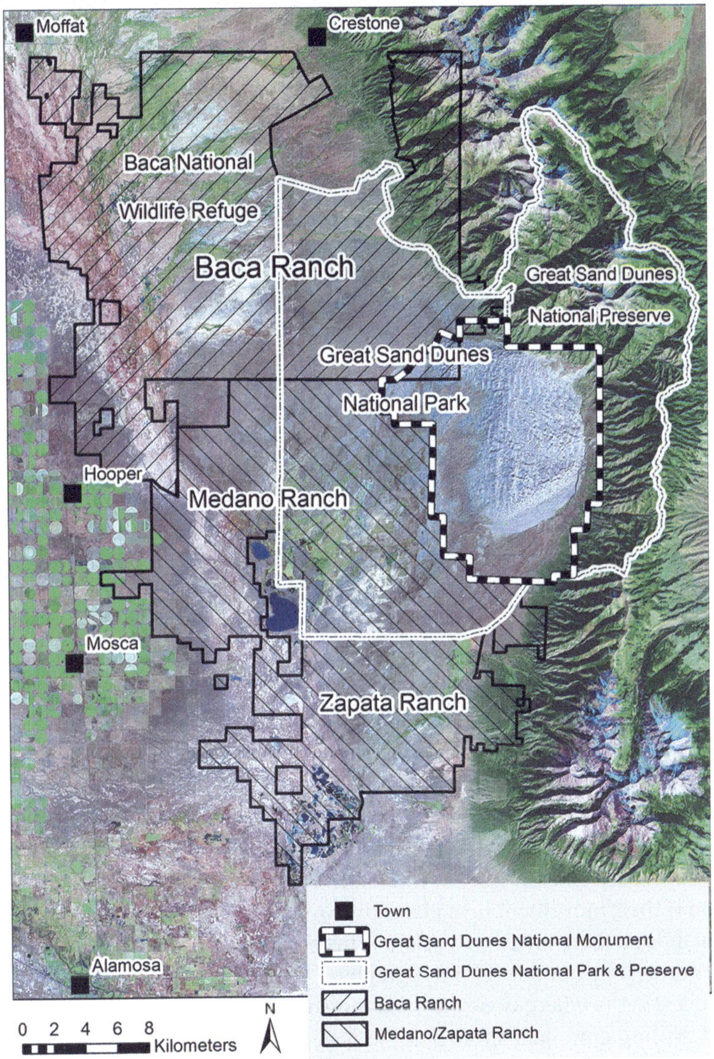


**Fig. 7.5** Land management near Great Sand Dunes. Prior to 2000, Great Sand Dunes was a national monument whose boundary is shown by the black and white dashed line. The Baca Ranch bordered on the northwest, the Medano Ranch on the west, and the Zapata Ranch on the south. The Great Sand Dunes National Park and Preserve Act of 2000 expanded the National Park Service boundary to the extent shown by the bold white line. The portion on the valley floor became a national park and the mountainous portion a national preserve so that hunting could continue there. The Baca Ranch was purchased by the federal government in 2004 and incorporated it into the national park. Areas of the Baca Ranch west of the park boundary became Baca National Wildlife Refuge. The mountainous portion of the Baca Ranch became part of Rio Grande National Forest

took center stage in the public forum. The NPS turned to a local geology professor for expertise and he developed the following theory; GSD was stable because of the moist sand that exists beneath the ground surface. That moisture comes from groundwater and if the groundwater is lowered, the dunes would dry up and blow away.

This theory was never properly tested by its proposer, but it became popular with the press, possibly because it was sensational and spoke of imminent danger. The GSDNM interpretation staff also started using it in the information they passed on to the public. Contradicting that theory was the AWDI expert witness, geologist Steve Fryberger who argued that the large dunes at GSD were 'stable' because of bimodal and complex wind regimes.

The dune behavior relationship to wind regime concept had been established for over a decade by that point and it was largely based on the work that Fryberger did at GSDNM in the 1970s, while he was part of a U.S. Geological Survey team lead by Ed McKee. Their combined work resulted in an iconic dune publication "A Study of Global Sand Seas" (McKee 1979b) and it included a method for calculating potential sand movement and effective wind regime from wind data (Fryberger et al. 1979), a method that is now commonly referred to as the Fryberger method. Fryberger also had an extensive list of dune-related research publications and his work is often cited, making him a well-qualified expert witness, especially for GSD.

The NPS expert witness, on the other hand, had a background in mineralogy. His theory on subsurface sand moisture stabilizing the dunes does not hold up well when applied to other dunefields or even dunes within GSD. First, what exactly does 'stable' mean in regard to dunes? In his theory it was based on the megadune at the south end of the dunefield looking similar today to what it did in historic photos. We know now that the ridges of these large dunes move several to tens of meters each year, so they are mobile and not stable in any sense. The direction and magnitude of the movement is dependent on wind regime as predicted by the Fryberger method, and discussed below. At GSD, the large dune forms have parallel dune ridges and if their movement is in phase, the overall appearance will be similar over time even though there is net movement. This movement is not likely to be apparent in a time series of ground level photographs.

Next question is where does the sand moisture come from, and will it go away? Well and drilling core data show that the water table drops quickly. At the base of the Star Dune, the water table is 40 m below ground level. The Star Dune rises 230 m above that ground level making a sand column 270 m high that in the stabilization theory it would be moistened by upward wicking from the water table. This concept was presented at a GRSA science conference and the hydrologists felt strongly that the idea was physically impossible. The most likely source of the sand moisture is precipitation that wets the surface and percolates downward. Unpublished density logs done on dry wells within the dunes suggest that moisture from large rain events may produce pulses of moisture that move down the sand column. Also the moist sand has field capacity, where water clings to the sand grains and gravity cannot pull it off (specific retention), so the dunes wouldn't 'dry up' regardless of where the water table was. Moisture can hinder sand movement if the surface is wet, but the aridity at GSD, makes the sand surface almost always dry.

Does subsurface moisture stabilize other dunes? Are dry dunes more mobile? Will the dunes blow away? There are coastal dunes in Oregon and Brazil; each is in a humid climate and is very likely to have moist sand below the surface, yet dune form and sand deposits indicate that the dunes are migrating from the coast inward. There are dunes in hyperarid climates, such as Algodones Dunes (CA). They seem to lack the subsurface moisture, yet have developed into a large, slow-changing megadune similar to GSD. Again the key is wind regime. Where the wind blows mostly in one direction, migrating dune forms develop. Where it blows in multiple directions, sand accumulates producing large dune forms. Even at GSD there are differing dune behaviors. East of Medano Creek are the Escape Dunes, a string of parabolic dunes that migrate up the lower slopes of the Sangres. West of Medano Creek is the dunefield. At creek level are small barchanoid ridge dunes that also migrate. With elevation gain, the dunes transition to reversing dunes that are the more stable form. Why wouldn't the subsurface moisture found in each dune stabilize all the dunes? The dunefield is the depocenter of the aeolian system because it's in a bimodal and complex wind regime zone. Unless that changes, why should the dunes blow away?

The point is, had the NPS included staff with a basic understanding of aeolian science, they would have been better prepared to select an appropriate expert witness. It was honorable of the local professor to try and help, but a better selection could have been made. The lingering effects are important, as it was difficult to convince the GSDNM interpretation staff that sand moisture doesn't stabilize GSD and many local residents still believe this to be true. It's hard to repair mis-information.

After scrambling to deal with the AWDI proposal, the NPS created a resource management division whose responsibility was to represent natural resource concerns and develop expertise about them. A planning effort brought in subject matter experts and produced a Resource Management Strategy that focused the division in defining important resources that include geologic, biologic, and cultural systems. In the geologic realm, aeolian geology was clearly very important at GSDNM. In the biologic realm, plants and animals adapted to sand, and in the cultural realm, human activity related to the dunes. The plan also brought to light that the dunes were a small part of a larger system. GSDNM also began monitoring hydrologic resources such as stream flow and groundwater levels. That type of data didn't exist previously, and it would be vital in evaluating the effects of groundwater pumping. A staff consisting of the division chief, a geologist, and eventually a biologist was hired. Cultural expertise is provided by NPS regional staff and consultants.

The ownership of the Baca Ranch following AWDI went to an investment group, whose on-site management, Stockman's Water Company (SWC), formulated another water development proposal. Project pumping was not clearly defined, but at meetings was described as an annual production around 125,000 acre-feet. The project included a political approach as supporters of SWC got 2 initiatives on the 1998 Colorado Election Ballot; both were defeated by a 3 to 1 margin (Geary, 2012) as election advertising described them as an effort to pump water out of the SLV. Of note is that the 1990s were a relatively wet decade in Colorado. Had 1998 seen a prolonged drought with Colorado cities facing water shortages, the SWC initiatives could have passed. SWC seemed less active in promoting their project after the 1998 election results.

### *7.3.3 National Park and Preserve*

As the 1990s came to an end, developments in and around GSDNM led to more fundamental changes. South of the Baca Ranch and west of GSDNM is another large ranch called the Medano/Zapata Ranch (Fig. 7.5). In 1999 it was purchased by The Nature Conservancy (TNC). By then GSDNM resource management staff was well established and had incorporated research done in the 1970s by the USGS McKee group. The NPS was actively thinking of GSD as part of a larger dune system that included a dry lakes province where sand originated, a vegetated sand province through which sand is transported, and a dunefield province as a depocenter (Fryberger 1990). Also, a new Park manager named Steve Chaney arrived; armed with the information about the greater dune system and having TNC, who commonly transfer their land to other conservation agencies, as a neighbor, he saw an opportunity to make management of the entire dune system a possibility. After getting approval from regional NPS management, he began to the pitch the idea in local communities. TNC was on board and together with the NPS, arranged for meetings with Colorado politicians who were also sold on the idea of boundary expansion. Representative Scott McGinnis led the way, and through no small effort, he brought a Bill to the House floor for a vote. The legislation passed through Congress and in November of 2000, President Clinton signed the Great Sand Dunes National Park and Preserve Act of 2000.

The primary purpose of the boundary expansion was to allow for the management of the greater dune system by the NPS (Fig. 7.5). The boundary was extended out to the alkali flats where the aeolian deposits originate. It included much of the sandsheet, and like the former monument boundary, the dunefield itself. In addition, it expanded into the mountains above the dunefield as they affect wind regime and supply the streamflow that modifies the edges of the dunefield and the recharge to aquifers. There are still extensive areas of playas and sandsheet outside of the NPS boundary to the south. The decision was made to stop expansion at Lane 6N, one of the access routes to GRSA, to keep management simple as well as to avoid conflict with the BLM, who had a great interest in maintaining their management of the playas.

The new boundary extended onto the Baca and Medano/Zapata Ranches, providing the federal government the authority to purchase land from willing sellers. The majority owners of the Baca Ranch were willing to sell. TNC negotiated and purchased the Baca Ranch in 2004 on behalf of the Federal government and shortly thereafter, was reimbursed, transferring ownership to the USA. The purchase of the Baca Ranch allowed for the park unit's name to officially become Great Sand Dunes National Park and Preserve (GSDNPP). The Baca Ranch covered a greater area than what the expanded NPS boundary covered and the mountainous portion on the east side of the ranch was added to Rio Grande National Forest. The western half of the ranch contains a complex of large meadows, and that area was used to establish Baca National Wildlife Refuge. In 2021, the federal government is expecting to purchase the Medano/Zapata Ranch property within the boundary from TNC and that will bring almost all of the lands in the expanded boundary under federal ownership.

### 7.3.4 *The Critical Role of Water*

An underlying interest in the boundary expansion was to end the decade of water development speculation in the SLV. Federal ownership of the Baca Ranch effectively does that. In addition, the expansion legislation required the NPS to apply for a water right from the State of Colorado that would protect the natural groundwater level beneath the Park. In 2008, the NPS applied for and received that right, which is believed to be the first such water right that protected in-place groundwater. There was objection to the water right application so the case went to Colorado Water Court. The NPS had to show that the water was available, that it would be put to beneficial use, and that it wouldn't impact existing water rights. Well data showed that the water is there. The beneficial use was the importance of the water to the natural processes and biologic diversity protected by the park. Since the right would be non-consumptive, there would be no impact on existing water right holders. The NPS also needed to define what such a water right was and how it was to be managed. It did so by selecting 10 sites along the park boundary where monitoring wells would track water pressure in the upper aquifer. Each well site has a defined water table elevation that resents a point of legal protection. Water levels in each well would continue to be monitored to ensure compliance with the water right and demonstrate to the State Engineer's office that the right is being exercised. This in-place water right is a very strong legal defense against future water development because they would have to address impacts to groundwater within GSDNPP.

Water is an important resource that is closely managed by the NPS at GSDNPP via an extensive network of stream gauges, monitoring wells, and weather stations. It is most fortunate to have a largely intact natural hydrologic system within GSDNPP due to the extensive aeolian deposits. The sand-dominated environment was very sparsely settled save for several large ranches that had minor development. The hydrologic system isn't completely intact as north of the aeolian deposits, the two largest streams in the system, Saguache and San Luis Creeks, are completely diverted for agriculture, preventing their flow from reaching the playas upwind of GSD. Lastly the NPS can now evaluate potential threats in a more defendable manner. If the water table under the park were lowered significantly, it would reduce streamflow as previously mentioned, but in addition, the lowering could impact shrubs on the sandsheet whose roots tap the ground water, and it could stop the evaporation process that produces salts on the alkali flats. It has the potential to alter processes important to the development of the aeolian system. It also would affect biodiversity should water features disappear. Water has culturally importance as springs and possibly a sipapu are important to affiliated Native American tribes.

## 7.4 Aeolian Features

The Great Sand Dunes aeolian system (Figs. 7.1 and 7.3) is composed of salt encrusted alkali flats, vegetated sand sheets, and active dunes. The main dunefield covers approximately 69 $km^2$ with a north-south extent of 12 km and 7 km east to west. Star and reversing dunes are common in the main dunefield where they reach heights up to 230 m. The sand sheet in surrounding the dunefield has an area of 405 $km^2$. Within the sandsheet is 21 $km^2$ of small active dunes. The sandsheet terminates on the east where it often laps onto mountain slopes as sand ramps. The alkali flats cover 192 $km^2$ on the western portion of the system. Thin and often dissected sandsheet deposits border the GSD system to the north, west, and south.

### *7.4.1 Sediments*

As with most aeolian systems, sand (64 μm to 2 mm in diameter) is the dominant grain size at GSD. It is compositionally immature and is generally subrounded. A rough estimate is sand makes up over 95% of the sediment in the sand sheet and dunefield. Clay and silt-sized materials (<64 μm diameter) are uncommon, but blown in by the wind, and can collect in standing water and trapped by vegetation on the sand sheet. Streams bring granules (2–4 mm) to cobbles (4–64 mm) to the margins of the dunefield. Boulders (>64 mm) can be found near the mountain front where they are brought in during flash flood events often interbedded with finer grained deposits. Seventy percent of the sand at GSD originates in the San Juan Mountains; the other 30% comes from the Sangre de Cristo Mountains (Madole et al. 2008).

### *7.4.2 Ripple Types*

The aeolian deposits on the dunes can display a range of ripple sizes. The smallest sand ripples are those formed by grains that are 'splashed' from the surface by impacting windblown sand grains; the low velocity of the splashed grains produces sand ripples 1–2 cm in wavelength that have been called reptation ripples (Anderson 1987). When there is an abundance of both saltating sand and coarse particles (>1 mm in diameter), a situation that occurs near the Medano Creek side of the main dunes in the park, ripples caused by the wind can grow to wavelengths of 1 to 3 m;

the size of such megaripples are much larger than those of typical sand ripples (see Planetary Analog section below). The Medano Creek coarse particles are quartz-rich, so megaripples on the south side of the main dune mass are consistently brighter (higher albedo) than the surrounding sand. However, granule ripples on the western half of the sandsheet are dark colored as the granules are from dark volcanic rocks.

### *7.4.3 Dune Types*

Sand dunes can be self-organizing or form in response to wind disturbances created by vegetation or topography (Mulligan and Tchakerian 2002). The GSD aeolian system is extensive enough that controls on dune formation vary, so it is not surprising that many dune types are present. The playas, sand sheet, dunefield, and mountain front include several characteristic dune types (Fig. 7.6).

On the playas and along the ephemeral Medano and Sand Creeks, water-deposited sand is deflated by wind and commonly trapped by vegetation at the margin of the water body. These materials produce lunettes downwind of playas and coppice dunes (nebkhas) along the banks of streams. The lunettes are sand-dominated and lack the silt cores common at other sites (Brunhart-Lupo 2011). Greasewood, four-wing saltbush, and rubber rabbitbrush are shrubs that commonly trap sand in the lunettes. Along the streams, narrow-leaf cottonwood and coyote willow trap sand, forming the coppice dunes.

On the sandsheet, sand mobility is directly related to vegetation cover. Measurements by the NPS indicate that a sparse grass cover (~10%) can reduce sand mobility by 50% relative to barren sand. An area with 50% vegetation cover has very little sand mobility with only trace amounts of the finest fraction of sand moving. Where sand is mobilized in quantities sufficient to form dunes, the results tends to be parabolic dunes with vegetation-stabilized arms. Some of the parabolic dune arms extend upwind for several kilometers, and they can be wide enough (400 m between arms) so that transverse dune forms can develop on the leading edge of the parabolic dune. Deflation on the backside of the parabolic dune is believed to serve as the sand supply for the parabolic dune, allowing it to build extensive arms. The sides of the parabolic dune are initially colonized by Indian rice grass and blowout grass followed by rubber rabbitbrush. The majority of the sandsheet is stabilized by vegetation but the hilly surfaces of the sandsheet are former dunes, classified as stabilized dunes.

Sand Creek can be a significant enough supplier of sand so that trains of barchans and transverse dunes develop along its course through the sandsheet. Blowouts are common erosional features on the sandsheet; they tend to form on elevated areas such as the arms of parabolic dunes.

The dunefield is an area with multidirectional winds where sand accumulates and has time to build large dune forms; the resulting composite feature can be classified as a draa (Wilson 1972). Within the draa are areas of sand accumulation where

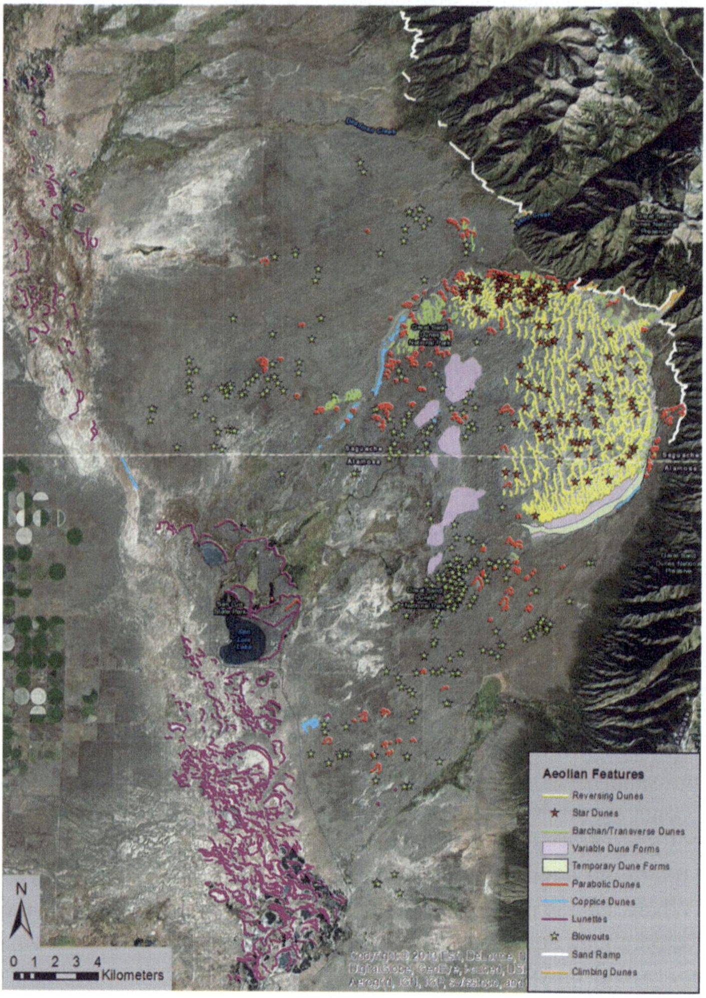


**Fig. 7.6** Dunes Types at Great Sand Dunes. Lunettes are common around playas. Parabolic dunes and blowouts are common in the sandsheet. Reversing and star dunes define the dunefield. Barchan and transverse dunes are common around the perimeter of the dunefield. Coppice dunes form adjacent to stream channels. Small dunes whose form can change types have been mapped as areas of variable dune forms. Intermittent steam channels allow seasonal formation of protodunes and have been mapped as areas with temporary dune forms. Sand ramps and climbing dunes are found on the lower slopes of the Sangre de Cristo Mountains

dunes are superimposed on larger dune forms, creating complex dunes or megadunes. In Fig. 7.7 individual megadunes are encircled by white dashed lines. On the southeast is the Medano Creek megadune, and the southwest has the Star Dune megadune. The east-central portion is the Castle Creek megadune, and the Cold Creek megadune is in the north-central portion of the dune field. The northwest arm

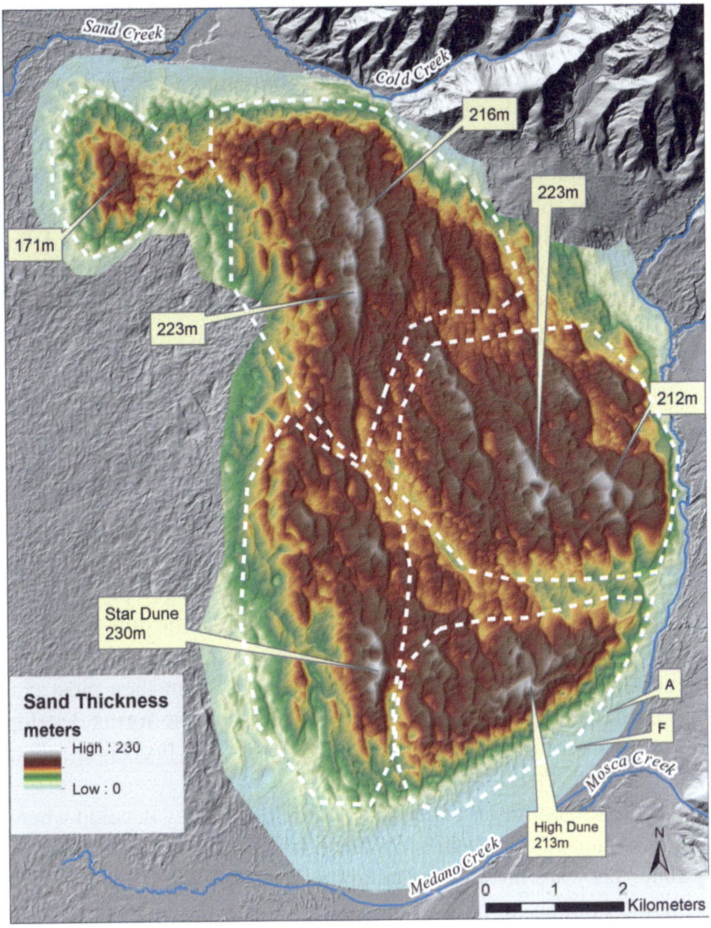


**Fig. 7.7** Sand thickness of the Great Sand Dunes dunefield. Created from elevation data from the 2011 USGS lidar survey of the San Luis Valley and a valley floor surface that was projected under the dunes. Sand thickness is the difference in elevation of the lidar survey and the projected valley floor. Selected dune heights are labeled. Megadunes are encircled by white dashed lines. 'A' and 'F' indicate the monitored dunes shown in Figs. 7.21 and 7.22, respectively

of the dunefield is the Sand Creek megadune, commonly referred to by GRSA interpretation staff as the Sand Creek star dune complex.

Figure 7.7 shows sand thickness in the dunefield as calculated by the NPS. A dune base was projected under the dunefield that matches the gradient of Sand and Medano Creeks with slopes generally parallel to the mountain front. The base also projects to the bedrock along the mountain front; it approximates what the slope of the valley floor would be if the dunefield was removed. Aeolian sand does extend below the projected base. In 1992 the NPS drilled into the valley floor at the base of the Star Dune. Cores were collected to a depth of 30 m and the entire core consisted of aeolian sand.

Measuring dune height can be difficult on megadunes. The summit is apparent, but identifying the base can be challenging on such a complex form. At GSD the dune height is defined as the difference between the summit elevation and the elevation of the projected base beneath it, which correlates to height of the dune above the valley floor. The tallest dune at GSD has the common name 'Star Dune' and based on the 2011 lidar data, is 230 m tall. The 'High Dune' is the highest dune near the dunes parking lot and it is 213 m tall. The full relief of those dunes is easy to see since they slope directly down to the stream where the dune base is defined. The large dunes in the central dunefield do not display their full relief as their ridges begin on surfaces already elevated by the buildup of sand on the megadune. The total volume of sand in the dunefield above the projected base is 6.4 billion cubic meters.

The most common dune type within the dunefield is the reversing dune (McKee 1979b), and they develop in areas with a bimodal or complex wind regime. They are ridge-shaped and somewhat symmetrical with each side typically having a slope of 15° to 30°. Near the top of the dune form is a smaller ridge with a slipface. The slipface position can be on either side of the ridge downwind of the most recent dune building wind event. Persistent southwesterly winds are common in the spring and early summer and if energetic enough, can move the ridgetops far enough to the east so that the slipface enlarges and reaches the bottom of the dune form. When this happens the reversing dune morphs into a transverse dune, but northeasterly winds rework them back into reversing dunes, which is the form they most commonly have. Only after extremely windy springs do the transverse forms develop. It has happened just a handful of times since the early 1990s, but Fryberger reports that it was common in the 1970s.

Star dunes are also common within the dunefield, and they develop where reversing dune ridges intersect or where secondary dune forms develop on the ridge of a reversing dune. They are most abundant at the northern end of the dunefield where the wind regime is likely to be most complex. Within the large dune troughs and near the base of the dunefield along the mountain front are migratory dunes such as barchan dunes and transverse dunes. The topography of the mountains and large dune forms may act as barriers, limiting the exposure to winds coming from their direction. Smaller dunes formed along Medano Creek and on the sandsheet can be reworked by large wind events and are mapped in Fig. 7.6 as having variable dune forms. East facing transverse dunes can be blown westward, transitioning to small reversing dunes, then to west facing transverse dunes.

On the braided channel sections of Medano and Sand Creeks is an area of temporary dune forms. Snowmelt runoff in the streams inundates and erodes previously established dune forms, re-working the sand into fluvial deposits. The streams typically dry in mid to late summer and the channel is once again exposed to the wind. Aeolian processes begin creating ripple beds, then protodune ridges, and occasionally small barchaniod ridge dunes. Drought years can leave these channel sections dry for multiple years, allowing the dunes to grow, but development is limited as sand supply gets cut off when lag surfaces develop in the interdunes. If the dunes get large enough they can deflect the stream around them, shifting the stream channel.

In general, though, they are eroded by streamflow, and preservation normally occurs when channel incision creates stream terraces, producing a more permeant shift in the channel.

Topography-related dunes can be found along the mountain front. Sand ramps form where sand migration laps onto the lower slopes of the Sangre de Cristo Mountains. Most sand ramps are currently stabilized by vegetation, but the sand ramp building process is active east of Medano Creek (Fig. 7.8). Lower Medano Creek's channel trends toward the southwest, parallel to southwesterly winds. The channel orientation changes to a north-south trend causing winds blowing up the lower channel to transport sand onto the east bank. When sand deposition is high enough to overcome vegetation, transverse and barchaniod dunes are generated that transition into parabolic dunes as they migrate upslope over vegetated areas. The migration pathway terminates where Castle Creek captures the sand back into the fluvial system. These dunes are called the 'escape dunes' by GRSA interpretation staff because former NPS staff described them as dunes that escaped from the dune-field in the 1950s during a drought. They are more likely spawned after wet periods of high streamflow with high sand transport. With an average migration rate of 10 m/year it would take a dune 600 years to travel from the edge of the creek to the mountain front.

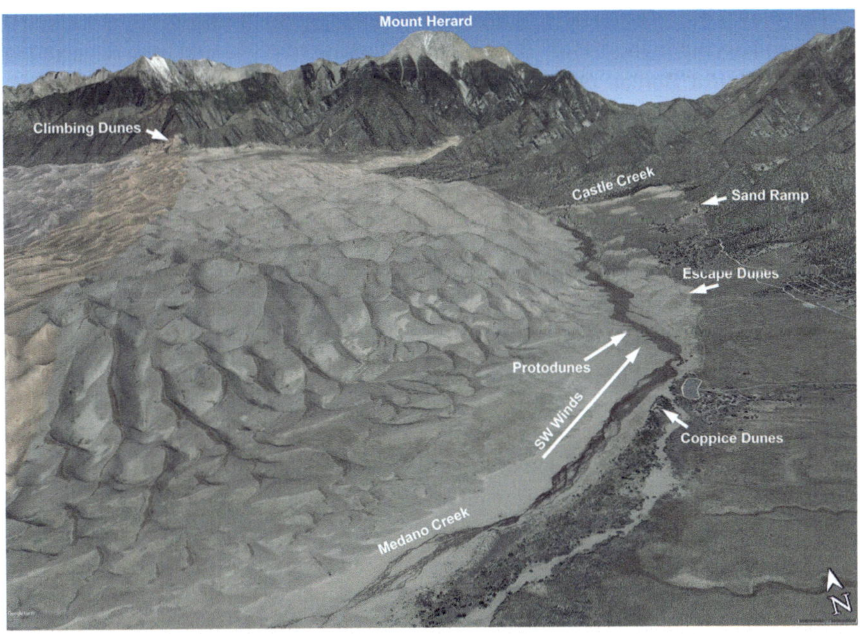


**Fig. 7.8** The Escape Dunes are a set of dunes developed east of Medano Creek. They begin on the bank of Medano Creek where southwest winds blow sand from the channel onto the bank. The dune migrate upslope forming sand ramps. The migration terminates at Castle Creek where sand is intercepted by streamflow. Also shown are coppice dunes, protodunes, and climbing dunes

Climbing dunes can be found on a mountain ridge north of the dunefield (Fig. 7.8). Here sand has migrated up slopes steep enough to form climbing dunes. The sand has reached the ridge summit, but has not gone over in large enough quantities to produce falling dunes on the eastern side of the mountains.

## 7.5 Climate Setting

### *7.5.1 Hydrology*

GSD is a product of climatic conditions. Precipitation produces streams flow that transports sediment from the surrounding mountains to the valley. Streams in the closed basin deposit the sediment in a trough along the lower drainage axis of the closed basin. This trough corresponds to the playa system and other low gradient areas adjacent to them. The aeolian system begins at this trough (Madole et al. 2016) where wind removes silt and clay, and sand bounces along the ground creating dunes and leaving sand deposits. Precipitation and streamflow are cyclical, suggesting repeated cycles of sediment input followed by deflation. Wind data are limited but seem to be consistent in regard to a wind regime that created zones of transport and areas of deposition.

The playas fill episodically, with historic accounts of filling in the 1920s and 1990s, and drying in the 1950s (Bunch 2018). Drying likely would have occurred in the 2000s, but a Bureau of Reclamation project now keeps several playas filled. Figure 7.9 shows the extent of standing water that would exist when the playas of the closed basin are filled to overflow. Aeolian sand surrounds the playa extent (Fig. 7.3) but is much thicker and more extensive on the east side, and the sand dominates the valley floor eastward to the mountain front.

Figure 7.10 shows a precipitation reconstruction for GRSA from 1035 to 1995. Reconstruction data came from tree ring cores collected in the Sangre de Cristo Mountains above GSD, and it is calibrated to local precipitation data from 1941 to 1967 (Grassino-Mayer 1998). Post-1995 precipitation data come from the GSD National Weather Service coop station. Also displayed is a 6 year moving average that highlights the cyclical nature of precipitation. Off scale and that the bottom of the graph is a 20-year average that further simplifies cycles and may be useful to infer when the playas are wet and dry. The 1920s and 1990s are peaks in the moving average; the 1950s is a low. Insert chart A shows how the annual discharge for Saguache Creek relates to precipitation. The yellow, polynomial curve shows a similar cycle with a peak in the 1920s and 1980s, and a low in the 1950s. Saguache Creek drains the San Juan Mountains and not the Sangres, but it was chosen for the comparison as it is the largest stream flowing into the closed basin, with the longest data record. Medano and Sand Creeks show an even more direct correlation to GSD annual precipitation.

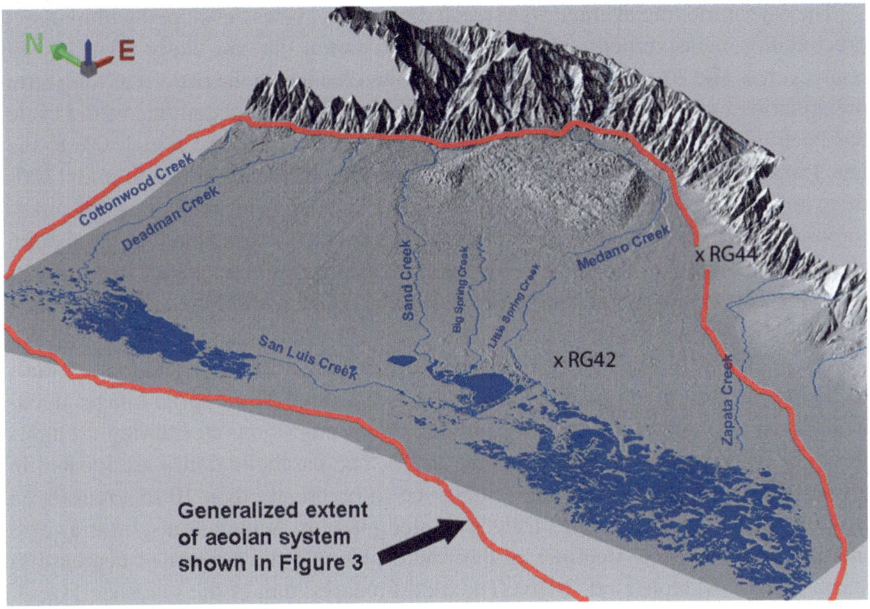


**Fig. 7.9** Hillshade topography showing Great Sand Dunes and filled playa basins. Elevation data comes from 2011 USGS lidar survey. This shows the extent of water when the closed basin fills and the position of aeolian sand deposits relative to them. Also shown is the location of wells RG42 and RG44

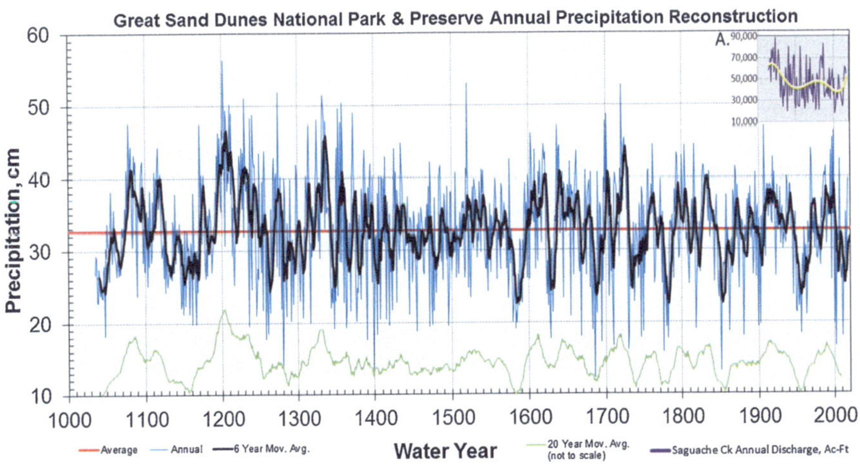


**Fig. 7.10** Great Sand Dunes, Colorado precipitation reconstruction. The blue line is annual data and black line is a 6 year moving average. The green line is a 20 year moving average and not on scale. Insert graph A shows annual discharge in acre-feet for Saguache Creek from 1910 to 2017. The yellow line is a polynomial curve fit to annual discharge

The 6-year moving average was selected based on cycles developed within direct precipitation measurements at the GSD NWS station that are likely related to El Nino cycles. The 20-year moving average shows changes in the pattern of long term precipitation cycles. From the late 1500s to the present, the magnitude of the cycle has been large, with a cycle length averaging 45 years. From the mid-1300s to the late 1500s, the variability in the cycle was less. From the mid-1000s to mid-1300s there appear to be longer cycles lasting around 125 years, with smaller cycles superimposed on them. The early 1200s was a wet period. It is possible that the pattern change in the late 1500s could be related to a change in tree species from limber pine pre-1600 and ponderosa pine post-1600 (Grassino-Mayer 1998; Smiley et al. 1991).

The sandsheet is especially sensitive to climatic variability as vegetation responds to changes in precipitation and groundwater levels. Analysis of aerial images document up to a six times increase in parabolic dunes migration rates during droughts as compared to wet years (Marin et al. 2005). The parabolic dunes are located in area where depth to groundwater is believed to be greater than 10 m, creating an assemblage of vegetation that survives on precipitation. Aeolian transport may also increase in response to declines in the water table and the associated decrease in shrub growth (Madole et al. 2016). The oldest mapped unit of the sandsheet, Qes3, was deposited between 7.2 and 3.5 ka when the water table was lower (Madole et al. 2016). Since then, the near surface water table has supported large shrubs that are most effective at stabilizing sand. The greasewood that grows there is indicative of a high water table.

Subsequent periods of declining water table may have led to widespread activation of the sandsheet, producing map unit Qes2. The deposition of the unit is believed to have occurred from 1.5–1.3 ka. The end date is based on the onset of the Little Ice Age and the return of wet conditions (Madole et al. 2016). Qes2 is characterized by precipitation-dependent vegetation, so perhaps the lowering water tables also indicated a dry period when sandsheet vegetation decreased because of decreased precipitation. There is some active sand movement on the sandsheet mapped as Qes1. Satellite and aerial imagery indicate that the extent of active vs stabilized sand varies on a decadal scale (Janke 2002).

Other clues of past climatic conditions come from the playa area. There are parallel sets of lunettes, some of which have stream incision. OSL dates from the lunettes suggest that 10–8 ka was a wet period when the outermost sets of lunettes formed. From 8–6 ka conditions were wet to semi-arid and the inner set of lunettes developed. From 6–4 ka it was arid to extremely arid, with deflation in surrounding areas. Fluvial dissection of lunettes occurred 4–2 ka when conditions were wet. From 2 ka to modern times it has been semi-arid with ephemeral playas (Brunhart-Lupo 2011). Sediment cores from these playas push back the climate record to 16.5 ka. For most of that record, San Luis Lake is hydrologically closed, but does fill and overflow in phase with other regional playa lake systems. The major control on wet vs dry conditions is the position of the intertropical convergence zone, and the wettest periods correspond to cold phases in the north Pacific. Results from the

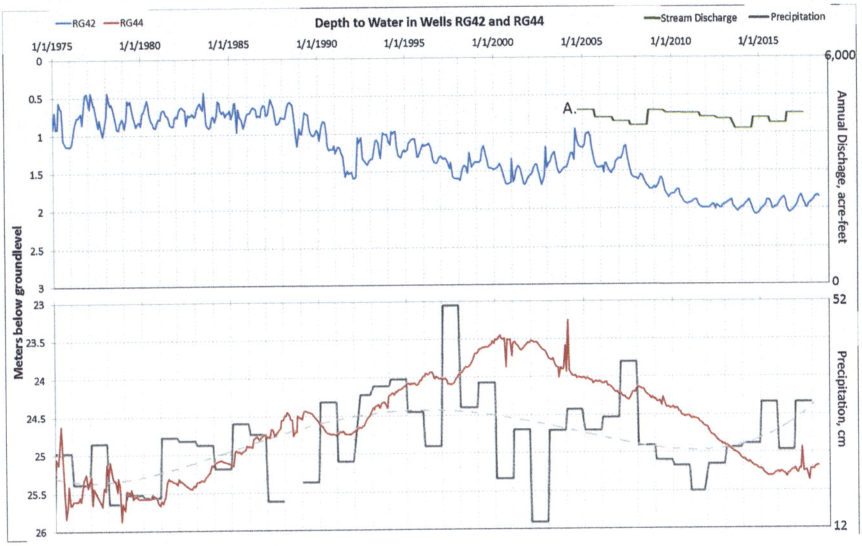


**Fig. 7.11** Water level in wells RG42 and RG44. Location of each well is shown in Fig. 7.9. RG42 is near the terminal end of the hydrologic system and groundwater is within a few meters of ground level. Big Spring Creek gains water from this part of the hydrologic system and its annual discharge is shown as line A. RG44 is near the mountain front the water table there is tens of meters below ground level. Water levels are cyclical and matches precipitation cycles measured at the Great Sand Dunes Visitor Center, gray line. The gray dashed line is a polynomial curve fit to annual precipitation

cores indicate that San Luis Lake was relatively dry 10.5–6.7 ka and wet 6.7–2.6 ka; from 2.6–0 ka the climate has been dry with low variability (Yuan et al. 2013).

Ground water data for the sandsheet is available starting in the mid-1970s for two wells that show how the water table varies based on location (Fig. 7.11). Well RG42 is at the low end of the hydrologic gradient and RG44 is higher, near the mountain front (Fig. 7.9). RG42 is in a setting similar to where unit Qes3 is found, and the near-surface water table promotes shrub growth and evaporation. The water table fluctuates on an annual cycle whose amplitude increases with proximity to the ground surface. There are stepdown-like drops around 1991 and 2008 that may be related to groundwater pumping by the Bureau of Reclamation. The current depth to groundwater is 2 m and the evaporite mineral trona is still actively forming on the ground surface near RG43. The water level does not seem to respond to climatic factors such as annual precipitation; this somewhat steady state groundwater level also explains the constant flow of the two streams that emerge from the sandsheet. Big Spring Creek has a year-round flow of 6.5 cubic feet per second (cfs) and Little Spring Creek's flow is 1.5 cfs. Aquifer residence time for water emerging in Big and Little Spring Creeks is at least 60 years (Rupert and Plummer 2004).

The water table in well RG44 ranges from 23 to 26 m below ground surface. The significant depth to groundwater, high permeability of the sand, and the outflow from mountain streams makes this area an effective groundwater recharge zone.

One large cycle is apparent, and it correlates well with annual precipitation at the nearby GSD NWS station. The 1970s were relatively dry, and then water level rises though the 1990s, the latest wet period, then drops again toward the last drought that on precipitation records was centered in 2011. The groundwater cycle lags behind the precipitation cycle, indicating about a 4-year response time between precipitation and the aquifer at this location.

Wells RG42 and RG44 illustrate the differences in the aquifer in highly variable areas of recharge vs more steady conditions in areas of discharge. Near surface and stable water tables at the lower elevation portions of the sand sheet my lead to long term sand stability as groundwater is readily available for shrubs. Once the water table drops below the root zone, sand mobility increases as vegetation density decreases. Along the mountain front, the position of the water table responds to decadal-level precipitation cycles.

Qes3 may be a key to understanding when new sand enters the GSD aeolian system, as it borders the playas and alkali flats where sand is sourced. It has been stable for thousands of years. This suggests that the multi-decadal filling and drying of the playas may play a limited role in sand supply as younger generations of sand input would have covered Qes3-age deposits. It is possible that significant inputs of new sand into the system occur with larger climatic cycles, such as times of glacial outwash and/or during glacial periods, when the cold temperatures inhibit vegetation growth.

### *7.5.2 Wind*

Wind is the primary agent to mobilize surface materials within the park. The dominant regional wind is from the west-southwest, which explains why the dunes are found along the east margin of the SLV. However, strong winds can blow from the east through the mountain passes, leading to a bimodal seasonal wind regime at the park (Fryberger 1979). There are several wind stations around the park, which can provide a good indication of regional wind conditions and can be correlated to nearby dune migration. The Remote Automated Weather Stations (RAWS; Zachariassen et al. 2003) is a nation-wide system maintained by the U.S. Forest Service and Bureau of Land Management; the GSD RAWS station located about 500 m south of the park visitor center. GSD RAWS wind records from 2004 to the present can be obtained over the internet (https://raws.dri.edu). A USGS Meteorological (MET) station is located on the sandsheet, near Indian Spring. It's a key location as the site is in-line with wind blowing over the dunefield and in and out of Medano Pass. Indian Spring MET data is also available online (https://waterdata.usgs.gov/co/nwis/uv/?site_no=374557105372401&PARAmeter_cd=00045). Figure 7.12 shows the location of each station and the sand drift potential wind regime calculated from 2017 data. Sand-moving wind events are common throughout the year but such events can be absent for week-long stretches during the winter (October–February).

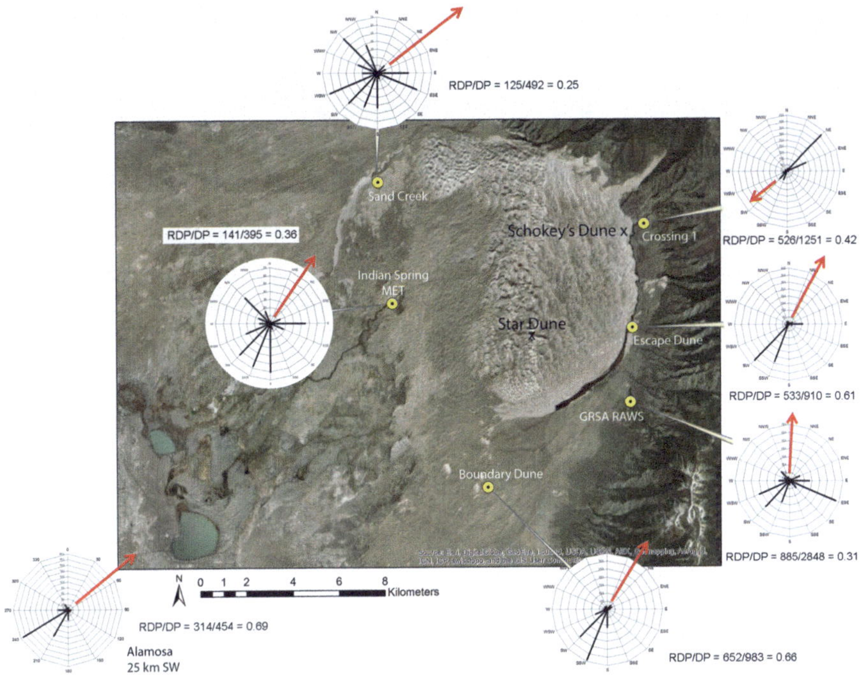


**Fig. 7.12** Great Sand Dunes wind rose diagrams for 2017. Wind data comes from the Alamosa airport, 25 km southwest of map area and weather stations within Great Sand Dunes National Park whose location is marked by a yellow dot. RDP/DP is the ratio of the resultant drift potential (red arrow) and the sum of the drift potentials (black lines)

The National Weather Service (NWS) station at the Alamosa airport is a good source for regional wind data (https://mesonet.agron.iastate.edu/request/download.phtml?network=CO_ASOS). It is 40 km southwest of GSD, has a record of 30 years of data, and is centrally located and on flat topography. In the 1970s, the USGS McKee group calculated an annual wind rose for Alamosa that showed a unimodal wind regime with a resultant pointing to the northeast (Fig. 7.13A). GRSA staff has repeated the calculation with data from 2007 to 2017 and got a similar, unimodal result. There was little variation in the wind rose from year to year and the average wind rose for those years is shown in Fig. 7.13. The total wind energy can vary as shown in Fig. 7.13B. The sand drift potential is unit-less and does not include winds of less than 5 m/s. The windiest year, 2011, has more than double the sand drift potential as the calmest year, 2016. Winds in the central portion of the SLV are unimodal from the southwest.

The various dune types found throughout the GSD aeolian system suggest that the wind regime is variable. GSD's location at the base of the Sangres and adjacent to mountain passes leads to a more complicated wind setting than what would be expected in Alamosa. Any longtime park employee knows there are many days when strong winds blow through the passes and over the dunefield while several

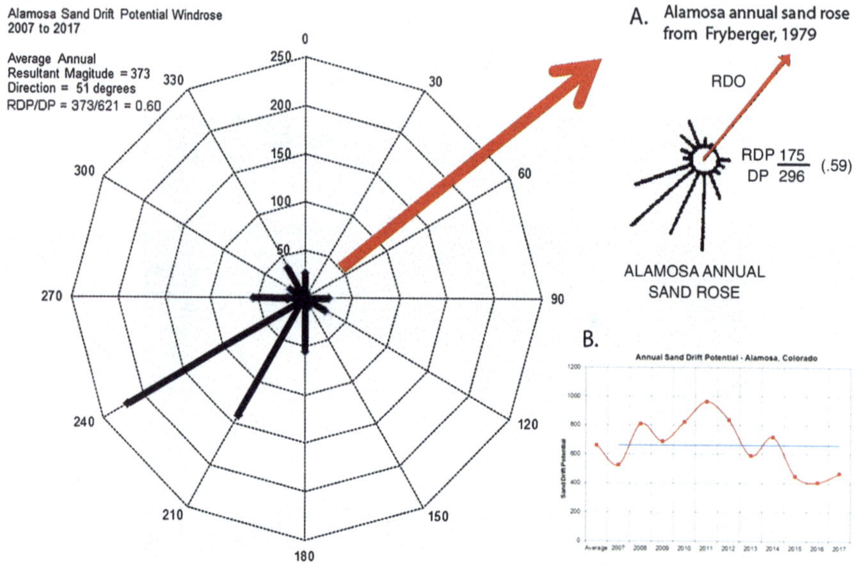


**Fig. 7.13** Alamosa average annual sand drift potential, 2007 to 2017. Wind direction is measured in 10° increments and for sand drift calculations were combined in 30° intervals. Wind regime is unimodal from the southwest which matches the wind regime calculated from 1970s data, insert A. Total sand drift potential for each year from 2007 to 2017 is shown in panel B

kilometers north or south of the dunefield, conditions are calm. Since the dunes are the primary resource at GSDNP&P, it is important that the NPS try to understand dune behavior and how it relates to wind. This effort included the installation of automated weather stations. The GSD RAWS was installed in 2004 and the Indian Spring MET station in 2007. The RAWS measures wind hourly while the MET does so every 15 min. Beginning in 2013, the NPS has installed portable weather stations near sand dunes whose position was monitored. They are based on the HOBO data logger and they collect data at 5 min intervals. The portable stations, while much less expensive, are not as reliable as the RAWS and MET stations, so data is occasionally lost from them.

The primary purpose of the GSD RAWS station is to determine fire weather in the pinon-juniper forest that is found on the lower slopes of the Sangres. The wind regime is bimodal with major wind vectors coming from the southwest and east. The eastern vectors are likely influenced by downslope winds from the Sangres. The resultant is northward at 13° (Fig. 7.14). Total sand drift potential shows some variation, insert A, but not to the same extent as seen at Alamosa.

Complex wind regimes have been measured on the sand sheet at the Indian Spring MET Station and the Sand Creek HOBO station. Each site has wind vectors that come from the northwest, east, and the southwest. The majority of wind comes from the southwesterly directions and the resultant is toward the northeast. 2014 to 2017 Indian Spring data from the MET station has been analyzed by month. Most months remain complex with the resultant trending toward the northeast. What did

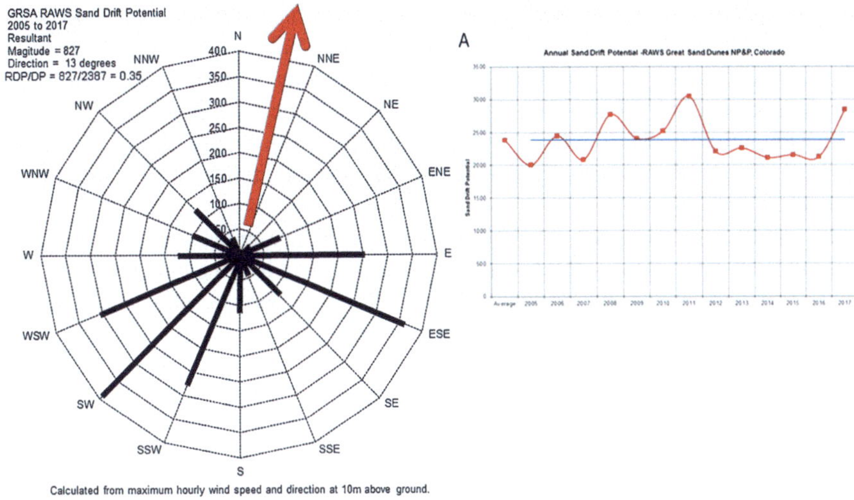


**Fig. 7.14** Great Sand Dunes Remote Automated Weather Station average annual sand drift potential, 2005 to 2017. Total annual sand drift potential is shown in panel A

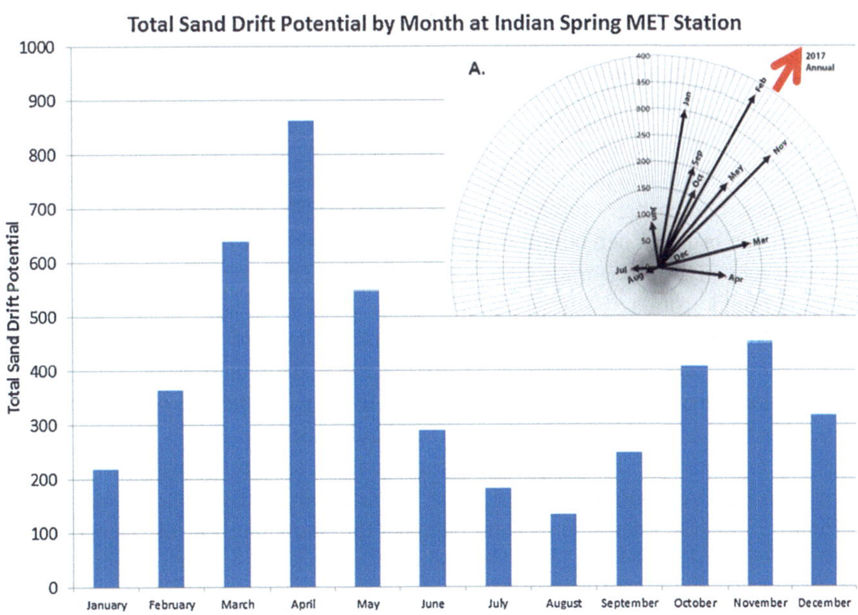


**Fig. 7.15** Average monthly sand drift potential at Indian Spring MET station, 2014 to 2017. Insert A shows the monthly resultants during 2017

become apparent from the monthly analysis is there is a bimodal distribution in wind energy. Figure 7.15 shows the average total sand drift potential for each month. The greatest wind energy is in the spring (March–May) with a peak in April. A lesser peak occurs in November. April has six times more wind energy then the calmest month, August. Figure 7.15A shows the distribution of resultants by month for 2017. The resultants with vectors greater than 250 are months that had unimodal regimes.

Unimodal wind regimes are present at the boundary dune, escape dune, and crossing 1 HOBO stations. The boundary dune site is south of the dunefield near the former national monument boundary. It is south of the corridor affected by winds blowing from Medano Pass and therefore lacks the northeast wind influence. The Escape dune site is at valley floor level and is likely below the elevation where northeasterly winds come into contact with ground level. The crossing 1 site is near the mouth of Medano Pass. It shows a strong unimodal pattern coming from Medano Pass. It is almost a certainty that winds blow into Medano Pass from the southwest making this area bimodal. The site is at valley floor level and the large dunes to the west may cause southwesterly winds to detach from the surface so as to not be registered at the crossing 1 site.

### *7.5.3 Dune Dynamics*

The density of wind data around the GSD allows for direct comparison of wind regime to dune form and dune behavior. Dune position is measured using 2 different data sources. Direct measurement using differential GPS provides sub-meter position of a dune brink. When height is needed, traditional surveys using total stations, and more recently, survey grade GPS have been used. The second method is digitizing dune brinks from aerial imagery. One meter aerial imagery is available via the U.S. Department of Agriculture National Agriculture Imagery Program (NAIP) for the following years; 2005, 2009, 2011, 2013, 2015, 2017, and 2019. When GPS measurements are made near the date of the NAIP imagery, they do overlay closely, suggesting that the imagery geo-referencing is quite accurate.

A few examples of GPS mapping of dunes will be shown starting with the Star Dune whose nearest weather station is the Indian Spring MET station 6 km to the E. Figure 7.16 shows periodic measurements of the Star Dune's brink from 1996 to 2017. The first year is a bold green line and the last is a bold red line to highlight the net change in position. White arrows indicate net change in position at selected sites. Insert A shows the wind rose and resultant from Indian Spring for 2006 to 2017. Insert B shows the migration of the dune ridges near the summit. In 21 years there has been a net migration of 64 m (3 m/year) and a total migration of 109 m (5.2 m/year). The net migration direction has a similar trend to the resultant at Indian Spring. The background image is from NAIP in 2011. The 2011–09 measurement was taken 2 days after the image and it is shown as a bold purple line for comparison to the background.

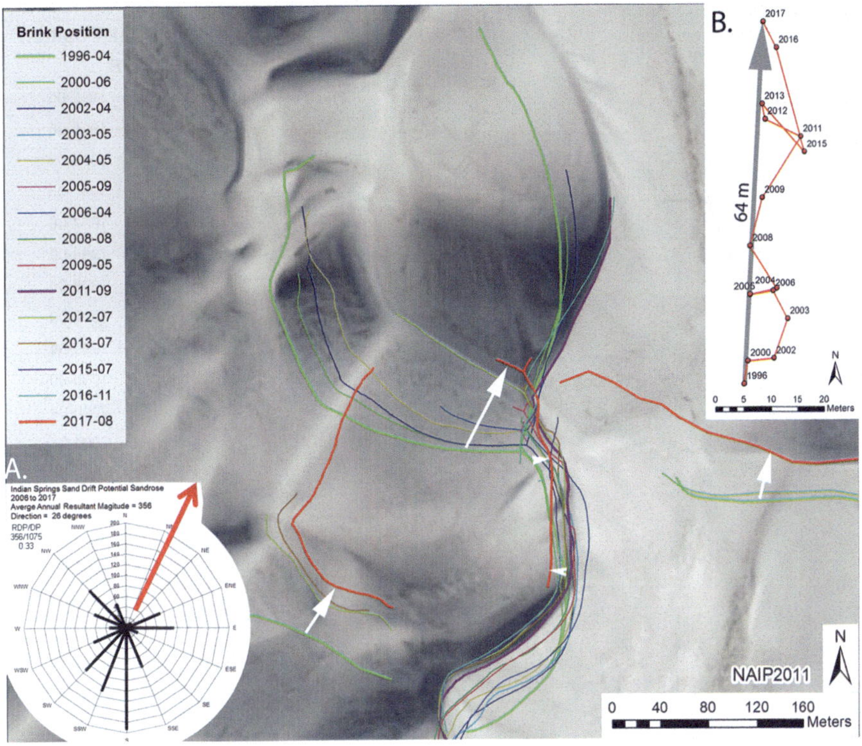


**Fig. 7.16** Position of the Star Dune, 1996 to 2017. Dune brinks were mapped by differential GPS. White arrows show net movement of brinks from 1996 to 2017. Insert A shows the sand rose for Indian Springs MET station which is 6 km E of the Star Dune and in line of sight. Insert B shows the position and net migration of the intersection of dune ridges at the top of the Star Dune

The main Star Dune ridge trends N-S and it shows little net change in position. The secondary, ridges are more perpendicular to the resultant and have migrated more. The migration along the crest is northward and that may result in elongation of the ridge, as is common on linear dunes. Another interesting trend is the Star Dune tends to lose a couple of meters in height when it shifts eastward. It gains the elevation back when it shifts westward. This is likely due to a difference in slope on either side of the Star Dune. Its NE facing slope is steeper (28° to 32°) than the SW slope (18° to 25°). The steeper north slope drops faster creating more accommodation space for sand when the dune shifts in that direction. Shifts back onto the gentler slopes, where there is less space, causes the sand to stack higher. Star dunes tend to grow vertically over time (McKee 1979b), but new growth requires the input of more sand. The process of getting more sand to the ridge top may be shown on Fig. 7.16. A small dune form has been mapped on the SW slope of the Star Dune and it is migrating upslope at a rate of 2.2 m/year. If the N-S ridge remains in place, then the small dune should reach the ridgetop in about 100 years, adding more sand and creating growth.

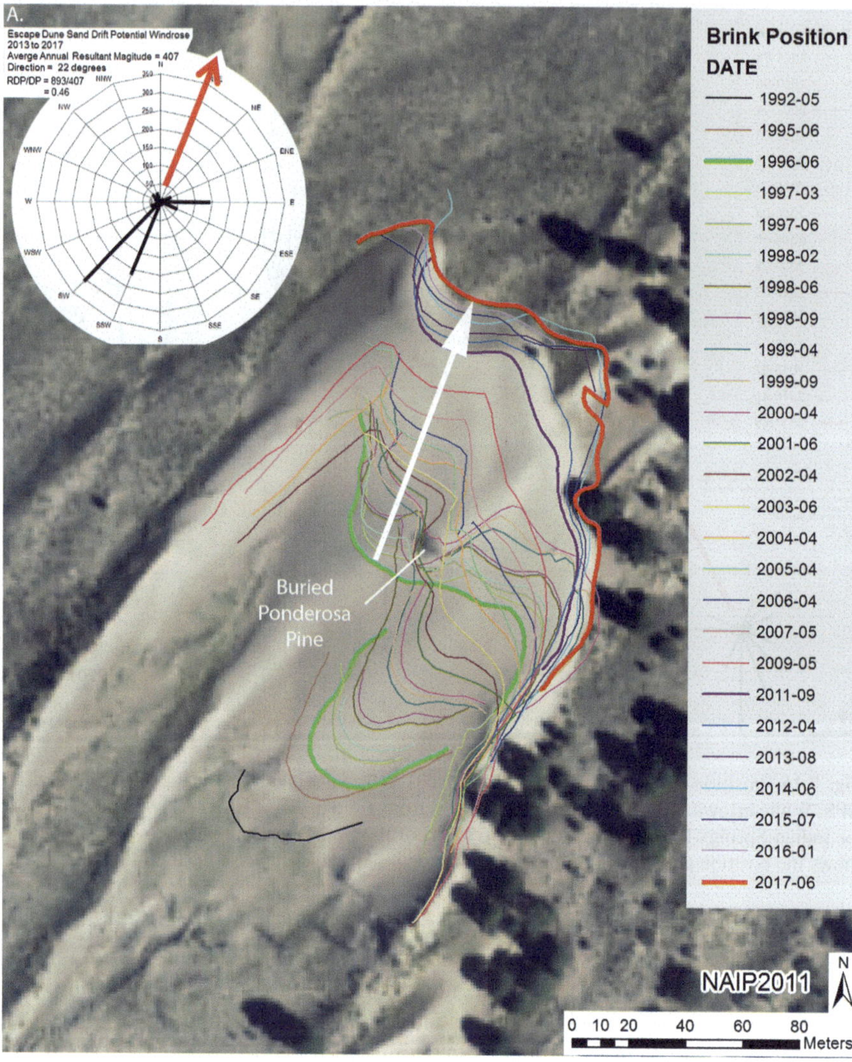


**Fig. 7.17** Position of the Escape Dunes, 1992 to 2017. Dune brinks were mapped using differential GPS. Insert A shows the wind rose for the escape dunes HOBO station located just off the map area west of the dune

The Escape Dune has also been mapped with GPS beginning in 1992 (Fig. 7.17). Overall it has a parabolic from, but barchan forms have developed on its backside and on the leading edge. The Escape dune HOBO station is 150 m SW of the dune brink. Figure 7.17A shows the 2013 to 2017 wind rose and resultant (data used are about 66% complete). The first mapping of the full extent of the brink occurred in June 1996 and is shown by the bold green line. The most recent was June 2017, the bold red line. The bold purple line is from September 12, 2011, 3 days before the underlying NAIP 2011 image was acquired. From 1996 to 2017 there was 97 m of linear migration, and both net and total migration average 4.6 m/year. The migration matches the wind rose resultant, which is mostly unimodal, bordering on bimodal with a downslope-wind-generated vector coming from the east. The progression of measurements shows modification of the dune form. The barchan present in 1992 (black line) migrated to the leading edge of the larger dune form and merged by

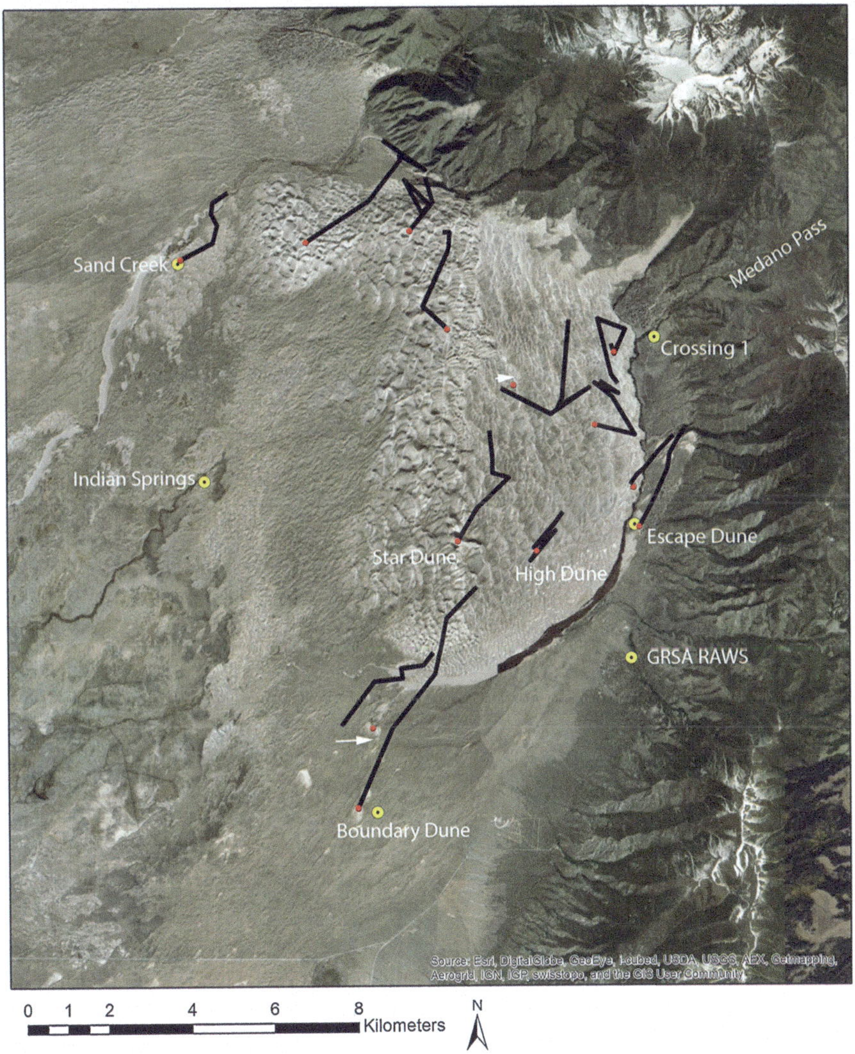


**Fig. 7.18** Dune migration patterns for select dunes at Great Sand Dunes, 2005 to 2017. Dune brink position was digitized from aerial imagery taken in 2005, 2009, 2011, 2013, 2015, and 2017. Dune location in 2005 is marked by a red dot. Length of migration is exaggerated 50× so that it is easily visible at map scale. Yellow dots are locations of weather stations

2004. There was also a large ponderosa pine tree that was ahead of the dune in 1996. It was buried by the dune and it caused a warp in the brink line until 2007.

Medano Creek is just west of this site and along the far side of the creek are small dunes that also show a unimodal form, mostly as barchanoid ridges. Beyond these dunes is the Medano Creek megadune and as the barchaniod ridges rise onto the megadune form, they transition to reversing dunes. This suggests that wind regime changes to bimodal with elevation at this location.

The NAIP imagery allows for tracking of dunes throughout the system on the odd years that the imagery is acquired. Figure 7.18 show selected dunes, (red dots) whose position was digitized in 2005, 2009, 2011, 2013, 2015, and 2017. It was difficult to find suitable dunes for tracking on the sand sheet as they tend to lack distinct, easily identifiable features and it is also common for an active dune to blowout in the listed time interval.

The dunes that were mapped had a distinct crest and/or a ridge intersection point that was tracked. The black lines show each dune's migration starting at the red dot (the dunes' location in 2005). The scale of the migration is 50 times greater than indicated by the background image, and a few paths are slightly offset so as to not overlap onto the results of a nearby dune. Some distinct patterns emerge. The dunes near the boundary dune and escape dune stations show uniform migration toward the NE. In the dunefield, migration becomes more complex with proximity to the mountain front.

This dune behavior, dune type, and wind regime data from the weather stations suggest that there are zones with distinct wind regimes at GSD (Fig. 7.19). Unimodal winds from the SW come from the SLV and that pattern persists south of the dunefield and at low elevations along the mountain front. The wind regime changes to complex in line with Medano and Music Passes as NE winds blow through the passes and out on the valley floor. The SW winds are a greater influence on the sand sheet and on the west side of the dunefield. Near the mountain front, the SW and NE wind seem to be more balanced. If this pattern persists, over time the extent of the draa may contract to its eastern half with larger dune forms holding the sand in a smaller area. North and northwest of the dunefield the wind remains complex, with net migration toward the NE. The parabolic dunes found east of the main dunes illustrate that they trend toward the NE, but their arms are irregular.

## 7.6 Relevance as Planetary Analogs

The setting for Great Sand Dunes makes it particularly relevant as an analog for the many sand dune fields that have been documented on Mars. Almost 2000 dune fields, each with an area greater than 10 km$^2$, have been cataloged on Mars, the vast majority of which are located on the floor of impact craters (Hayward et al. 2007, 2010, 2014). The largest accumulation of dunes in the park is adjacent to the western flank of the Sangre de Cristo Mountains, where the mountain topography produces a natural barrier to the eastern progression of the sand across the San Luis

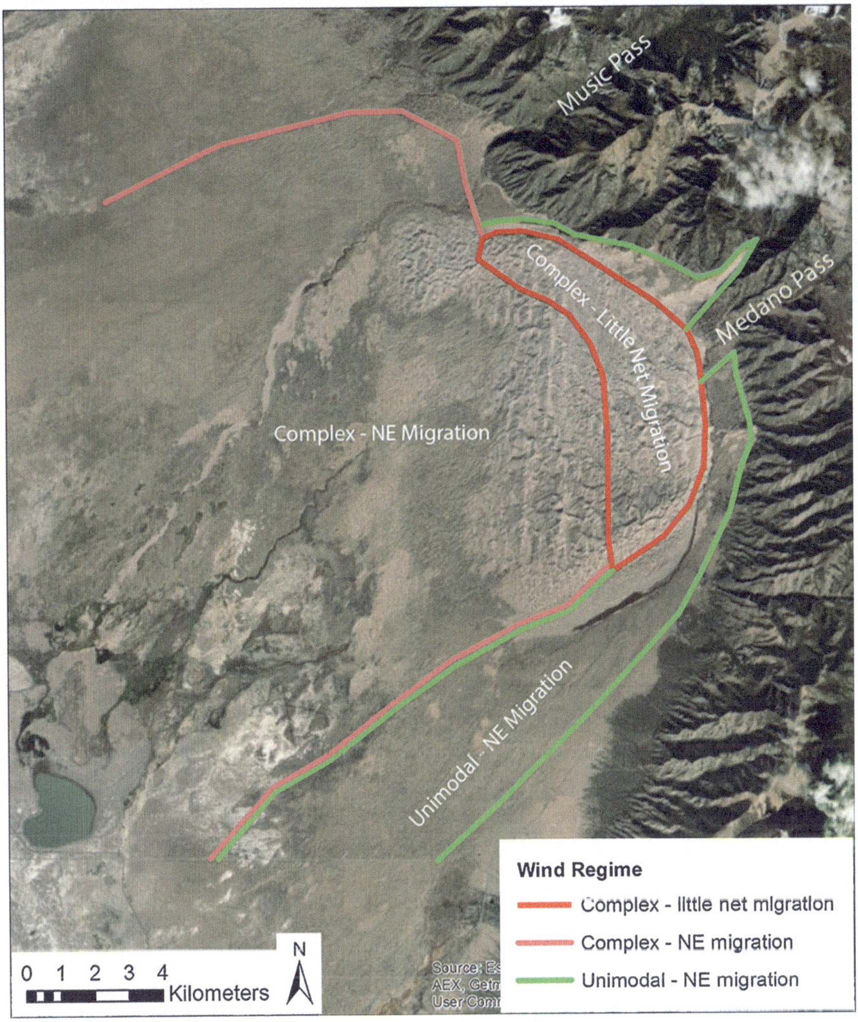


**Fig. 7.19** Wind regime areas at GRSA. Unimodal Wind regimes exist south of the dunefield and along the mountain front east and north of the dunefield. Complex wind regimes are developed southwest of Music and Medano Passes

Valley, much like dune fields on Mars where the sand has become confined by crater topography. Dune type within the park is dependent upon the local wind regime, which can be quite variable (e.g., see Fig. 1 of Marin et al. 2005), but the confinement of the sand by the regional topography is certainly relevant to many places on Mars. Of course, Mars has no vegetation to anchor the arms of parabolic dunes like those that are prevalent in the San Luis Valley upwind of the main dune field.

Transverse and reversing dunes are common along the southern edge of the main dunefield (McKee 1979a), the area most often visited by the public. The public

**Fig. 7.20** (**a**) Differential Global Positioning System (DGPS) survey perpendicular to the crest of Dune A. This survey-grade Trimble R8 system is capable of measuring positions in three dimensions to <2 cm horizontal and <4 cm vertical. JRZ photo, 7/30/09. (**b**) Trimble GeoXH survey of the crest of Dune F. With subscription post-processing using fixed base stations in the area, positions are measured in three dimension to <15 cm horizontal and vertical. The GeoXH system is far easier to use among the dunes than is the R8 system, showing crest position changes to much better than one meter. (JRZ photo, 6/8/16)

accessibility to this part of the dunefield facilitates repeated visits to monitor the movement of some of dunes, information that should prove useful for evaluating similar dunes seen on Mars where confined by topography. Initial efforts focused on obtaining precise topographic profiles across some of the dunes, data that documented the very symmetric topography evident on reversing dunes (Fig. 7.20). The profile information confirmed that many of the symmetric reversing dunes still migrated several meters per year toward the east; this result held even though a reversing dune is the product of a bimodal wind regime (McKee 1979a). We changed the monitoring effort to document repeatedly the crest location on several dunes (Figs. 7.20, 7.21 and 7.22), something more easily accomplished using a handheld GPS receiver (Fig. 7.22) capable of making differential corrections using information obtained from several permanent base stations throughout the area.

Documented movement of the crests of reversing dunes at Great Sand Dunes shows the general eastward movement of the dune forms, but in a way that is more complex than what was originally thought. Crest migration direction and magnitude is inconsistent for individual dunes (Figs. 7.21 and 7.22), and in some cases is highly variable over length scales of tens of meters. One dune showed a consistent extension up the steep slope of the main dune field even while undergoing migration to the east. The complex behavior of the crest migrations indicates that wind conditions at a micro scale may be more important than can be inferred from regional wind data, a situation that is also likely be true for dunes on other planets. Great Sand Dunes provides an easily accessible location to test the quantities and types of wind data records that may be needed in order to understand the complex crest migration behavior of individual dunes, information that would be very helpful in assessing how best to interpret the motions of individual dunes on other planets.

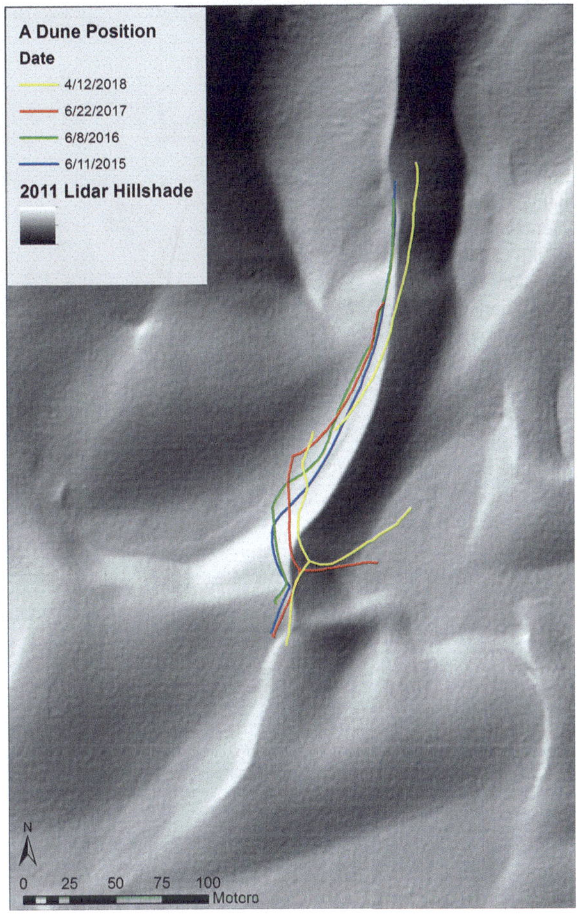


**Fig. 7.21** Trimble GeoXH surveys of the crest and/or brink of Dune A (see Fig. 7.20a). The reversing dune arm extending north from the summit of Dune A shows relatively little E-W change over four years whereas the dune summit (an incipient star dune) shows consistent NNE movement of several meters per year. Survey lines are overlaid on a georeferenced LIDAR hillshade image from 2011, indicating summit movement consistent with the survey results, along with confirmation that the N arm shows limited E-W movement (consistent with the reversing dune profile illustrated in Fig. 7.20a) but considerable N movement, which is up the steep face of the southern edge of the main dune mass

Some unique aeolian bedforms that are common at Great Sand Dunes are megaripples that are covered with particles considerably coarser than medium sand. Sharp (1963) described such features in the Mojave Desert, referring to them as granule ripples because many of the coarse particles were in the granule size range of 2–4 mm in diameter. The coarse-grained megaripples at Great Sand Dunes (Fig. 7.23) have a surface cover of angular to subrounded, mostly quartz grains in the coarse sand to granule size range (1 to >2 mm in diameter) (Fig. 7.24). The coarse particles are too large to be moved in saltation, and are instead induced to roll or bounce across the surface through the impact of saltating sand grains through a process called 'creep' (Bagnold 1941; Sharp 1963; Greeley and Iversen 1985). The coarse particles originate in the mountains and are transported from there in Medano Creek, where they become remobilized by impacting sand grains when the creek surface dries (even if water continues moving beneath the surface). The granites in the mountains provide abundant quartz particles (Fig. 7.23), which imparts a relatively high albedo (bright, as compared to nearby sand; Fig. 7.22) to accumulations

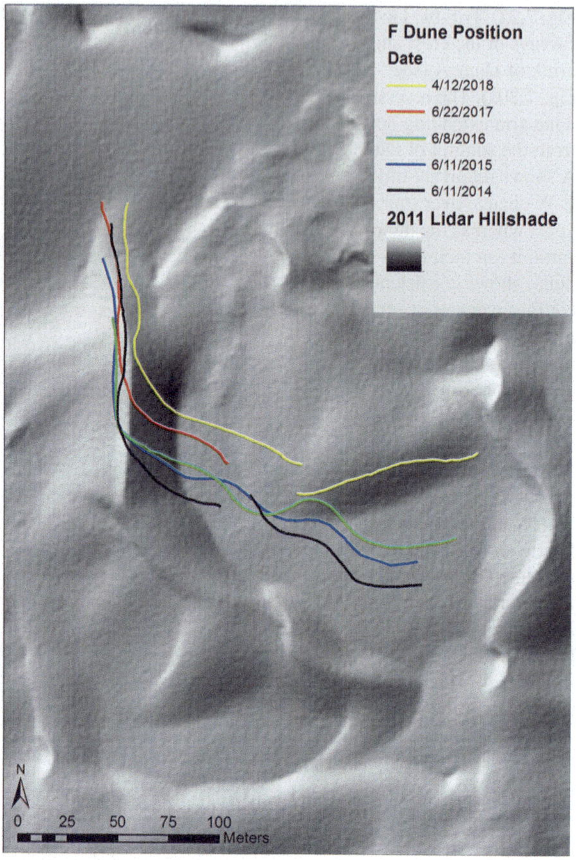


**Fig. 7.22** Trimble GeoXH surveys of the crest and/or brink of Dune A (see Fig. 7.20b). As with Dune A (Fig. 7.21), the reversing dune arm extending north shows relatively little E-W change over 5 years whereas the E-W-oriented portion of the crest shows consistent movement of several meters per year, consistent with the motion of the summit of Dune A. Survey lines are overlaid on a georeferenced LIDAR hillshade image from 2011, indicating overall movement consistent with the survey results

**Fig. 7.23** Dark sand ripples (background) merging into bright granule ripples (foreground). (JRZ photo, 4/11/18)

**Fig. 7.24** Large subrounded grains (ranging from coarse sand to granule size) on the surface of a granule ripple. (JRZ photo, 9/15/05)

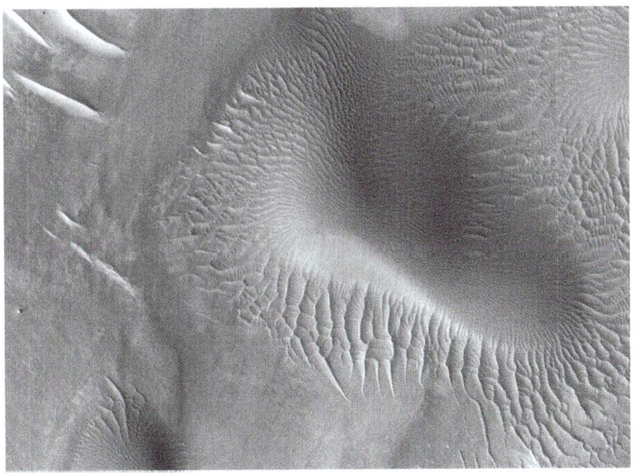

**Fig. 7.25** Portion of HiRISE frame PSP_002721 showing ripples and megaripples on Mars. Large sand ripples (on the dark sand dunes, center and lower left) merge into megaripples. Some megaripples merge with small TARs (upper left side of central sand dune); isolated individual TARs are at upper left. Scene width is 460 m

of these particles, even from a considerable distance (Fig. 7.8). Measurements of the movement of crests of megaripples at Great Sand Dunes lead to the conclusion that 25-cm-tall megaripples on Mars (typical of the size of megaripples traversed by the Opportunity rover; Sullivan et al. 2005) could need up to 2000 Earth years to move 1 cm under wind conditions as observed by Viking (Zimbelman et al. 2009).

Megaripples are in fact quite common on Mars (Wilson 2004; Balme et al. 2008) (Fig. 7.25), where their symmetric profiles (Zimbelman 2010) are distinct from the uneven profiles of typical sand ripples (Zimbelman et al. 2012). The term 'Transverse Aeolian Ridge' (TAR) has been applied to such features on Mars, to preserve them having formed as either ripples or dunes; TARs tend to be brighter than their surroundings, as are the megaripples at the park, and so far there is no evidence of TAR movement under current martian conditions (Balme et al. 2008; Berman et al. 2011; Bridges et al. 2011). Observations by rovers on Mars show coarse particles (Fig. 7.23) on the surface of features at the small end of the size range of TARs

**Fig. 7.26** Dingo Gap bedform on Mars before it was crossed by the Curiosity rover. The feature has been interpreted to be a small TAR (Zimbelman and Foroutan 2017, in review). Navcam image mosaic, looking south

(Fig. 7.26), leading to the proposal that small TARs (those with heights <1 m) can now be called coarse-grained megaripples (Zimbelman and Foroutan 2017). Great Sand Dunes is one of the best places in the world to conduct field investigations of coarse-grained megaripples.

Great Sand Dunes provides unique conditions to investigate the interaction between fluvial and aeolian processes (as along the banks of Medano Creek) or how snow might interact with aeolian sand. The high altitude of the park means that snow is frequent in the winter and early spring, allowing the chance to observe how snow can affect sand deposits (Fig. 7.26). The cold winters at Great Sand Dunes have been used to study the dunes as analogs for sand movement, or the lack thereof, on the cold surface of Mars (e.g., Dinwiddie et al. 2012). The park has also been the site for field tests of both NASA and student rovers for possible application to rovers on Mars (Valdez and Zimbelman 2016).

## 7.7 Conclusions

The wind did it! (Bunch 2018), along with rifting, stream flow, vegetation, and evaporation.

The setting for Great Sand Dunes is created by the Rio Grande Rift. Subsidence of the San Luis Basin creates a depositional basin that can supply enough sediment to feed aeolian activity. The aeolian processes begin where rift related subsidence is greatest and the internally drained Closed Basin has developed. The water that periodically drains to the sump of the closed basin ends up in playas which dry leaving sand and silt exposed to the wind. The fine sediment can be blown out of the basin leaving the sand that sources aeolian deposits found downwind (Fryberger 1990). Winds from the southwest blow across the San Luis Valley floor transporting sand to the Sangre de Cristo mountain front where the wind regime becomes complex and vertically growing dune types develop. Stream flow around the margin of the dunefield, modifying the edges, first by eroding sand and truncating the dunefield, then by depositing sand upwind of the dunefield, creating an ample sand supply. The result is large dune forms developed on the edge of the dunefield that highlights Great Sand Dunes National Park and Preserve's claim to fame; big dunes. Much of the aeolian surface is vegetated which hinders sand movement and active dune development. The sand can also be stabilized by evaporate minerals in areas around the playas and where deflation has lowered the ground surface to the capillary fringe of the water table, inducing evaporation.

These processes produce distinct areas within the Great Sand Dunes aeolian system. At the lowest end topographically, evaporation is the dominant process creating salt encrusted playas and alkali flats. As the land surface rises above the capillary fringe, the sand remains loose and vegetation growth becomes the important process. This broad area lacks active dunes and has been classified as a sandsheet. Vegetation cover on the sandsheet can quickly change and is a control on aeolian activity. Near the mountain front, dune activity overcomes vegetation and a dunefield of draa size has developed.

Water plays an important role as a hydrologic system is superimposed on the aeolian system. Aeolian transport is generally from the southwest going up the topographic gradient from the lowest portion of the valley floor to the mountain front. Fluvial sand transport moves in the opposite direction. Equilibrium between the two transport mediums can be seen by the deflection of streams flowing around the dunefield and when they enter the sandsheet. Water also adds to the biological diversity of the park and to its recreational value.

Water is a resource that is subject to anthropogenic threats and perhaps what needs to be managed most closely at Great Sand Dunes National Park and Preserve. It is a valuable commodity in the arid and semiarid western U.S. and the San Luis Valley has seen efforts to extract this valuable resource. The agency that manages water resources in Colorado is the Colorado Division of Water Resources and they have determined that in the San Luis Valley water use has historically been greater than water supply so they are in the process of trying to reach sustainable water use levels. Great Sand Dunes has played a role in that effort by objecting to water development efforts and showing that such projects would impact water rights held by the then national monument. This threat forced the National Park Service to more actively manage natural resources and to collect data that helps understand the processes important for dune formation. In doing so, Great Sand Dunes was seen as more than just the dunefield. It was understood to be part of a larger system where the dunefield was just the 'tip of the iceberg'. That knowledge led to a boundary expansion in 2000 that allowed the National Park Service to manage most of that larger system. The boundary expansion permitted lands that were once the site of water speculation to be incorporated into federally managed lands, removing the threat from so near the dunes. The increased knowledge of the aeolian system has also improved the story that the National Park Service passes on to the visiting public. Their most common question is, 'How did these dunes get here?' The goal of the

Park's interpretation staff is to help visitors understand what they are looking at and why Great Sand Dunes is special.

Great Sand Dunes is valuable to more than just the visiting public. The variety of dune types present, the elevation of the park, the large volume of the sand accumulations, and their proximity to the adjacent mountains all can serve as analogs to sand dunes on Mars. The sediment contribution from the mountain streams that flow near the dunefield bring in particles in the coarse sand to granule size range, along with abundant sand; the coarse particles can be mobilized by the wind (through creep induced by saltating sand) to form megaripples up to 3 m in wavelength and tens of centimeters in height, much larger than typical sand ripples but comparable to some large aeolian ripple-like features on Mars. Study of both the dunes and the megaripples at Great Sand Dunes makes this a particularly valuable location for improving our understanding of how wind-related features might develop on Mars.

## References

Anderson RS (1987) A theoretical model for aeolian impact ripples. Sedimentology 34(5):943–956

Andrew RL, Ostevick KL, Ebert DP, Rieseberb LH (2012) Adaptation with gene flow across landscape in a dune sunflower. Mol Ecol 21:20178–22091

Bagnold RA (1941) The Physics of Blown Sand and Desert Dunes. Chapman and Hall, London, 265 p

Balme M, Berman DC, Bourke MC, Zimbelman JR (2008) Transverse Aeolian Ridges (TARs) on Mars. Geomorphology 101(4):703–720

Berglund HT, Sheehan AF, Murray MH, Roy M, Lowry AR, Nerem RS, Blume F (2012) Distributed deformation across the Rio Grande Rift, Great Plains, and Colorado Plateau. Geology 40(1):23–26. https://doi.org/10.1130/G32418.1. Boulder: Geological Society of America

Berman DC, Balme MR, Rafkin SCR, Zimbelman JR (2011) Transverse Aeolian Ridges (TARs) on Mars II: Distributions, orientations, and ages. Icarus 213(1):116–130

Bridges NT, Bourke MC, Geissler PE, Banks ME, Colon C, Diniega S, Golombek MP, Hansen CJ, Mattson S, McEwen AS, Mellon MT, Stantzos N, Thomson BJ (2011) Planet-wide sand motion on Mars. Geology 40(1):31–34

Brunhart Lupo ME (2011) Paleoclimate and Paleoenvironment conditions of the great sand dunes National Park and Preserve – 10,000 BP to present. Dissertation, Colorado School of Mines

Bunch FB (2018) Oral communication with Fred Bunch, chief of resource Management at Great Sand Dunes National Park & Preserve, Colorado

Carter JC (1978) Pike in Colorado, the explorations of Zebulon Montgomery pike in the San Luis Valley of Colorado. The Old Army Press, Fort Collins, p 18

Dickenson WR (1981) Plate tectonic evolution of the Southern Cordillera. In: Dickenson WR, Payne WD (eds) Relations of tectonics to ore deposits in the Southern Cordillera, Digest 14. Arizona Geological Society, Tucson, pp 113–135

Dinwiddie CL, Michaels TI, Hooper DM, Stillman DE (2012) Environmental conditions and meteorological context for modification of the Great Kobuk Sand Dunes, Northwestern Alaska. In: Third international Planetary Dunes workshop: remote sensing and data analysis of planetary dunes, LPI contribution no. 1673. Lunar and Planetary Institute, Houston, pp 36–37

Drenth BJ, Turner KJ, Thompson RA, Grauch VJS, Cosca MA, Lee JP (2011) Geophysical expression of elements of the Rio Grande rift in the Northeast Tusas Mountains-preliminary interpretation. In: Koning DJ, Karlstrom KE, Kelley SA, Lueth VW, Aby SB (eds) Geology of the Tusas Mountains and Ojo Caliente. New Mexico Geological Society, Socorro, pp 165–176

Drenth BJ, Grauch VJS, Thompson RA, Rodriguez BD, Bauer PW, Turner KJ (2016) Preliminary 3D model of the San Luis Basin, northern Rio Grande rift, Colorado and New Mexico, Abstracts with program 48(7). Geological Society of America, Boulder. https://doi.org/10.1130/abs/2016AM-281204

Fryberger SG (1979) Dune forms and wind regime. In: McKee ED (ed) A study of global sand seas, U. S. Geol. Survey prof. Paper 1052. U. S. Gov. Printing Office, Washington, DC, pp 137–169

Fryberger SG (1990) Chapter 6: Depositional models for modern eolian sand seas. In: Fryberger SG, Krystinik LF, Schenk CJ (eds) Modern and ancient eolian deposits: hydrocarbon exploration and production. Rocky Mountain Section Society of Economic Paleontologists and Mineralists, Denver

Fryberger SG, Ahlbrandt TS, Andrews SA (1979) Origin, sedimentary features, and significance of low-angle eolian "sand sheet" deposits, great sand dunes National Monument and vicinity, Colorado. J Sediment Petrol 49(3):0733–0176

Geary MM (2012) Sea of sand: a history of Great Sand Dunes National Park and Preserve. University of Oklahoma Press, Norman, pp 126–139

Gonzales DA, Karlstrom KE (2011) A legacy or mountains past and present in the San Juan Mountains. In: Blair R, Bracksieck G (eds) The eastern San Juan Mountains. University Press of Colorado, Boulder, pp 3–16

Grassino-Mayer HD (1998) A multycentry reconstruction of precipitation for Great Sand Dunes National Monument, Sourthwestern Colorado. Final report. Submitted to Mid-continent Ecological Science Center, Fort Collins, Colorado, for the National Park Service, 40p

Greeley R, Iversen J (1985) Wind as a geological process on Earth, Mars, Venus and Titan. Cambridge Univ. Press, Cambridge, UK, 333 p

Hayward RK, Mullins KF, Fenton LK, Hare TM, Titus TN, Bourke MC, Colaprete A, Christensen PR (2007) Mars Global Digital Dune Database and initial science results. J Geophys Res 112(E11)

Hayward RK, Fenton LK, Tanaka KL, Titus TN, Colprate A, Christensen PR (2010) Mars Global Digital Dune Database: MC1. U. S. Geol. Survey Open-File Report 2010–1170, pubs.usgs.gov/of/2010/1170/

Hayward RK, Fenton LK, Titus TN (2014) Mars Global Digital Dune Database (MGD3): Global dune distribution and wind pattern observations. Icarus 230:38–46

Hearne GA, Dewey JD (1988) Hydrologic analysis of the Rio Grande basin north of Embudo, New Mexico, Colorado and New Mexico. U.S. Geological Survey Water-Resources Investigations Report 86–4113, 120 pp

HRS (1999) Hydrogeologic investigation Sand Creek and Indian Springs Area, Great Sand Dunes National Monument, Colorado. HRS Water Consultants, Inc., Lakewood

Hutchison DM (1968) Provenance of sand in the Great Sand Dunes National Monument. Morgantown, West Virginia University, Ph.D. dissertation, Colorado, 132 p

Janke JR (2002) An analysis of the current stability of the dune field at Great Sand Dunes National Monument using temporal TM imagery (1984–1998). Remote Sens Environ 83:488–497

Jodry MA (1999) Folsom technological and economic strategies: views from Stewart's Cattle Guard Sites and the Upper Rio Grande Basin. Ph.D. dissertation, Department of Anthropology, The American University, Washington, DC

Jodry MA, Stanford DJ (1996) Changing hydrologic regimes and prehistoric landscape use in the Northern San Luis Valley, Colorado. In: Thompson RA, Hudson MR, Pillmore CL (eds) Geologic excursions to the Rocky Mountains and beyond, Geological Society of America Special Publication 44. Geological Society of America, Denver

Johnson RB (1967) The Great Sand Dunes of Southern Colorado. US Geol Surv Prof Pap 575-C:C177–C183

Kirkham RM, Magee AW (2020) Chapter 1: Geologic history of the San Luis Valley. In: Beeton JM, Saenz CN, Waddell BJ (eds) The San Luis Valley: its geology, ecology, and human history. University of Colorado Press, Boulder

Lindsey DA (2010) The geologic story of Colorado's Sangre de Cristo Range. US Geol Surv Circ 1349, 14 p
Lipman PW, Mehnert HH (1975) Late Cenozoic basaltic volcanism and development of the Rio Grande depression in the Southern Rocky Mountains. In: Curtis BF (ed) Cenozoic history of the Southern Rocky Mountains, Memoir 144. Geological Society of America, Boulder, pp 119–154
Lorenz RD, Zimbelman JR (2014) Dune Worlds: How windblown sand shapes planetary landscapes. Praxis/Springer-Verlag, Berlin, 308 p
Machette MN, Marchetti DW (2006) Pliocene to middle Pleistocene evolution of the Rio Grande, Northern New Mexico and Southern Colorado. Geol Soc Am Abstr Prog 38(6):36
Machette MN, Thompson RA, Marchetti DW, Smith RSU (2013) Evolution of ancient Lake Alamosa and integration of the Rio Grande during the Pliocene and Pleistocene. In: Hudson MR, Grauch VJS (eds) New perspectives on Rio Grande rift basins: from tectonics to groundwater, Special paper 494. Geological Society of America, Boulder, pp 1–20. https://doi.org/10.1130/2013.2494(01)
Madole RF, Romig JH, Aleinikoff JN, VanSistine DP, Yacob EY (2008) On the origin and age of the Great Sand Dunes, Colorado. Geomorphology 99:99–119. https://doi.org/10.1016/j.geomorph.2007.10.006
Madole RF, VanSistine DP, Romig JH (2016) Geologic map of Great Sand Dunes National Park, Colorado, Scientific investigations map 3362. U.S. Geological Survey, Reston. https://doi.org/10.3133/sim3362
Marin L, Foreman SL, Valdez A, Bunch F (2005) Twentieth century dune migration at Great Sand Dunes National Park and Preserve, Colorado, relation to drought variability. Geomorphology 70(2005):163–183
Martorano MA (1999) Chapter 7: Culturally peeled Pondersosa pine trees. In: Colorado prehistory: a context for the Rio Grande Basin. State Historical Fund, Colorado Historical
McCalpin JP (2006) Active faults and seismic hazards to infrastructure at Great Sand Dunes National Monument and Preserve. Final report. Submitted to National Park Service, Great Sand Dunes National Park and Preserve, Colorado
McKee ED (ed) (1979a) A study of global sand seas, U.S. Geol. Survey prof. Paper 1052. U.S. Gov. Printing Office, Washington, DC, 429 p
McKee ED (1979b) Introduction to a study of global sand seas. In: A study of global sand seas, U.S. Geol. Survey prof. Paper 1052. U.S. Gov. Printing Office, Washington, DC, pp 1–19
Mulligan KR, Tchakerian VP (2002) Toward a genetic classification of Aeolian Sand Dunes. Poster presented at international conference on Aeolian Research, Lubbock
Murray KD (2015) GPS measurements of Rio Grande rift deformation. M.S. thesis. New Mexico Institute of Mining and Technology, Socorro
Naranjo A (2016) Email correspondence with the National Park Service resource management specialist Fred Bunch
Rowlands PG, Geary MM (1998) The great sand dunes Rephotographic project. Historic images of a changing landscape. Submitted to the National Park Service
Ruleman CA, Machette MN, Thompson RA, Miggins DP, Goehring BM, Paces JB (2016) Geomorphic evolution of the San Luis Basin and Rio Grande in Southern Colorado and Northern New Mexico. In: Keller SM, Morgan ML (eds) Unfolding the geology of the west, Field Guide 44. Geological Society of America, Boulder, pp 291–333
Rupert MG, Plummer LN (2004) Ground-water flow direction, water quality, recharge sources, and age, Great Sand Dunes National Monument and Preserve, South-Central Colorado. 2000–2001: U.S. Geological Survey Scientific Investigations Report 2004–5027, 32 p
Sharp RP (1963) Wind Ripples. J Geol 71(5):617–636
Smiley TL, Bryson RA, King JE, Kukla GJ, Smith GI (1991) Quaternary paleoclimates. In: Morrison, E. B., ed., quaternary nonglacial geology: conterminous U.S., The Geology of North America, vol K2. Geological Society of America, Boulder, pp 13–44
Spencer FC (1925) The story of the San Luis Valley. Alamosa J, 83 p

Sullivan R, Banfield D, Bell JF, Calvin W, Fike D, Golombek M, Greeley R, Grotzinger J, Herkenhoff K, Jerolmack D, Malin M, Ming D, Soderblom LA, Squyres SW, Thompson S, Watters WA, Weitz CM, Yen A (2005) Aeolian processes at the Mars Exploration Rover Meridiani Planum landing site. Nature 436(7047):58–61

Thompson RA, Shroba RR, Machette MN, Fridrich CJ, Brandt TR, Cosca MA (2015) Geologic map of the Alamosa 30′ x 60′ Quadrangle, South-Central Colorado, Scientific investigations map SIM-3342. U.S. Geological Survey, Reston

Toll RW (1931) United States Department of the Interior, National Park Service, Report on Great Sand Dunes National Monument. Unpublished report in Great Sand Dunes National Park archives, p 47

Trimble S (1975) A sketch on the history of the establishment of great sand dunes NM. Great Sand Dunes National Park archives, p 11

Valdez A, Zimbelman J (2016) The quest to explore other planets. Ranger 32(3):6–9

Watkins TA (1996) Geology of the Northeastern San Luis Basin, Saguache County, Colorado. In: Thompson RA, Hudson MR, Pilmore CL (eds) Geologic excursions to the Rocky Mountains and beyond. Field trip guidebook for the 1996 annual meeting, Geological Society of America, Special publication 44. Colorado Geological Survey. CD-ROM, Denver

Webster RJ (2000) Kinematic analysis and structural style of the Mosca Creek window, west side of the Sangre de Cristo Mountains. Wichita, Wichita State University, M.S. thesis, Alamosa County, 164 p

White D (2005) Seinanyedi an ethographic overview of Great Sand Dunes National Park and Preserve. Submitted to the National Park Service

Whitmeyer SJ, Karlstrom KE (2007) Tectonic model for Proterozoic growth of North America. Geosphere 3(4):220–259. https://doi.org/10.1130/GES00055.1. Boulder: Geological Society of America

Wiegand JP (1977) Dune morphology and sedimentology at Great Sand Dunes National Monument. MS thesis, Colorado State University, Fort Collins, 165 pp

Wilson IG (1972) Aeolian bedforms: their development and origins. Sedimentology 19:1973–1210

Wilson SA (2004) Latitude-dependent nature and physical characteristics of transverse aeolian ridges on Mars. J Geophys Res 109(E10)

Yuan F, Koran MR, Valdez A (2013) Late glacial and Holocene record of climate change in the Southern Rocky Mountains from sediments in San Luis Lake, Colorado, USA. Paleogeogr Paleoclimatol Paleoecol 392(2013):146–160

Zachariassen J, Zeller KF, Nikolov N, McClelland T (2003) A review of the Forest Service Remote Automated Weather Station (RAWS) network. Gen. Tech. Rep. RMRS-GTR-119. U.S. Department of Agriculture, Forest Service, Rocky Mountain Research Station, Fort Collins, 153 p. PDF available from https://www.fs.fed.us/rm/pubs/rmrs_gtr119.pdf

Zimbelman JR (2010) Transverse Aeolian Ridges on Mars: First results from HiRISE images. Geomorphology 121(1–2):22–29

Zimbelman JR, Foroutan M (2017). Transverse aeolian ridges: Mars spacecraft data analyses and a new Earth analog. Fifth Int. Planet. Dunes Workshop, LPI Contrib. No. 1961, Abs. 3037

Zimbelman JR, Williams SH (2007) Eolian dunes and deposits in the western United States as analogs to wind-related features on Mars. In: Chapman M (ed) The geology of Mars: evidence from earth-based analogs. Cambridge University Press, Cambridge, pp 232–264

Zimbelman JR, Irwin RP, Williams SH, Bunch F, Valdez A, Stevens S (2009) The rate of granule ripple movement on Earth and Mars. Icarus 203:71–76. https://doi.org/10.1016/Icarus.2009.03.033

Zimbelman JR, Williams SH, Johnston AK (2012) Cross-sectional profiles of sand ripples, megaripples, and dunes: a method for discriminating between formational mechanisms. Earth Surf Process Landf 37(10):1120–1125

# Chapter 8
# Sand Dunes, Modern and Ancient, on Southern Colorado Plateau Tribal Lands, Southwestern USA

**Margaret H. Redsteer**

**Abstract** A mantle of both active and stable aeolian sand covers approximately 34,000 km$^2$ of northern Arizona, western New Mexico and southern Utah on the southern Colorado Plateau. From west to east, these deposits can be subdivided into the Kaibab-Moenkopi dunes, Chinle Valley dunes, and Chaco dunes, all of which include relict, partly stable and mobile aeolian sand. Locally, these deposits have distinct compositional characteristics. An examination of previous studies into disparate aspects of Colorado Plateau dunes, taken in the context of local geology, Quaternary landscape history and geomorphic processes, provides new insights into interpretation of this regional aeolian sedimentary record. Additional new data about the characteristics of the deposits, and an assessment of present-day climatic conditions enhances our ability to interpret the relative influences of ecosystem and geomorphologic processes with climate variability that continue to influence both new dune formation and reactivation of older deposits. Taken as a whole, the data emphasizes the role that local landscape conditions and history play in providing the context for correctly interpreting aeolian activity and depositional environments, and whether sediment supply or climate play a dominant role in sand dune formation. This is particularly true in the Little Colorado River Valley of northeastern Arizona, where Quaternary volcanic activity has significantly influenced the local landscape processes, deposit characteristics, and dune paleohistory.

**Keywords** Sand dune mobility · Drought · Navajo · Hopi · Paleoclimate · OSL · Dune vegetation

---

M. H. Redsteer (✉)
School of Interdisciplinary Arts & Sciences, University of Washington Bothell, Bothell, WA, USA
e-mail: mredst@uw.edu



N. Lancaster, P. Hesp (eds.), *Inland Dunes of North America*, Dunes of the World, https://doi.org/10.1007/978-3-030-40498-7_8

## 8.1 Introduction

Nearly a third of Navajo and Hopi Tribal lands of the Southern Colorado Plateau, in the southwestern USA, are mantled with an extensive and diverse array of active and partly active sand dunes, as well as poorly developed aeolian-derived soils. Covering parts of northeastern Arizona, western New Mexico and southern Utah, these substantial aeolian deposits overly gently dipping Paleozoic, Mesozoic, and Cenozoic sedimentary strata and are flanked to the southwest by Tertiary and Quaternary volcanic rocks of the San Francisco Volcanic Field. The boundaries of this region of aeolian deposits are generally defined by the Colorado River and San Juan Rivers to the west and north, the Little Colorado River that follows a northwest trending strike valley to the south, and by the mountainous region including Largo Canyon and Huerfano Mountain to the east in New Mexico (Fig. 8.1). Currently, dunes exhibit a broad spectrum of mobility and morphology that reflects local variations in climate, vegetation, sediment supply and the complex dynamics of a windy, semi-arid to arid landscape. Moreover, the dormant and relict dune deposits of this

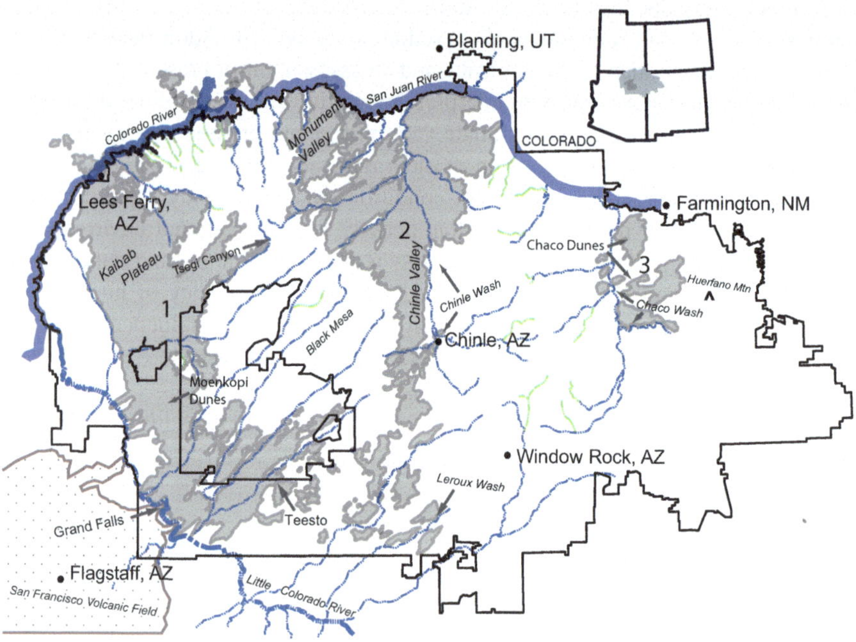


**Fig. 8.1** Map showing the spatial distribution of dune deposits on southern Colorado Plateau tribal lands, within three main areas, as well as the location current ephemeral (blue dashed line) and past ephemeral (dashed green line) fluvial sources. The location of the San Francisco Volcanic Field and lava dams are shown by stipled area. Larger rivers are shown with wide blue line. Solid borders indicate boundaries of the Navajo (larger) and Hopi (smaller inset) tribal lands. Arrows point to specific sites discussed in text

province document a longer perspective of Holocene and Pleistocene landscape history in the southwestern United States.

Active and semi-stable dunes deposits (mappable at 1:250,000) can be divided into three major areas (Fig. 8.1). The westernmost area of aeolian deposits (1) is the largest, and extends from the western reach of the Little Colorado River northward (covering approximately 18,000 km$^2$) across the Moenkopi and Kaibab Plateaus, and includes dunes near Lee's Ferry, House Rock Valley and the Paria Plateau in addition to more well-known dunes of the Moenkopi Plateau. The central area of aeolian deposits (2-covering approximately 12,000 km$^2$) includes scattered deposits in the Painted Desert near Leroux Wash that thicken northward and become more continuous through the Chinle Valley, Ventana Mesa and Rock Point, covering extensive areas of Monument Valley near its northern limit. The easternmost dune field group (3) is smaller, with many deposits modified by land use, but still significant in size (covering approximately 3600 km$^2$). It is comprised of deposits to the north and east of the Chaco Wash in New Mexico and is referred to as the Chaco Dunes (Fig. 8.1).

## 8.2 Dune Morphologies

Aeolian landforms in all three regions of dune deposits include linear, transverse, barchan, parabolic dunes, falling dunes, echo dunes, sand ramps, zibars and sand sheets that frequently overly one or more generations of poorly to well indurated relict dunes. Variations of dune morphology on Navajo and Hopi tribal land were studied by Hack (1941) who developed a ternary classification diagram of three principal dune types, parabolic, barchan and linear dunes, as well as describing conditions of local vegetation and sediment supply. In addition to observations about dune morphology, this early work also included a highly generalized map of those dune types and their occurrence (Hack 1941). Newer, higher resolution imagery provides an opportunity to revisit the diversity of dune morphologies and the landscape conditions where they occur.

Since the middle of the twentieth century, study of available temporal aerial photos series has shown that in some areas dunes have migrated, and new dune fields have formed downwind of riparian areas (Redsteer et al. 2010a; Bogle et al. 2015). Changes to streamflow dynamics have altered rates of sediment delivery because of regional declining snowfall and increasingly ephemeral streamflow (Redsteer et al. 2010b; Draut et al. 2012; Redsteer et al. 2013). In addition, some changes to dune mobility and morphology have occurred as a result of drought and land use disturbances (Redsteer and Block 2004).

The Moenkopi Plateau, situated within the westernmost dune area, is a well-known locality of linear dunes (Breed and Breed 1979; Greeley and Iverson 1985; Billingsley 1987; Stokes and Breed 1993; Lancaster 1995, and others). The size and spacing of dunes here are not as well ordered as linear dunes in other areas of the southern Colorado Plateau. Typically 2–10 km long, their crests are moderate to low relief (3–12 m high). Dune spacing is somewhat irregular, from 50 m to over 300 m apart, dependent on dune size. The largest linear dunes on this plateau are more widely-spaced complex forms (≥ 250 m apart) with superimposed parabolic and barchan morphologies on dune crests (Figs. 8.1 and 8.2a). Dune heights and sediment mobility on dune crests both decrease away from the plateau margins, and downwind. Differences in size and spacing may be due to complex variations in sediment supply from proximity to climbing dunes along the Plateau escarpment, variable wind energy due to topographic irregularities, and the nature of underlying mesa geology. More regularly spaced linear dunes occur to the south within kilometers of the Little Colorado River where topography and sediment supply are less spatially variable. Here, larger linear dunes are 100 m apart and alternate with smaller, lower (1–2 m) crested dunes that bisect the interdune corridors at 50 m (Fig. 8.2b). Frequently, linear dunes exhibit Y-branching morphologies (e.g. Tsoar 2001) such as those located south of Blanding, UT (Fig. 8.2c). Some Y-branching linear dunes occur downwind from or adjacent to linear sequences of parabolic dunes. Where closely spaced, these two dune types can be hard to distinguish and closely resemble braided dunes observed by Tsoar and Møller (1986). Tsoar (2008) discusses the braided morphology as a consequence of herbivory by livestock, but land use practices vary in the vicinities with braided morphologies on the southern Colorado Plateau, and braided and non-braided morphologies coexist in areas with the same land use. These spatial relationships suggest that vegetation decline in general may lead to alteration of linear morphologies to braided morphologies. Currently, most areas of linear dunes exhibit braided morphologies, possibly as a consequence of land use stresses and increased aridity in recent decades. In areas where vegetation cover is sparse to non-existent, the vegetated linear morphology is replaced completely by bare sand that forms active transverse ridges (Fig. 8.2d).

The Chaco dune field (3) of southern Colorado Plateau dunes was studied in detail by Wells et al. (1990) who described the stratigraphy of aeolian derived soils. These deposits are mantled by active compound parabolic dunes and linear dunes, as well as parabolic and barchan dunes along the margins of Chaco Wash (Fig. 8.2e). Their orientation suggests that the predominant direction of wind drift potential in this easternmost area of dunes tends to be in a more westerly (WSW) direction, in contrast to dunes in the other two sub-regions, where the predominant wind energy is from the southwest.

Both parabolic and barchan dunes are found on mesa tops in areas with thicker sediment, dipping topography, or where they overly abandoned channels (Fig. 8.2f). However, these dune forms most commonly occur near riparian areas, associated with dry or ephemeral stream beds that act to supply sediment. Parabolic dunes are usually found within broad channels that are oriented in the direction of the wind (Fig. 8.2g). Barchans dunes most commonly occur as source-bordering dunes that are downwind of straight channel segments oriented perpendicular to the wind, above areas where bedrock controls have lowered stream gradient or restricted channel adjustments (Fig. 8.2h). Barchan dunes tend to be small in scale (frequently several meters high and ~100 m across) and thus, able to migrate at discernible speeds of about 1 m in a typical wind event (Bogle et al. 2015).

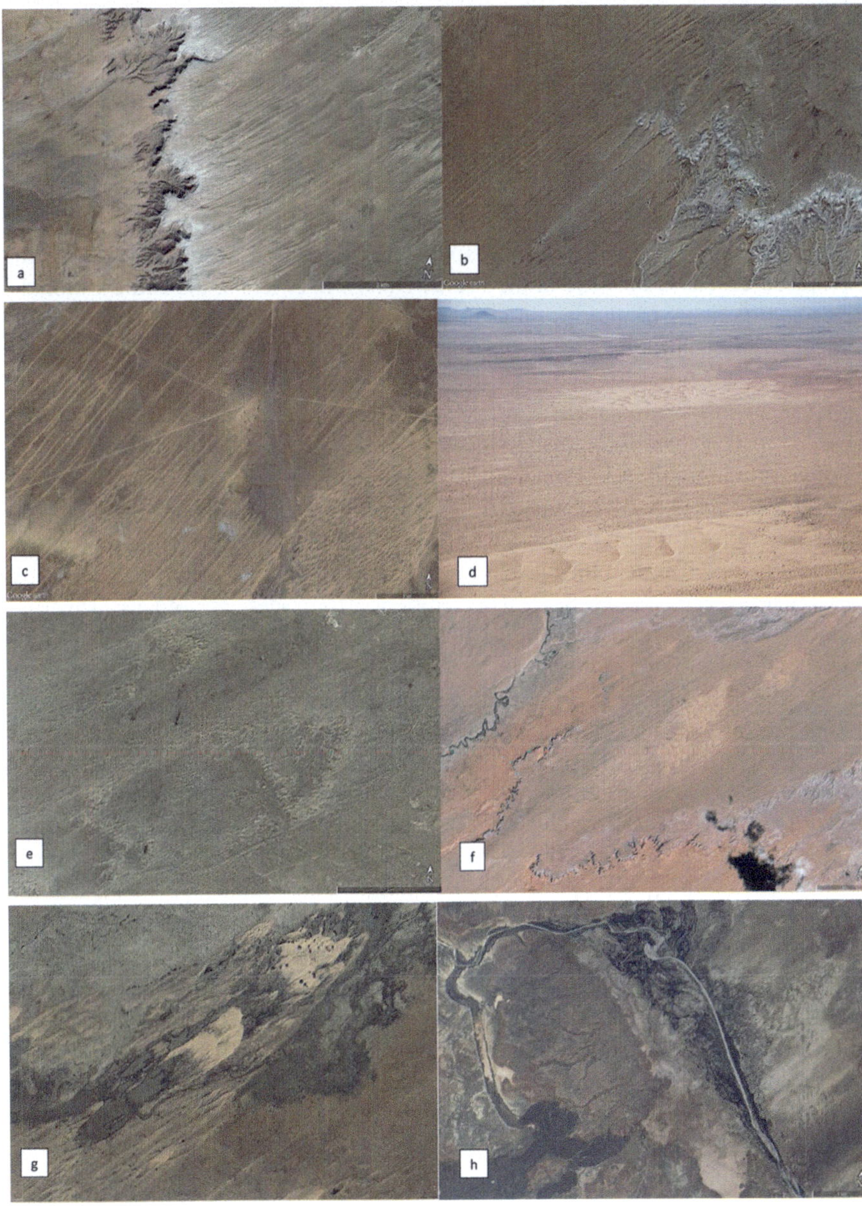


**Fig. 8.2** Imagery and oblique aerial photo showing dune morphologies including (**a**) Moenkopi Plateau complex linear dunes, (**b**) regularly-spaced linear dunes near the Little Colorado River, (**c**) Y-branching and braided linear dunes in southern Utah, (**d**) active transverse dunes adjacent to linear vegetated dunes near Teesto, AZ (**e**) compound parabolic dunes at the Chaco Dune Field, NM, (**f**) variations in dune morphology along a NE dipping mesa, (**g**) parabolic dunes in wash aligned in direction of sediment transport, (**h**) source-bordering barchan dunes on downwind side of Little Colorado River where river diversion from lava dams have altered stream gradient. Current channel is cutting through lake deposits (satellite imagery courtesy of Google Earth)

## 8.3 Sediment Supply Characteristics

Sediment availability is reflective of local Quaternary geologic history, the ephemeral and "flashy" nature of streamflow and floods, as well as the age and state of dune mobility. In addition to variations in sediment supply and aridity, topographically controlled variations in wind energy via "wind gaps" may be an important predictor in the location of active dunes (Billingsley et al. 2013). Sediment characteristics are highly dependent on local sources, and are often closely associated with fluvial sources, as is the case in the Chaco Dunes. In Arizona, the Little Colorado River and its many tributaries on the northeast side of the river are an important source of sediment for extensive aeolian sand sheet and sand dune deposits that are transported by southwesterly winds to the northeast towards Black Mesa (in Area 1). Changes in river dynamics during the Holocene and Pleistocene are likely to have led to considerable influence on sediment supply to nearby dune fields. As eruptive activity in the San Francisco Volcanic Field migrated to the northeast during the Pleistocene and Holocene, it caused multiple channel abandonments and diversions of the Little Colorado River (Fig. 8.2h). Eruptions produced lava flows that periodically dammed the current river channel three times, leading to flooding and aggradation upstream. Several generations of terraces, as well as lacustrine deposits, are evidence of flooding and channel adjustment to within-channel lava flows and lava dams in the Little Colorado River and its tributaries (Billingsley et al. 2014). A well-documented example occurred at Grand Falls about 19.6 ± 0.14 ka (basalt flow age was determined by four techniques: optical luminescence, cosmogenic $^{3}$He, $^{40}$Ar/$^{39}$Ar, and magnetic secular variation (Esser 2003; Duffield et al. 2006). Basaltic tephra also blocked tributaries and streams, and blanketed parts of the Little Colorado River Valley with ash deposited during local strombolian eruptions as late as 900 years ago (Fig. 8.3).

In areas close to the Little Colorado River, dune mineralogy and chemical composition reflect variations in local geology, with a myriad of eroding sedimentary sources and local input from ash-fall (Fig. 8.4). Windblown sediments generally consist of angular clear quartz, red, yellow and milky quartz, chert, biotite, iron-oxides and feldspars, in addition to vesicular basalt glass (Billingsley 1987; Redsteer and Hayward 2015). Therefore, the characteristics of dune sediments vary considerably depending on location and exposures of local sedimentary strata, including the Navajo Sandstone and Entrada Formations, two extensive deposits of ancient sand seas that are exposed 45 km north of the Little Colorado River Valley, as the cap rock of the Moenkopi Plateau. In the regions to the north (on the Kaibab Plateau, Monument Valley and Chinle Valley) where outcrops of Navajo and De Chelly Sandstone are common, aeolian sediment is mineralogically mature, consisting primarily of frosted, well-rounded quartz with iron-oxide coatings (Fig. 8.5). In the central dune area (2), aeolian sediment has accumulated and recycled through the Chinle Valley and Monument Valley, where the De Chelly and Navajo Sandstones, as well as other sedimentary strata, are uniform, ubiquitous outcrops that contribute sediment to Chinle Wash.

**Fig. 8.3** Tephra dam in abandoned Little Colorado River channel. Similar deposits dammed the river and its tributaries as late as 900 years ago

Some dunes, such as those high on the Moenkopi Plateau tend to be fairly uniform fine-and very fine-grained sand, with sorting characteristics that reflect the position on the dune (e.g. crest vs. plinth). In contrast, other dunes can be coarse-grained and poorly-sorted, especially those in the south where aeolian sandstones are not present as bedrock, areas close to fluvial sources, or in areas containing coarse-grained sand derived locally from volcanic tephra. For example, saltating sand trapped in 1 m height BSNEs at Grand Falls can often include very coarse tephra (> 2 mm). These dune sediments tend to become better sorted downwind, in the direction of sediment transport, on the downwind side of dune fields (Fig. 8.5).

## 8.4 Current Climate Setting of the Southern Colorado Plateau

Precipitation occurs bi-modally on the southern Colorado Plateau, during winter storms (December–March) and the North American monsoon season (July–September). Locally, approximately 45% of annual precipitation occurs during the summer monsoon season (Redsteer et al. 2010b). These two wet seasons are separated by a dry, windy spring (April–June) when most aeolian activity occurs,

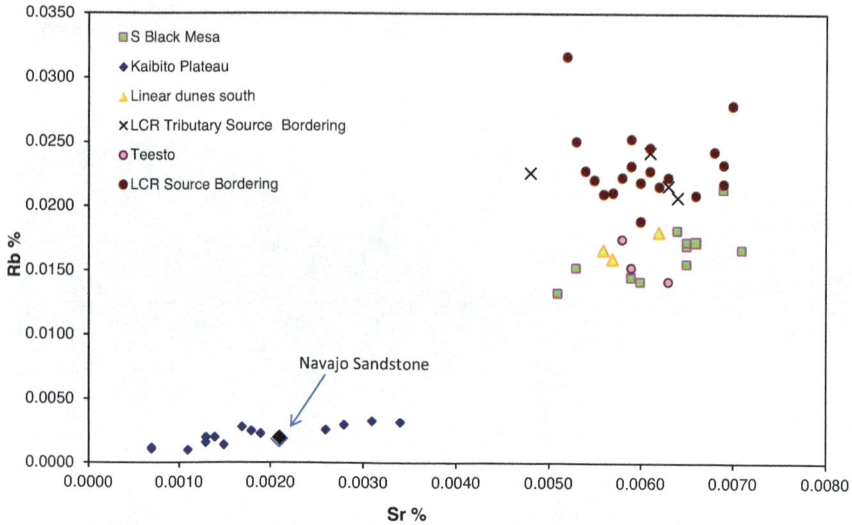


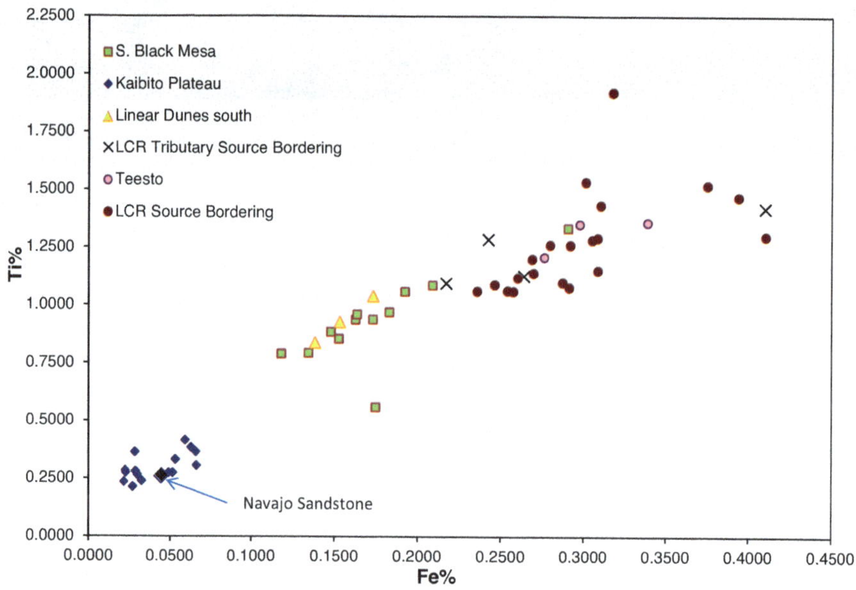


**Fig. 8.4** Plots showing compositional differences among sediment sources that reflect local sediment supply inputs to dunes on the southern Colorado Plateau. Larger black diamond denotes composition of the Navajo sandstone, whereas "Linear dunes south" show compositions of well-sorted linear dune sands that occur south of the southern limit of Navajo Sandstone outcrops

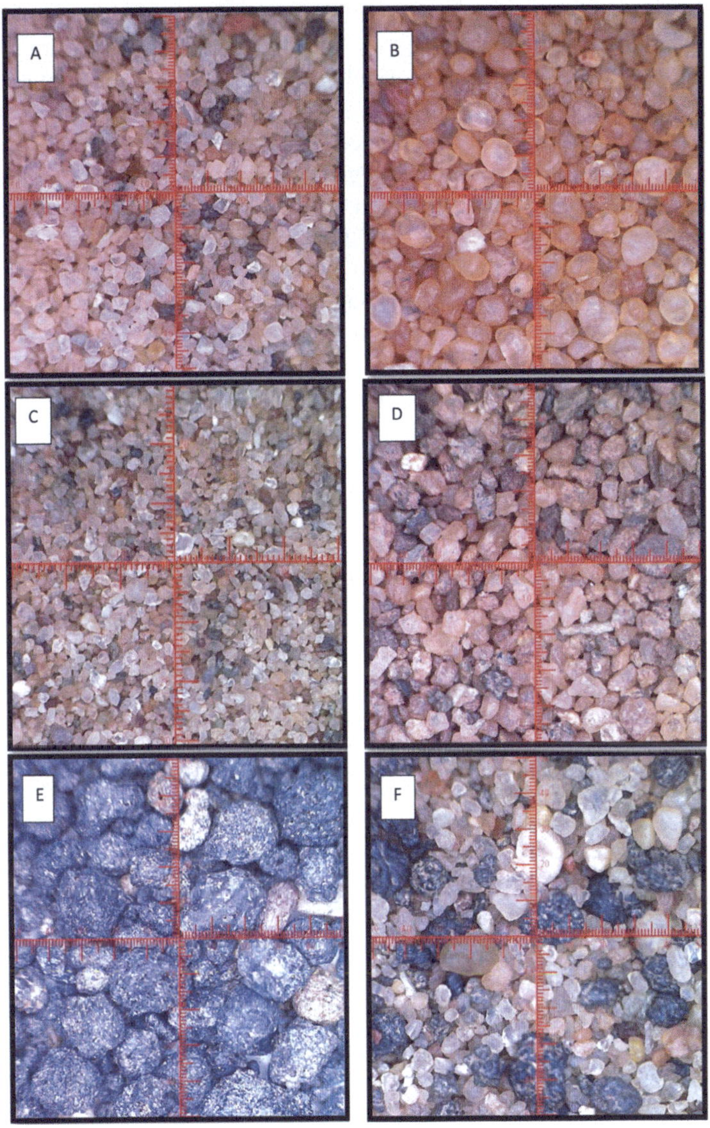


**Fig. 8.5** Photomicrographs showing grain size, composition and sorting characteristics of different dune sands (with analyses in Fig. 8.4) and local volcanic tephra

because the regional timing of the dominant wind and precipitation seasons are out of phase with each other (Bogle et al. 2015; Draut et al. 2012).

Meteorological documentation within the region is difficult because it is topographically and climatically variable, and large portions of the region are poorly monitored, with a density of operating National Weather Service COOP (Cooperative Observer Program) monitoring stations of one for every 6400 km$^2$. Assessing the relation between dunes and climate variability on the southern Colorado Plateau is difficult because there are few continuous, extended instrumental records, and especially records of wind speed. The US Geological Survey has four additional meteorological monitoring sites in northeastern Arizona, co-located with dune study sites that record temperature, rainfall, wind speed and direction, and soil moisture.

Moisture availability varies by sub-region: lowlands in the western and central dune areas at 1200–1500 m in elevation are the hottest and driest, with average annual precipitation from 100–150 mm during recent decades (1980–2010). The eastern Chaco Dune area in New Mexico is wetter with annual precipitation from 200–250 mm. The Lukachukai Mountains, aligned N-S along the Arizona, New Mexico border (between areas 2 and 3) are 3000 m high at the crest, and receive the greatest proportion of the regions' snowfall and rainfall, followed by Black Mesa with an average of 250–300 mm annually. There are also large seasonal and diurnal variations in temperature, with average annual temperatures from 11.0 °C in areas of higher altitude, to 15.0 °C in the valleys and lowlands.

### *8.4.1 Utilization of Climatic Factors to Assess State of Dune Mobility*

Drought conditions that have persisted since 1996 on the southern Colorado Plateau Tribal lands have produced significant changes in dune mobility in recent decades (Redsteer and Block 2004; Redsteer et al. 2010b). Related sand and dust storms can damage rangeland and cause dangerous travel conditions. These storms can also generate economic, cultural, and health consequences for the Navajo and Hopi people (Redsteer et al. 2013). Declining precipitation and warming temperatures in the southwestern U.S. have also resulted in ecological changes (Westerling et al. 2006; Seager et al. 2007; Weiss and others 2009). However, assessing how much the current state of dune mobility is directly linked to climatic conditions has been a challenge due to the historic dependence of dune ecosystems for grazing livestock.

Many studies of dune stability and movement in arid and semi-arid environments have sought to quantify links between climate variables and the observed conditions of sand dunes. Lancaster (1988) proposed the use of the ratio of P (precipitation) to the total annual PE (potential evaporation) as an indicator of the amount of stabilizing vegetation on dunes, used by many subsequent workers (Muhs and Maat 1993; Lancaster 1995; Wolfe 1997; Lancaster and Helm 2000). This ratio (P/PE) is a measure of effective precipitation. Winds capable of transporting sand occur frequently on the southern Colorado Plateau. Therefore, the ratio P/PE has been shown to be a critical factor controlling dune mobility because of its direct link to the amount of stabilizing vegetation, unless local pulses of sediment supply or other dune disturbances have occurred (Lancaster 1994; Tsoar 2005; Barchyn and Hugenholtz 2015).

In order to generally depict the spatial relationship of effective moisture to mapped southern Colorado Plateau dunes, P/PE was calculated from all available meteorological data and compared to dune field distributions (Fig. 8.6). Based on

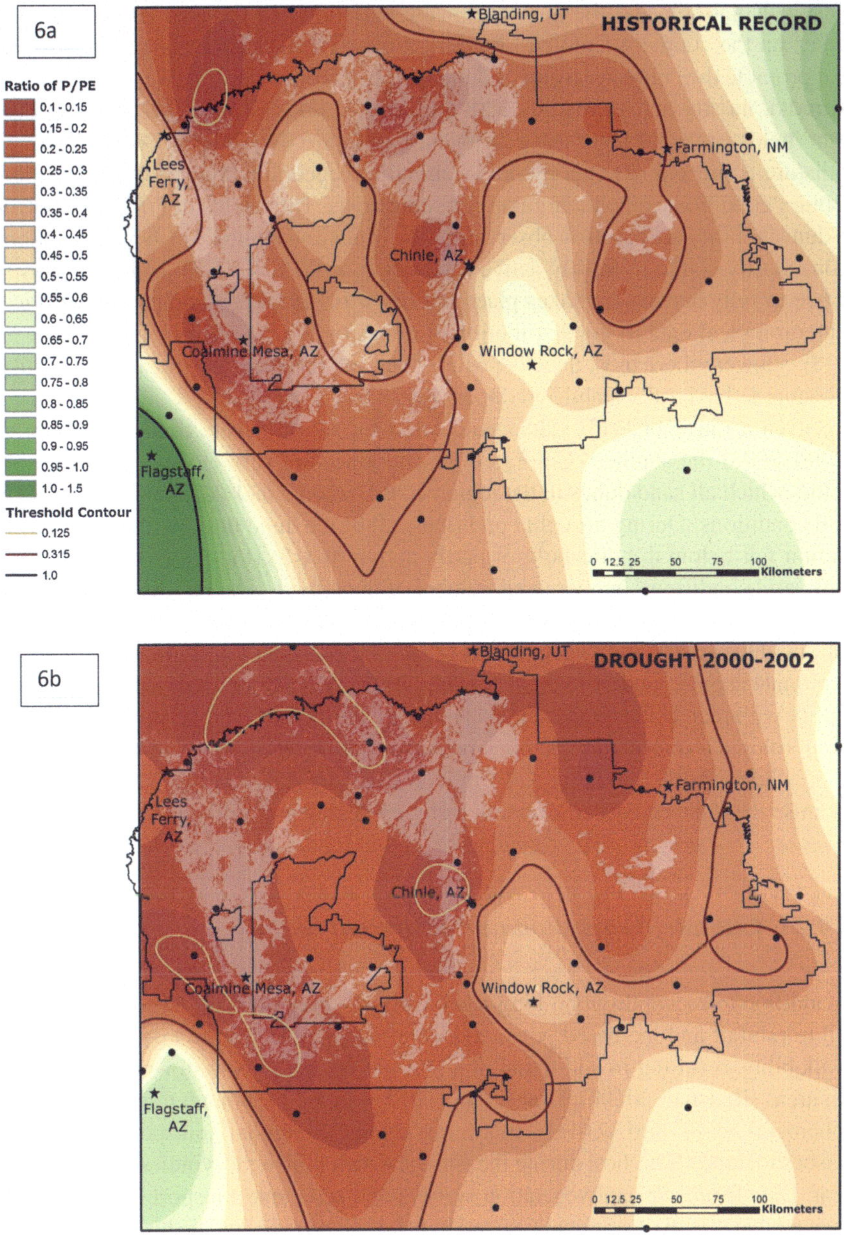


**Fig. 8.6** (**a**) Map showing the distribution of normal average ranges of P/PE, calculated from all the years of record for each station (locations indicated as points on map). White areas are the mapped distribution of aeolian deposits. Where white areas overlap with red regions of low P/PE, sand dune movement is likely to occur. (**b**) Map showing spatial distribution of P/PE for the drought years 2000–2002. Threshold lines indicate moisture conditions related to vegetation required for stability (> 0.315) and moisture deficits where vegetation is unlikely to survive on dunes (< 0.125)

these calculations, and the known climatic variability of the past 100 years, sand dunes on the southern Colorado Plateau currently exist under climatic conditions that promote the entire spectrum of dune mobility, from mostly stable (during humid periods), mostly active, to fully active (during periods of drought). Although the dune fields on tribal lands have an overall moisture deficit (P/PE < 1.0), enough moisture is present to support some stabilizing vegetation under normal twentieth century conditions, albeit marginally in some areas. Thresholds in P/PE for changes in dune mobility are based on observations by Muhs and Holliday (1995): Transitions from mostly stable to mostly active sand occur at P/PE = 0.315, and from mostly active to fully active sand takes place at P/PE = 0.125. When comparing historical meteorological data to the distribution of dunes on the southern Colorado Plateau, P/PE is lower where most dune deposits occur, but there are also areas with enough moisture to have fully stabilized dunes (Fig. 8.6a).

Given a range of P/PE values calculated for recent severe drought conditions (2000–2003) the southern Colorado Plateau dunes are very close to the threshold below which all sand dunes in the region could become mobile, due to increasingly arid conditions. During drought conditions, all of the dune fields within the Navajo Nation fall below the threshold of partly active (0.315) (Fig. 8.6b). With increasingly arid conditions, an increased number of dune fields also fall below the threshold of P/PE for mostly active (0.125) indicating that not enough moisture is available for any dune stabilizing vegetation to grow. The continuation of drought projected in climate change models by Seagar et al. (2007) is a major concern, as is increased aridity with increasing temperatures. Holocene climate variability as well as current meteorological information demonstrates that prolonged drought would likely alter regional dune fields to a perpetually mobile state (e.g. Yizhaq et al. 2009), and compromise the dune ecosystems upon which people depend for their livelihood.

### *8.4.2 Climate, Vegetation and Sediment Supply*

In addition to recent drought, long-term declining precipitation, a shift from snowfall to rainfall and increased potential evapotranspiration lead to less surface water availability, resulting in additional dune formation and growth downwind of riparian areas (Redsteer 2010b). Draut et al. (2012) assessed sediment transport from an ephemeral wash into actively migrating dunes. Sediment transport from local washes to dunes is highest during the spring, which is also the windiest period of the year. Data from seasonal vegetation surveys and sediment transport measurements showed that stabilizing annual vegetation is not present during the highest mobility-prone seasons. Multi-year seasonal surveys also found that short pulses of moisture favor invasive annual plant growth, whereas multiple years of above average precipitation would be needed for perennial vegetation recovery from drought. Therefore, seasonal variations in plant growth and the relative composition of perennial vs. annual plant cover may become increasingly important in evaluating susceptibility to wind erosion as invasive plants become more dominant in dune ecosystems.

Numerous studies have provided varying estimates of the amount of vegetation cover above which dune deposits are stable, and wind erosion is minimal. The relationship between vegetation cover and sand mobility is nonlinear and varies by habitat type (Wiggs et al. 1995; Lancaster and Baas 1998). In these studies, sand mobility was apparent when vegetation cover was 14–16% or less. Additional studies have shown that dunes are stable when vegetation cover exceeds 30% (Ash and Wasson 1983). In addition to total cover, the spatial arrangement of vegetation may also be important (Okin 2008). Therefore, although P/PE (effective moisture) provides a proxy for vegetation, complications arise from empirical measurements about the characteristics of vegetation that constitutes a stabilized dune.

Hack (1941) provided the first description on the southern Colorado Plateau dune vegetation, and observed a potential relationship of plant rooting type with sand mobility. He mapped dune types across Navajo County and developed a classification of six inter-merging dune zones with their component plant species characterized by presumed rooting depth and water utilization traits. He observed that the traits of plants occurring on stable dunes had shallower roots, utilized water close to the surface, and were intolerant of root erosion by sand. On the active dunes, plants had deeper roots, utilized deeper soil moisture, and had roots tolerant of sand abrasion. He also suggested that the shallow-rooted species do not occur on dune surfaces with moving sediment because of their intolerance for sand erosion or burial. Likewise, the deep-rooted plants would be less likely to occur on the stable dunes because the shallow-rooted plants utilize water before it is stored more deeply in the sand.

Thomas and Redsteer (2016) reviewed Hack's observations of the plant species occurring on the dunes of the southern Colorado Plateau tribal lands and hypothesized that vegetation characteristics would vary depending on dune stability. They examined how rooting traits reflect their capacity to tolerate sand abrasion and burial or to survive under conditions of limited precipitation. The relative amount of sand vs. silt was utilized as a proxy for surface stability on 37 plots of partially vegetated dunes near Teesto, AZ. Dunes with higher surface mobility, as evidenced by coarser sand composition, were dominated by shrubs with deeper taproots than grasses and forbs. The positive association of shrub cover with percentage of sand is consistent with Hack's hypothesis that more active dunes favor taprooted species. Grass cover was not significantly associated with sand composition, but increased with the amount of shrubs (possibly serving as protection from the wind). The relationship of vegetation type with sediment mobility provides information about the nature of vegetation recruitment that will be required to mitigate wind erosion and the migration of sand dunes.

### 8.4.3 Sand Dune Migration

Although sand dune mobility on southern Colorado Plateau tribal lands is increasing, associated with a general increase in aridity, understanding the factors that lead to dune migration and assessment of current dune mobility is complex. Declining precipitation, a shift from snowfall to rainfall and increased potential evapotranspiration contribute to reduced surface water on southern Colorado Plateau tribal lands. At least 30 streams and lakes that were perennial in the 1920s are now dry or ephemeral (Redsteer et al. 2010b). Much of the recent dune activity has resulted from an increase in dry fluvial sediment and the formation of source-bordering dunes such as those found at Grand Falls (Redsteer et al. 2011). Sand and dust movement in the region is also closely linked to regional aridity, yearly variations in wind energy, ephemeral flood events along the riparian-aeolian interface, and regional synoptic weather events (Redsteer 2016).

Current dune migration and growth were documented through the study of source-bordering dunes on the Little Colorado River at Grand Falls. These dunes are very young, with the aerial imagery record indicating that they formed sometime after 1935 but before 1953 (Redsteer et al. 2011). The Grand Falls dune field is located upstream from a lava dam that diverts river flow and currently exerts control on streamflow adjustment to variable discharge. Therefore the location of these dunes also coincides to a segment of the river where overbank deposits will accumulate during periods of high river discharge. Dune development may have been initiated by a substantial amount of overbank sediment deposition as a result of very large spring floods in the 1940s, typical of wet years of the early twentieth century (Hereford 1984). These extremely wet years were followed by dry, windy conditions during the 1950s drought when river sediment was drier, and therefore more likely to be mobilized by wind and form dunes downwind of the riverbed (Redsteer et al. 2010a). The dune field has expanded rapidly over the past 60 years, growing in area from approximately 0.65 $km^2$ to approximately 2.25 $km^2$.

Seasonally repeated surveys were used to track the location of migrating sand dunes, and compared to in-situ meteorological data on wind speed and direction, temperature, precipitation, soil moisture, and vegetation in order to examine climatic parameters and seasonal variations that affect dune mobility. Three years of data on migration rates and weather information were collected, and provide a detailed snapshot of the seasonal and annual variability of active sand dune movement in the Grand Falls Dune Field (Bogle et al. 2015). Recent average migration rates for dunes in the dune field, ranged from 25 m/year to 43 m/year, changing by as much as 48% annually, at rates directly proportional to local wind energy.

A comparison and evaluation of current methods used to quantify dune mobility indicate that Grand Falls dunes, under current arid and sparsely vegetated conditions, have acquired little measurable stabilization from precipitation and therefore, plant growth. During the period of observation, the lowest annual precipitation rate was less than 50% of the local historical average of 160 mm, and the peak precipitation for all 3 years was only 75% of the long-term average. These low precipitation amounts, combined with increasing temperatures, may lead to less availability of effective moisture needed for plant growth. The location of the dunefield in a channel segment prone to increased sediment supply during floods, and a seasonal climate with bimodal precipitation that is out of phase with periods of high wind energy, are conditions that are highly conducive to continuing sand dune mobility even while surrounding areas are mantled with semi- stable vegetated dunes.

## 8.5 The History of Dune Activity

Throughout the Quaternary history of the American Southwest, periods of aeolian sediment mobility have occurred in response to changes in climate and variations in sediment supply. The oldest known dunes on the southern Colorado Plateau occur near Teesto, AZ, where they occur as relict deposits preserved below a mantle of colluvium that defines a regional pediment. These Pleistocene deposits were first described by Sutton (1976) who referred to them as Dilkon deposits. They are a well-indurated, structureless aeolian deposit (Fig. 8.7). A mid- Pleistocene (Irvingtonian) age was assigned to the mammalian fossils collected from fluvial sediment underlying this deposit (Colbert and Marshall, unpubl. data). Attempts at OSL age determination yielded an imprecise age of > 100 ka.

In sediment transport applications, the optically stimulated luminescence (OSL) dating technique measures the length of time since a grain was buried. OSL dates in periodically active dune forms often reflect only the most recent episodes of deposition and the timing of stabilization. Dating Quaternary episodes of activity becomes challenging because exposure to sunlight during periods of dune activity reset the OSL ages. Falling sand dunes form in canyons and in the lee of other topographic obstructions such as mesas and buttes. These dunes may retain the history of aeolian activity in some circumstances, because they are less vulnerable to reworking during periods of increased dune activity inherent to periodically active dune systems. In an attempt to provide a more complete aeolian history of the southern Colorado Plateau, falling dunes preserved in areas 1 and 2, within deep canyons oriented perpendicular to the predominant wind, were sampled using an auger (Manning 2010; Rittenour unpubl. data). Final OSL age determination of a sample from Tsegi Canyon yielded an age of 92.3 ± 7.4 ka (MIS 5c). The next oldest deposits were more significant in volume, and four samples (two each from Tsegi Canyon and Canyon de Chelly) provided ages that span a period of early MIS 3 from 57–51 ka (Fig. 8.8). Although local paleoclimatic information is lacking for MIS 5 and MIS 3 on the southern Colorado Plateau, Glover et al. (2017) provide a multi-proxy overview of regional climate history through an analysis of lake cores and paleoclimate sites. They find that MIS 5c coincides with increased insolation and lowstand sea level conditions, suggesting that warm, dry conditions prevailed. During MIS 3, a period with data from across the southwest, they found widespread southwestern aridity was common to those sites where Dansgaard-Oeschger interstadials were identified. For example, Lake Manley in Death Valley transitioned from a mudflat to a saltpan around 59–57 ka (Forester et al. 2005).

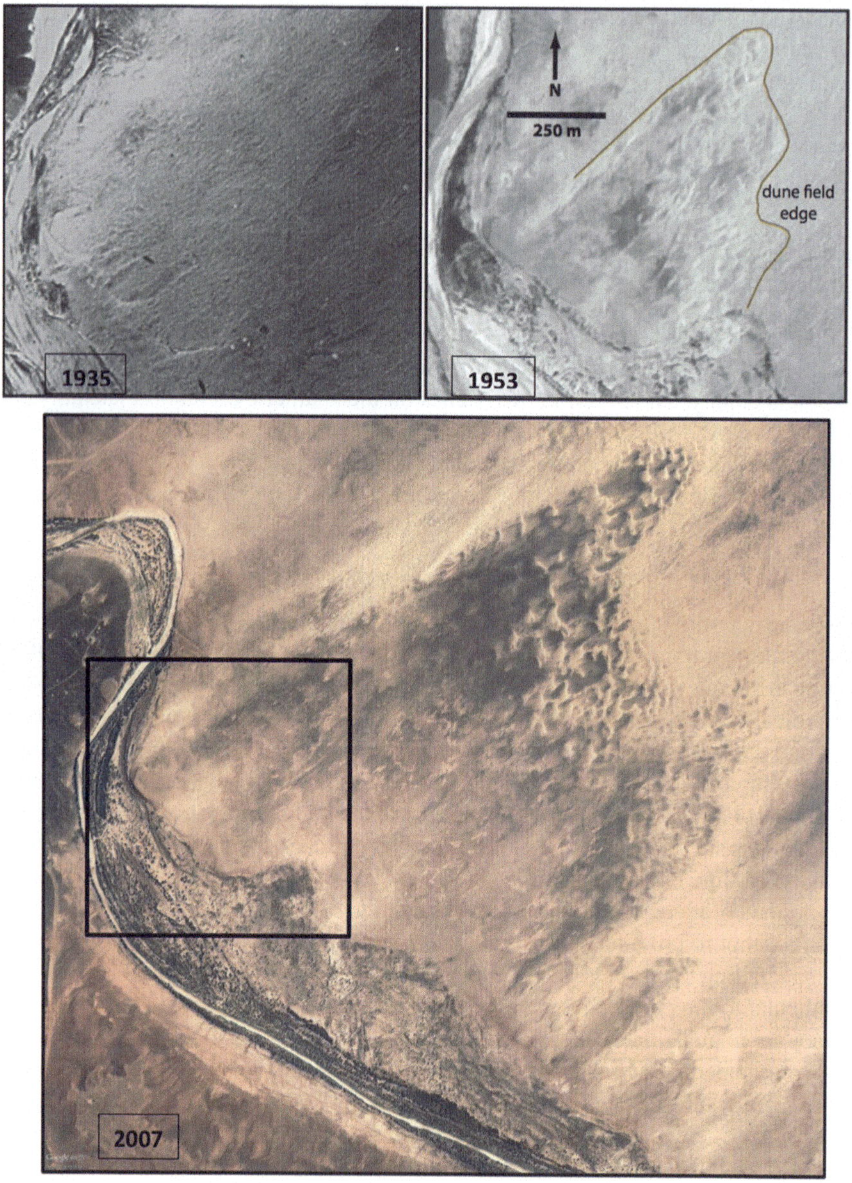


**Fig. 8.7** Aerial images from the Grand Falls Dune field location in 1935 and 1953 showing pre and post-dune field formation (after Bogle et al. 2015), with more recent 2007 color imagery of area that the dune field now encompasses. (Inset area on 2007 image shows the reference location of 1935 and 1953 imagery)

**Fig. 8.8** Outcrop photo showing the location of OSL sample site in Pleistocene Dilkon aeolian deposit, near Teesto Arizona. Contact with overlying mantle of carbonate-rich colluvium is immediately above shovel handle. Red tape marks on shovel are 30 cm apart for scale

The oldest dune age from Black Mesa is roughly 30 ka (Ellwein et al. 2016) and corresponds to a period of aridity, when lakes on the Mogollon Rim had been dry for up to 5000 years (Anderson et al. 2000). Additional evidence also points to complex hydrologic conditions during MIS 2, with drought recorded in southern California around 27.5–25.5 ka (Glover et al. 2017). In contrast to dune ages from other areas of the Colorado Plateau, OSL ages from the Black Mesa area overlap in age with the Jeddito Formation, first described by Hack (1942) near the area studied. The Jedditio formation is a thick alluvial sequence that occurs in many canyons across the southern Colorado Plateau, suggesting that it was deposited during a period of widespread aggradation and sediment availability. Stratigraphic studies of the Jeddito Formation by Sutton (1976) in the Black Mesa area include a $^{14}C$ age of ~40 Ka on detrital wood from the middle- lower section, and mammalian fossils indicating that upper sections are Late Pleistocene.

Ellwein et al. (2011) note that most of the dune deposition on Black Mesa occurred before 20 ka, corresponding closely to the age of lava damming the Little Colorado River at Grand Falls. Damming of the river and lake formation may have temporarily curtailed local aeolian sediment transport from the river northward. The youngest dune ages from Black Mesa are sand sheets from 12-8 ka and occur during the Pleistocene-Holocene transition, a period of relatively wet conditions (Weng and Jackson 1999). Latest Pleistocene aeolian deposits from the Chaco Dune field also fall within this period (Wells et al. 1990; based on $^{14}C$ ages by Simmons 1983). An assessment of dune activity in southwestern deserts found that dune construction occurred during the early Holocene from (11.8- 8 ka) as a result on enhanced sediment supply from fluvial and lacustrine sources (Halfen et al. 2016). Although >50 ka deposits from the Black Mesa are absent, the region of Black Mesa is higher in elevation (2600 m), receives more precipitation, and is cooler than surrounding dune-mantled areas. Additionally, there are no deeply incised canyons oriented perpendicular to the wind, where optimal dune preservation would be expected. Black Mesa has sparser aeolian deposits today than most of the southern Colorado Plateau, and these current conditions may have also occurred in the past (Fig. 8.9).

Holocene dune ages have been documented by Wells et al. (1990) and Stokes and Breed (1993). New OSL ages from Tsegi Canyon, Canyon de Chelly, and Leroux Wash (Rittenour, unpubl. data) provide additional context to understanding regional conditions. Wells et al. (1990) interpreted multiple radiocarbon ages in aeolian deposits from Chaco and surmised that aeolian deposition there was initiated after 5.6 ka with the bulk of Holocene dune activity occurring at Chaco from 4.0 to 2.8 ka. A mid-Holocene age of 4.7 ± 0.40 ka from the Moenkopi Plateau roughly overlaps with this broad period of dune deposition, and also with sediment accumulated in a falling dune from Canyon de Chelly with an age of 5.28 + 0.49 ka. Similarly, an OSL age of 3.60 ± 0.4 from the Moenkopi Plateau corresponds to basal aeolian deposits from the middle unit at Chaco from 3.7–3.9 ka (derived from $^{14}C$ ages). Additionally, late Holocene ages overlap in several localities at 2.2–2.8 ka (Moenkopi Plateau, Canyon de Chelly, and Chaco Dunes) and 1.5 ka (Tsegi Canyon, Leroux Wash, Chaco Dunes).

Holocene dune ages fall within distinct periods, and suggest that dune activity occurred in response to Colorado Plateau climatic fluctuations. The mid-Holocene accumulation of dust in the southern Rockies (Muhs and Benedict 2006) may have been a precursor to the 5.6 ka onset of dune formation (and stabilization) at the Chaco dunefield. Broader correlations also include Holocene ages of 1.5 ka at three locations that coincide with a period of dune construction in western North America noted by Halfen et al. (2016). Because the nature of the deposits sampled for age comparison provides information of both sediment accumulation and stability (sand dunes and sheets) and sediment transport and accumulation (falling dunes) it is important to distinguish what overlapping ages from different deposits might imply. Because of the limitations in age resolution to at least the relative errors assigned to OSL and $^{14}C$ ages, it is possible that when ages from falling dunes and sand sheets overlap they may indicate that dune activity was followed by dune stabilization within a period of decades or centuries. The age overlaps that include both sand sheets and falling dunes, therefore, might imply that drought was followed by abnormally wet and stable conditions.

Aeolian processes are also a landscape response to wind energy, for which other paleo-proxy comparisons are difficult, although fire-scar studies could be relevant. Many researchers as well as this study note the absence of dune activity during the well-documented 1300s drought that led to widespread migration and abandonment

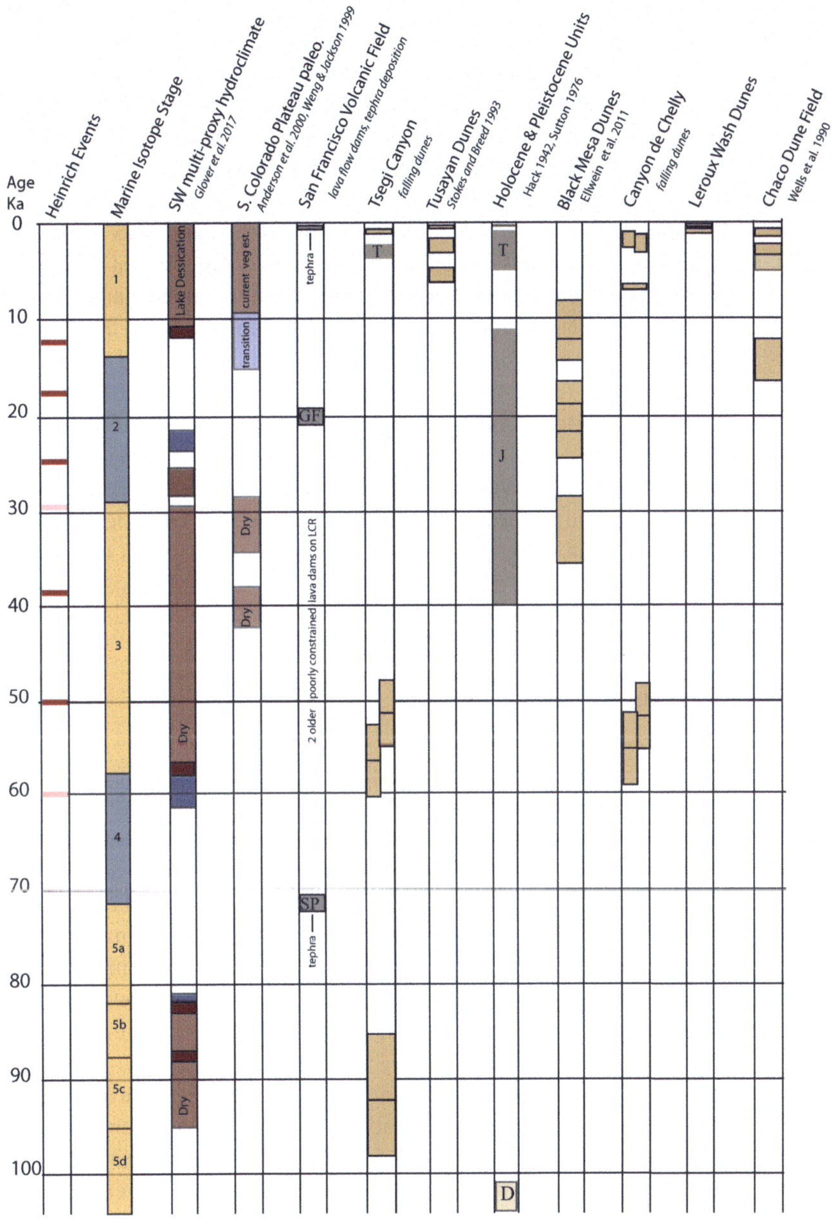


**Fig. 8.9** Dune age correlation chart with chronology of local geologic deposits and events, Marine Isotope Stages and Heinrich events (red for N. American dominance, pink for European dominance), and paleo-proxy records from Glover et al. (2017), Anderson et al. (2000), and Weng and Jackson (1999). Blue indicates significantly wetter conditions, red indicates warmer, drier conditions

of well-known southwestern archaeological sites (Benson et al. 2007; Halfen et al. 2016). One possible explanation may be that the period in question was accompanied by changes to wind circulation patterns and lower wind energy.

Pleistocene dune ages can be more difficult to interpret, especially before 50 ka due to the lack of local paleoclimatic proxies and limits to $^{14}C$ dating (Coats et al. 2008). Deposits from MIS 5 may have resulted from aeolian activity during a response to a warm global climate. Deposits of this age are unusual, possibly due to their rare preservation. More ubiquitous deposits occur during early MIS 3, from 57 to 50 ka, at both Canyon de Chelly and Tsegi Canyon, and suggest that this was a period of widespread dune activity on the southern Colorado Plateau, and therefore less likely a result of local landscape perturbations in sediment supply. This observation is supported by an assessment of paleoclimate sites scattered across the southwest that exhibited dry interstadials during MIS 3 (Glover et al. 2017).

Unfortunately, the role of seasonality during the Holocene and Pleistocene is not well delineated. Glover et al. (2017) noted low seasonality during MIS 3 in southern California, but it does not necessarily reflect the spatial distribution of monsoon climates. Coats et al. (2008) note an expansion of monsoonal climatic influence throughout the Wisconsinan, and suggest this period coincided with a temporally variable climate.

## 8.6 Conclusions

The myriad diversity of dune forms, ages and states of dune mobility present on the southern Colorado Plateau tribal lands provides an excellent opportunity for studying aeolian processes. Sand and dust movement across the region is mostly linked to periods of regional aridity. However, unraveling the history of dune activity is a complex undertaking because local geologic history and landscape changes have significant influence on sediment characteristics and supply, complicating an evaluation of direct and indirect climatic influences. Moreover, it is clear from studies of sediment transport, vegetation composition and cover, as well as migration of currently active dunes that seasonal variations are additionally important in assessing climatic influences on dune mobility.

In this climatically variable region, dune morphologies and the relative amount of dune activity varies, although dune activity has increased in recent decades. Based on seasonal vegetation surveys and an evaluation of dune mobility, drought impacts to dune ecosystems may require successive years of above average moisture for any substantive vegetation recovery to occur. The current increase in aridity, coupled with active dune movement in the southern Colorado Plateau, may hinder the recruitment and establishment of plants needed to stabilize mobile dunes in this region. Thomas and Redsteer (2016) demonstrate that only species with specific rooting traits may be successful pioneers into areas with active sand movement.

A comparison and evaluation of current methods used to quantify dune mobility indicate that dunes in drier areas, and especially areas with an augmented sediment supply from ephemeral stream beds, have begun migrating at rates directly proportional to local wind energy. Within the multiple possible states of dune stability described by Yizhaq et al. (2009) and Hugenholtz and Wolfe (2005) that are dependent upon the degree of plant cover, locations with high sediment supply, such as the fully-mobile dunes at Grand Falls, could require a significant change to wetter and calmer conditions in order to re-couple the state of dune activity to what current climatic (P/PE) conditions would predict. Once brought to a highly active state, via the absence of stabilizing cover eliminated by repeated perturbations, normal climatic regimes are likely insufficient to reverse activity. It is highly likely that increases to P/PE over decades or centuries could be required and/or significant decreases to wind drift potential are necessary to stabilize fully mobile dunes.

Changing seasonal and inter-annual climate variability, changing run-off as snowpack declines and glaciers recede, and prolonged and more intense droughts are expected to be part of our future as climate change occurs (Prein et al. 2016). These changes also dramatically affect the amount and composition of vegetation and aeolian geomorphologic processes, but the details of the connections between the two are not well understood. The southern Colorado Plateau dune ecosystems are highly vulnerable to degradation due to climate change and human activities. The Lancaster mobility index used over a period of years or decades serves as a valuable indicator of perturbations that force dune fields into new mobility states. Monitoring the trend and spatial variability in P/PE over shorter time scales provides insight into the relative influence of drought conditions, and the duration of dry conditions that lead to changes in vegetation and dune mobility.

Currently, understanding and evaluating periods of dune stability and aeolian sediment movement within the context of regional paleoclimatic records have become important for evaluating future responses to drought and global climate change. Because of topographic complexities and changes to sediment supply and transport that can occur during discrete geologic or climate-related events, an examination and understanding of deposits in the context of local geology and landscape altering geologic events are critical for interpretation of controls on dune mobility. Climatic variations and droughts can be localized or regional, and changes to mobility may vary depending on where reactivation of older dunes and/or renewed sediment supply occurs (Muhs and Holliday 2001). However, correlating across different dune areas of the southern Colorado Plateau suggest that discrete periods of dune activity have occurred in during MIS 3 and MIS 5 interglacials as well as during the mid-Holocene, as a response to the warm, dry conditions. These correlations suggest periods of aridity control regional responses in dune ecosystems, whereas less correlative ages are more likely reflective of local basin attributes. Future studies could work toward further refinement of comparisons between falling dune ages (periods of sediment transport) and ages of fixed dunes (periods of stabilization). As yet, the role of seasonality in paleodune and paleodrought history is still unclear, although current seasonality imparts significant influence on dune mobility today.

## References


Anderson RS, Betancourt JL, Mead JI, Hevly RH, Adam DP (2000) Middle- and late Wisconsin paleobotanic and paleoclimatic records from the southern Colorado Plateau, USA. Palaeogeogr Palaeoclimatol Palaeoecol 155:31–57

Ash JE, Wasson RJ (1983) Vegetation and sand mobility in the Australian desert dunefield. Z Geomorphol Suppl 45:7–25

Barchyn TE, Hugenholtz CH (2015) Predictability of dune activity in real dune fields under unidirectional wind regimes. J Geophys Res 120:159–182

Benson LV, Berry M, Joliec EA, Spanglerd JD, Stahle DW, Hattori EM (2007) Possible impacts of early-11th, middle-12th- and late-13th-century droughts on Western Native Americans and the Mississippian Cahokians. Quat Sci Rev 26:336–350

Billingsley GH (1987) Geology and geomorphology of the southwestern Moenkopi Plateau, and southern Ward Terrace, Arizona, U.S. Geological Survey Bulletin 1672. U.S. Government Printing Office: Washington, DC, 18

Billingsley G H, Block D L, and Redsteer MH (2013) Geologic map of the Winslow 30'× 60′ quadrangle, Coconino and Navajo Counties, AZ; U. S. Geological Survey Scientific Investigations Map 3247, 1:100,000

Billingsley GH, Block DL, Redsteer MH (2014) Geologic map of the eastern quarter of the Flagstaff 30'× 60′ quadrangle, Coconino County, Northern Arizona; U. S. Geological Survey Scientific Investigations Map 3279, 1:50,000

Bogle R, Redsteer MH, Vogel J (2015) Field measurements and analysis of climatic factors affecting dune mobility in the Grand Falls area of the Navajo Nation of the Colorado Plateau, southwestern United States. J Geomorphol 228:41–51

Breed CS, Breed WJ (1979) Windforms of Central Australia, and a comparison with some linear dunes on the Moenkopi Plateau, Arizona. In: El-Baz F (ed) Scientific results of the Apollo-Soyuz Mission, National Aeronautics and Space Administration (NASA) special paper, vol 412, pp 319–358

Coats LL, Cole KL, Mead JI (2008) 50,000 years of vegetation and climate history on the Colorado Plateau, Utah and Arizona, USA. Quat Res 70:322–338

Draut, AE, Redsteer MH, Amoroso L (2012) Vegetation, substrate, and eolian sediment transport at Teesto Wash, Navajo Nation, 2009–2012; U.S. Geological Survey Scientific Investigations Report 2012–5095, 72p

Duffield W, Riggs N, Kauffmann D, Champion D, Fenton C, Forman S, McIntosh W, Hereford R, Plescia J, Ort M (2006) Multiple constraints on the age of a Pleistocene lava dam across the Little Colorado River at Grand Falls, Arizona. Bull Geol Soc Am 118:421–429

Ellwein AL, Mahan SA, McFadden LD (2011) New optically stimulated luminescence ages provide evidence of MIS3 and MIS2 eolian activity on Black Mesa, northeastern Arizona, USA. Quat Res 75:395–398

Ellwein AL, Mahan SA, McFadden LD (2016) Impacts of climate change on the formation and stability of late Quaternary sand sheets and falling dunes, Black Mesa region, southern Colorado Plateau, USA. Quat Int 362:87–107

Esser RP (2003) $^{40}Ar/^{39}Ar$ Geochronology results from volcanic rocks from Grand Falls, Arizona. Socorro, New Mexico Bureau of Mines and Mineral Resources, New Mexico Geochronological Research Laboratory Internal Report NMGRL-IR-227, 5p

Forester RM, Lowenstein TK, Spencer RJ (2005) An ostacode based paleolimnologic and paleohydrologic history of Death Valley 200-0 ka. Geol Soc Am Bull 117:1379

Glover KC, MacDonald GM, Kirby ME, Rhodes EJ, Stevens L, Whitaker A, Lydon S (2017) Evidence for orbital and North Atlantic forcing in alpine southern California between 125 and 10 ka from multi-proxy analysis of Baldwin Lake. Quat Sci Rev 167:47–62

Greeley R, Iverson JD (1985) Wind as a Geological Process on Earth, Mars, Venus and Titan. In: Axford WI, Hunt GE, Greeley R (eds) . Cambridge Planetary Science Series, Cambridge/London/New York/New Rochelle/Melbourne/ Sydney, pp 168–169

Hack JT (1941) Dunes of the Western Navajo country. Geogr Rev 31:240–263
Hack JT (1942) The erosion and sedimentation in the Jeddito Valley and other valleys of Western Navajo country. In: The changing physical environment of the Hopi Indians of Arizona: reports of the Awatovi expedition. Peabody Museum, Harvard University report no. 1, Cambridge Massachusetts, USA, pp 45–69
Halfen AF, Lancaster N, Wolfe S (2016) Interpretations and common challenges of aeolian records from North American dune fields. Quat Int 410:75–96
Hereford R (1984) Climate and ephemeral-stream processes: Twentieth-century geomorphology and alluvial stratigraphy of the Little Colorado River, Arizona. Geol Soc Am Bull 95:654–668
Hugenholtz CH, Wolfe SA (2005) Recent stabilization of active sand dunes on the Canadian prairies and relation to recent climate variations. Geomorphology 68:131–147
Lancaster N (1988) Development of linear dunes in the southwestern Kalahari, southern Africa. J Arid Environ 14:233–244
Lancaster N (1994) Controls on aeolian activity: some new perspectives from the Kelso Dunes, Mojave Desert, California. J Arid Environ 27:113–125
Lancaster N (1995) Geomorphology of desert dunes. Routledge, p 290
Lancaster N, Baas A (1998) Influence of vegetation cover on sand transport by wind: field studies at Owens Lake, California. Earth Surf Process Landf 23:69–82
Lancaster N, Helm P (2000) A test of a climatic index of dune mobility using measurements from the southwestern United States. Earth Surf Process Landf 25:197–207
Manning JC (2010) Falling dunes in northeastern Arizona: an evaluation of Quaternary aeolian history using optically stimulated luminescence, M.S. Thesis. Northern Arizona University, Flagstaff, AZ, 102 p
Muhs DR, Benedict JB (2006) Eolian additions to late Quaternary alpine soils, Indian Peaks Wilderness Area, Colorado Front Range. Arct Antarct Alp Res 38:120–130
Muhs DR, Holliday VT (1995) Evidence for active dune sand on the Great Plains in the 19th century from accounts of early explorers. Quat Res 43:198–208
Muhs DR, Holliday VT (2001) Origin of late quaternary dune fields on the Southern High Plains of Texas and New Mexico. Geol Soc Am Bull 112:75–87
Muhs DR, Maat PB (1993) The potential response of aeolian sands to green house warming and precipitation reduction on the Great Plains of the United States. J Arid Environ 25:351–361
Okin GS (2008) A new model of wind erosion in the presence of vegetation. J Geophys Res 113:F02S10. https://doi.org/10.1029/2007JF000758
Prein AF, Holland GJ, Rasmussen RM, Clark MP, Tye MR (2016) Running dry: the U.S. Southwest's drift into a drier climate state, Geophys. Res Lett 43:1272–1279. https://doi.org/10.1002/2015GL066727
Redsteer MH (2016) Characterizing Dust Sources on Colorado Plateau Tribal Land. National Atmospheric Deposition Program Annual Conference, Santa Fe, NM
Redsteer MH, Block D (2004) Drought conditions accelerate destabilization of sand dunes on the Navajo Nation, southern Colorado Plateau. Geol Soc Am Abstr Programs 36(5):171. Paper No 66–8
Redsteer MH, Hayward RK (2015) A field comparison of basalt vs. quartz sediment transport in the Grand Falls Dune Field, northeastern Arizona, USA. International Planetary Dunes Workshop, Boise ID
Redsteer MH, Bogle R, Vogel J, Block D, Middleton B (2010a) The history and growth of a recent dune field at Grand Falls, Navajo Nation, NE Arizona. Geol Soc Am Abstr Programs 42(5):416. Paper No. 170–5
Redsteer MH, Kelley KB, Francis H, Block D (2010b) Disaster risk assessment case study: recent drought on the Navajo Nation, southwestern United States *in* Annexes and Papers for the 2011 Global Assessment Report on Disaster Risk Reduction, United Nations, 19p. Available at http://www.preventionweb.net/english/hyogo/gar/2011/en/what/drought.html

Redsteer MH, Bogle RC, Vogel JM (2011) Monitoring and Analysis of Sand Dune Movement and Growth on the Navajo Nation, Southwestern United States; U.S. Geological Survey Fact Sheet 2011–3085. Available at http://pubs.usgs.gov/fs/2011/3085/

Redsteer MH, Bemis K, Chief KD, Gautam M, Middleton BR, Tsosie R (2013) Unique challenges facing Southwestern tribes: impacts, adaptation and mitigation. In: Garfin G, Jardine A, Overpeck J (eds) Assessment of climate change in the Southwest United States: a technical report prepared for the U.S. National Climate Assessment. Island Press, pp 385–404

Seager R, Ting M, Held I, Kushnir Y, Lu J, Vecchi G, Huang H, Harnik N, Leetmaa A, Lau N, Li C, Velez J, Naik N (2007) Model projections of an imminent transition to a more arid climate in Southwestern North America. Science 316(5828):1181–1184

Simmons AH (1983) Archaeological investigations into the prehistory of northwestern New Mexico: data recovery on Block IV and V of the Navajo irrigation project. Final report submitted to ESCA-Tech. Corp., Costa Mesa, CA

Stokes S, Breed CS (1993) A chronostratigraphic re-evaluation of the Tusayan Dunes, Moenkopi Plateau and southern Ward Terrace, Northeastern Arizona. In: Pye K (ed) The dynamics and environmental context of Aeolian sedimentary systems, Geological society special publication, vol 72, pp 75–90

Sutton RL (1976) Cenozoic chronostratigraphic studies, Black Mesa region of Arizona: Geological Survey Research. US Geol Surv Prof Pap 1000:75–76

Thomas K, Redsteer MH (2016) Vegetation of semi-stable rangeland dunes of the Navajo Nation, southwestern USA. Arid Land Res Manag 00:1–12

Tsoar H (2001) Types of aeolian sand dunes and their formation. In: Balmforth NJ, Provenzale A (eds) Geomorphological fluid mechanics. Springer, Berlin/Heidelberg/New York, pp 403–429

Tsoar H (2005) Sand dune mobility and stability in relation to climate. Physica A 357:50–56

Tsoar H, Møller JT (1986) The role of vegetation in the formation of linear sand dunes. In: Nickling WG (ed) Aeolian geomorphology. Allen and Unwin, Boston, pp 75–95

Tsoar H, Blumberg DG, Wenkart R (2008) Formation and geomorphology of the north-western Negev sand dunes. In: Breckle S-W, Yair A, Veste M (eds) Arid dune ecosystems, Ecological Studies, vol 200. Springer, Berlin/Heidelberg, pp 25–48

Weiss JL, Castro CL, Overpeck JT (2009) Distinguishing pronounced droughts in the southwestern United States: seasonality and effects of warmer temperatures. J Clim 22:5198–5932

Wells SG, McFadden LD, Schultz JD (1990) Eolian landscape evolution and soil formation in the Chaco dune field, southern Colorado Plateau, New Mexico. Geomorphology 3:517–546

Weng C, Jackson ST (1999) Late Glacial and Holocene vegetation history and paleoclimate of the Kaibab Plateau, Arizona. Palaeogeogr Paleoclimatol Paleoecol 153:179–201

Westerling AL, Hidalgo HG, Cayan DR, Swetnam TW (2006) Warming and earlier spring increase western U.S. forest wildfire activity. Science 313:940–943

Wiggs GF, Thomas DSG, Bullard E (1995) Dune mobility and vegetation cover in the Southwest Kalahari Desert. Earth Surf Process Landf 20:515–529

Wolfe SA (1997) Impact of increased aridity on sand dune activity in the Canadian Prairies. J Arid Environ 36:421–432

Yizhaq H, Ashkenazy Y, Tsoar H (2009) Sand dune dynamics and climate change: a modeling approach. J Geophys Res 114(F1):F0102

# Chapter 9
# Dunefields of the Southwest Deserts

**Nicholas Lancaster**

**Abstract** The southwestern parts of the USA and adjacent areas of northern Mexico are the most arid part of North America, and include the Great Basin, Mojave, Colorado, Sonoran, and parts of the Chihuahuan deserts. The region contains many small and several larger dune fields, spanning a range of states of activity from vegetation-stabilized to largely vegetation-free or active dune fields. Dunes have accumulated on the floors and margins of paleolake basins; adjacent to major perennial, seasonal, or ephemeral streams and in areas of groundwater discharge. This chapter provides a regional survey of dunefield occurrence and characteristics, sediment sources, and wind regimes, followed by in-depth discussion of dunefield characteristics and boundary conditions for selected dune fields, and concludes with an assessment of the role of past climates and environments on dunefield origins and development.

**Keywords** Basin and Range Province · Dune fields · Wind regime · Sediment supply · California · Arizona · Mexico · Nevada

## 9.1 Introduction

The southwestern parts of the USA and adjacent areas of northern Mexico are the most arid part of North America, and include the Great Basin, Mojave, Colorado, Sonoran, and parts of the Chihuahuan deserts, located in the Basin and Range physiographic province. The region contains many small and several larger dunefields (Fig. 9.1), spanning a range of states of activity from vegetation-stabilized to largely vegetation-free or active dune fields. Many of these dune areas are important centers of biodiversity and host endemic sand-obligate plants and insects (Epps et al. 1998; Pavlik 1989), as well as being the location of important archaeological sites (e.g. Quade 1986).

---

N. Lancaster (✉)
Desert Research Institute, Reno, NV, USA
e-mail: nick.lancaster@dri.edu



N. Lancaster, P. Hesp (eds.), *Inland Dunes of North America*,
Dunes of the World, https://doi.org/10.1007/978-3-030-40498-7_9

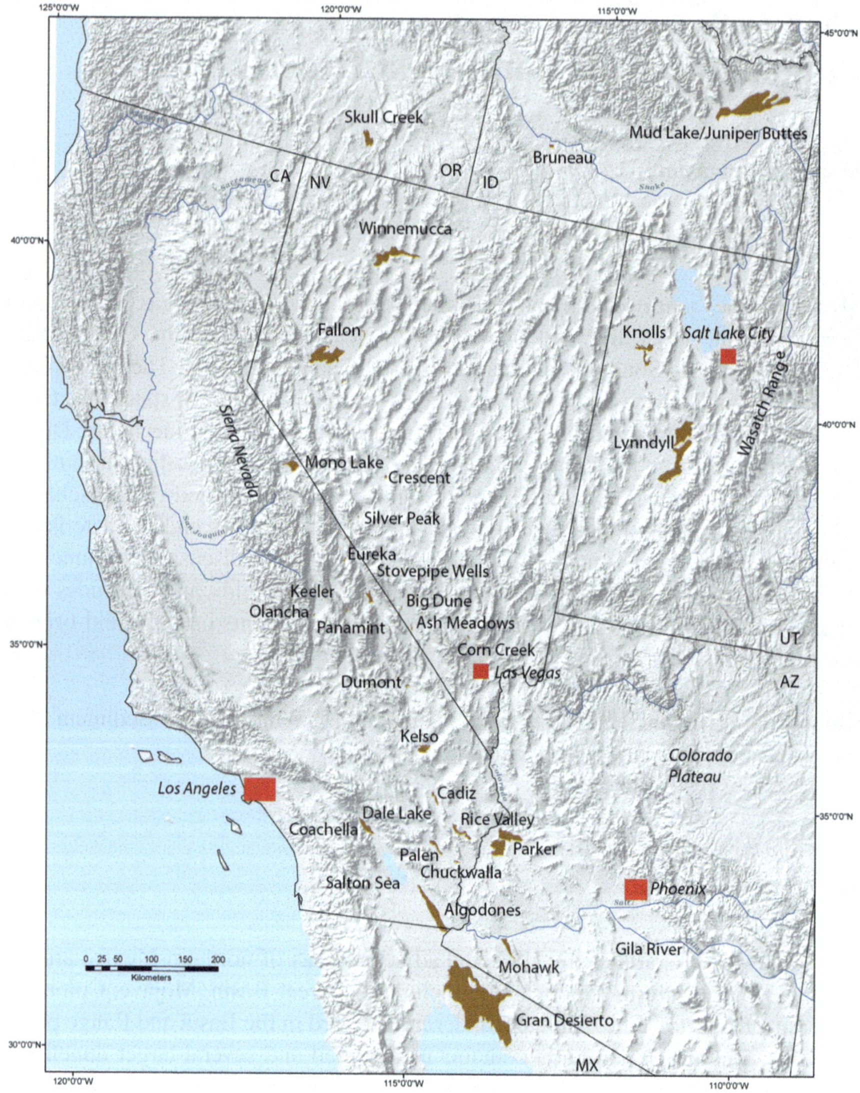


**Fig. 9.1** Dune fields of the Basin and Range Province, USA and Mexico

Although generally accessible, many dunefields in the region have not received scientific study, while others (e.g. Kelso, Algodones, Gran Desierto) have been extensively researched over many decades. Smith (1982) listed many, but not all, of the region's dune areas and provided a summary of the then-available information. Southern California dune fields are briefly described in Dean (1978). Sand sources are discussed by Muhs (2017) and Muhs et al. (2003). Wind regime characteristics for the Great Basin have been discussed by Jewell and Nicoll (2011) and for the Mojave Desert by Bach et al. (1996) and Laity (1987).

This chapter provides a regional survey of dunefield occurrence and characteristics, sediment sources, and wind regimes, followed by in-depth discussion of dunefield characteristics and boundary conditions for selected dunefields; and concluding with an assessment of the role of past climates and environments on dunefield origins and development. This chapter concentrates on dunes in the Sonoran, Colorado, Mojave, and Great Basin ecoregions. Colorado Plateau dunes are discussed by Margaret Hiza (Chap. 8).

## 9.2 Regional Setting

Most dune areas in the western deserts of North America are located in the Basin and Range physiographic province, which extends from northern Mexico to the Columbia Plateau in Idaho; and from the Sierra Nevada and Transverse Ranges in the west to the Wasatch Range and Colorado Plateau in the east. The region has a distinctive topography consisting of wide north-south trending valleys separated by narrow fault-bounded ranges and exhibits a wide range of elevations from −86 m in Death Valley to 4421 m at Mount Whitney. Mountain fronts are typically fringed by coalescing alluvial fans. Valley floors are flat to gently undulating and include ephemeral streams draining from the adjacent ranges to local playas. Generally, valleys are wider and ranges smaller in the southern Basin and Range (e.g. southern Arizona), where relief is lower, compared to those in central Nevada and eastern California.

Much of the region is part of the hydrologic Great Basin and is internally drained. Major internal drainages include the Mojave River, the Owens River, the Amargosa River draining to Death Valley, the rivers draining to the Lahontan Basin in northern Nevada; and those draining to the Bonneville Basin in Utah and adjacent areas of Nevada. Most drainages are ephemeral or seasonal; only rivers draining from the mountain ranges fringing the region (e.g. Sierra Nevada, Wasatch Range, Uinta Mountains) are perennial. Some areas of the eastern Mojave, Sonoran, and Colorado deserts drain to the Colorado River and its tributaries, which comprise the only external drainage for the region.

The region exhibits a wide range of climatic conditions, from the hot, arid lower elevations of the Sonoran, Mojave, and Colorado Deserts to the cold (semi-) arid high-elevation deserts of the central Great Basin. Precipitation mainly falls in the winter months in the Great Basin and Mojave Deserts; the Sonoran and Colorado deserts experience a summer rainfall maximum.

### 9.2.1 *Wind Regimes*

Wind regimes of the region reflect its mid-latitude location, with prevailing winds from NW to SW directions, which reach their maximum strength in the late winter-early spring (February – May) period. In the eastern Mojave, Sonoran, and Colorado deserts, summer SE winds associated with the North American monsoon circulation also occur, but they are generally of low velocity. Topographic effects are locally important, for example in the Owens Valley of eastern California, and the Coachella Valley of southern California, leading to locally increased wind speeds.

Analyses of wind regimes for the Mojave Desert (Bach 1995; Bach et al. 1996; Laity 1987), indicate the importance of the position and strength of the Pacific high-pressure cell wind velocity and direction. During the spring and early summer months, there is a thermal low in the interior of the desert, while the Pacific high is strong, resulting in prevailing westerly winds, with some southerly flow from the Gulf of California. In autumn and winter, the pressure gradient is weaker, resulting in lower wind speeds, except when frontal systems pass through or near the area (Shiyuan et al. 2008). In the Mojave Desert, western areas are dominated by westerly winds, whereas the eastern Mojave experiences more variable winds. In all areas of the Mojave, the direction of winds is strongly controlled by the topography, so that winds are funneled parallel to valley axes; this includes a prominent north-south element to the winds in the Colorado River Valley and immediately adjacent areas. In the Great Basin, Jewell and Nicoll (2011) point to the importance of the passage of extra-tropical cyclones and dry cold fronts, which reach maximum frequency in the February-May period. Wind directions are influenced by topography, although west, northwest, and southwest directions dominate.

Estimates of sand transport (drift) potential (DP) using the Fryberger and Dean (1979) approach by Jewell and Nicoll (2011) and Muhs et al. (2003) (Table 9.1), show that sand moving wind regimes are generally stronger in the western and eastern Great Basin, with high sand transport potential (DP); and of intermediate DP throughout most of the Mojave and Colorado deserts, except where topographic funneling of winds occurs (e.g., in the Coachella and Owens valleys). DP is low in the Sonoran Desert (Yuma, Phoenix). Resultant sand transport directions (RDP) are towards the E-NE in the western Great Basin and to the NE in the southern and eastern Great Basin (Jewell and Nicoll 2011). In the Mojave Desert, sand transport potential decreases from west to east, and resultant directions are towards the SE or E (Muhs et al. 2003).

## 9.3 Dunefield Distribution and Characteristics

Dune fields (mostly small) are distributed throughout the region. Table 9.2 summarizes the location, area, and dune types of the major dune fields in this region as shown in Fig. 9.1. In addition to the listed dune areas, there are many small dune

**Table 9.1** Wind regime characteristics of the basin and range

| | DP | RDP | RDP/DP | Resultant direction (°) | Energy Class |
|---|---|---|---|---|---|
| Bishop | 419 | 91 | 0.22 | 358 | High |
| China Lake NAS | 364 | 254 | 0.70 | 59 | Intermediate |
| Reno | 264 | 155 | 0.59 | 43 | Intermediate |
| Fallon NAS | 158 | 89 | 0.56 | 78 | Low |
| Winnemucca | 144 | 71 | 0.49 | 76 | Low |
| Elko | 106 | 65 | 0.61 | 66 | Low |
| Ely | 385 | 222 | 0.58 | 10 | Intermediate |
| Milford | 687 | 477 | 0.69 | 27 | High |
| Tonopah | 370 | 133 | 0.36 | 85 | Intermediate |
| Las Vegas | 475 | 251 | 0.53 | 53 | High |
| Salt Lake City | 282 | 157 | 0.56 | 4 | Intermediate |
| Cedar City | 419 | 202 | 0.48 | 28 | High |
| Daggett | 673 | 609 | 0.90 | 101 | High |
| Blythe | 203 | 77 | 0.38 | 94 | Intermediate |
| Indio | 114 | 103 | 0.90 | 135 | Low |
| Twentynine palms | 212 | 152 | 0.72 | 102 | Intermediate |
| El Centro | 392 | 350 | 0.89 | 80 | Intermediate |
| Yuma | 102 | 27 | 0.26 | 17 | Low |
| Phoenix | 21 | 5 | 0.24 | 63 | Low |

Data provided by D. Muhs, and from Jewell and Nicholl (2011)

areas in the valleys of the Great Basin and in the Mojave Desert that have yet to be systematically mapped and described.

Analysis of the topographic and geologic setting of these aeolian sand deposits indicates three main settings (Halfen et al. 2015): (1) the floors and margins of paleolake basins, especially the Bonneville and Lahontan basins; (2) adjacent to major perennial, seasonal, or ephemeral streams such as the Colorado, Gila, and Mojave rivers; and (3) areas of groundwater discharge (e.g. Ash Meadows, Nevada and California).

Dune fields in the northern Great Basin are characterized by vegetated linear and parabolic dunes. Where active dunes occur, they are generally of crescentic form. In the west-central Great Basin and northern Mojave, dune fields are comprised of small crescentic dunes and larger star and reversing dunes. Active dunes in the central and southern Mojave and adjacent areas are mostly crescentic in form, apart from the large complex ridges at Kelso. Vegetated linear and parabolic dunes are locally important in the eastern Mojave and adjacent areas of the Sonoran Desert. Most dune fields in the Colorado Desert are active and comprised of crescentic dunes. In the Sonoran Desert, dune fields are mostly small, with the exception of the Gran Desierto, which is comprised of a mosaic of star, crescentic, reversing, and vegetated linear dunes.

**Table 9.2** Dunefield characteristics

| | Center Lat (°) | Center long (°) | Area (Sq km) | Primary dune types |
|---|---|---|---|---|
| **Great Basin** | | | | |
| Silver peak | 37.66 | −117.627 | 2.84 | Crescentic, star |
| Fallon | 39.572 | −118.784 | 763.00 | Parabolic, linear |
| Winnemucca | 41.017 | −118.047 | 556.00 | Parabolic, crescentic, linear, barchan |
| Big Dune (Amargosa) | 36.649 | −116.583 | 3.74 | Star |
| Olancha | 36.291 | −117.968 | 7.18 | Crescentic, reversing |
| Keeler | 35.5 | −117.893 | 0.99 | Crescentic, parabolic |
| Skull Creek | 42.432 | −118.99 | 194.18 | Parabolic |
| Sand Mountain | 39.309 | −116.387 | 10.08 | Star |
| Ash Meadows | 36.431 | −117.351 | 11.20 | Crescentic |
| Corn Creek | 36.461 | −115.385 | 9.03 | Nebkha |
| Crescent Dunes | 38.239 | −117.326 | 12.13 | Star |
| Lynndyl (Little Sahara) | 39.4 | −112.462 | 1323.14 | Parabolic, linear, crescentic |
| Knolls dunes | 40.518 | −113.334 | 222.12 | Crescentic, reversing |
| **Sonoran** | | | | |
| Gran Desierto | 31.87 | −114.09 | 4816.65 | Sand sheet, crescentic, linear, star |
| Mohawk | 32.654 | −113.783 | 146.48 | Crescentic |
| Parker Dunes | 33.969 | −114.157 | 741.16 | Parabolic, linear |
| Gila River | 33.245 | −112.126 | 2.97 | Crescentic |
| **Mojave** | | | | |
| Stovepipe Wells | 36.683 | −117.137 | 75.63 | Crescentic, star |
| Eureka | 37.086 | −117.67 | 8.09 | Star |
| Dumont | 35.683 | −116.218 | 13.70 | Star |
| Panamint | 36.454 | −117.454 | 7.94 | Star |
| Kelso | 35.02 | −115.88 | 122.36 | Crescentic, star, parabolic, linear |
| Cadiz | 34.30 | −115.33 | 603.32 | Crescentic, parabolic |
| Dale Lake | 34.12 | −115.70 | 545.32 | Sand sheet |
| Palen Valley | 33.73 | −115.23 | 86.39 | Crescentic, parabolic |
| Chuckwalla | 33.59 | −114.84 | 17.19 | Sand sheet |
| Rice Valley | 33.98 | −114.88 | 136.02 | Linear, parabolic |
| **Colorado** | | | | |
| Algodones | 32.88 | −115.06 | 1696.85 | Crescentic |
| Salton Sea barchans | 33.192 | −115.855 | 8.91 | Barchan |
| Coachella Valley | 33.871 | −116.38 | 164.84 | Crescentic, parabolic |

### 9.3.1 Colorado and Lower Sonoran Deserts

Aeolian deposits in this region include sand sheets, nebkhas, and small parabolic dune fields in the Coachella Valley (Beheiry 1967; Griffiths et al. 2002; Katra et al. 2009; Wasklewicz and Meek 1995), many of which have now been destroyed by urban development. Sediment sources for these dunes are dominated by the Whitewater River, draining the San Bernardino Mountains, although local sources in ephemeral washes are also important (Griffiths et al. 2002).

In the Salton Trough, the major dunefield is the active compound crescentic Algodones Dunes of southeastern California (Derickson et al. 2008; Sweet et al. 1988), sourced from quartz-rich sediments deposited during periods when the Colorado River avulsed to the central Salton Trough (Muhs et al. 2003). In addition, small areas of barchans occur on the western margin of the Salton Sea (Haff and Presti 1995; Pelletier 2013).

In the Sonoran Desert, fluvial source-bordering dunes occur in the Gila River valley (Wright et al. 2011) and adjacent to the Colorado River in the Parker dunefield, which comprises mainly vegetation-stabilized linear and parabolic dunes (Muhs et al. 2003). The Gran Desierto sand sea of northern Mexico lies between the Colorado River delta, the Gulf of California and the Pinacate volcanic complex. It is comprised of extensive sand sheets and active and relict crescentic dunes that surround the core of chains and clusters of star dunes (Beveridge et al. 2006; Blount and Lancaster 1990; Scheidt et al. 2011).

### 9.3.2 Mojave Desert

Aeolian sediments and landforms are widespread throughout the Mojave Desert of southern California and adjacent areas of Nevada (Fig. 9.2). Major dunefields include those at Dumont Kelso, Cadiz, Palen, and Rice Valley. Smaller areas of dunes also occur in Death Valley at Stovepipe Wells (Edgett and Blumberg 1994); and Ash Meadows (Lancaster and Mahan 2012). Sand ramps – topographically-controlled accumulations of aeolian sand interstratified with alluvial and colluvial deposits are common and preserve a long history of aeolian accumulation (Lancaster and Tchakerian 1996). Topographic control of winds, and therefore sand transport, results in a series of more or less well-defined aeolian sediment transport corridors that are characterized by areas of active and inactive (sparsely vegetated) dunes and sand sheets, together with sand ramps (Muhs et al. 2003; Zimbelman et al. 1995). Mineralogical and geochemical analyses point to the existence of local sand sources, which suggests that many of these sand transport corridors may be more apparent than real (Muhs et al. 2003; Zimbelman and Williams 2002). Currently, the most active of these transport corridors extend from the terminal fan-delta of the Mojave River to the Kelso Dunes; and in the Palen area of the southeastern Mojave.

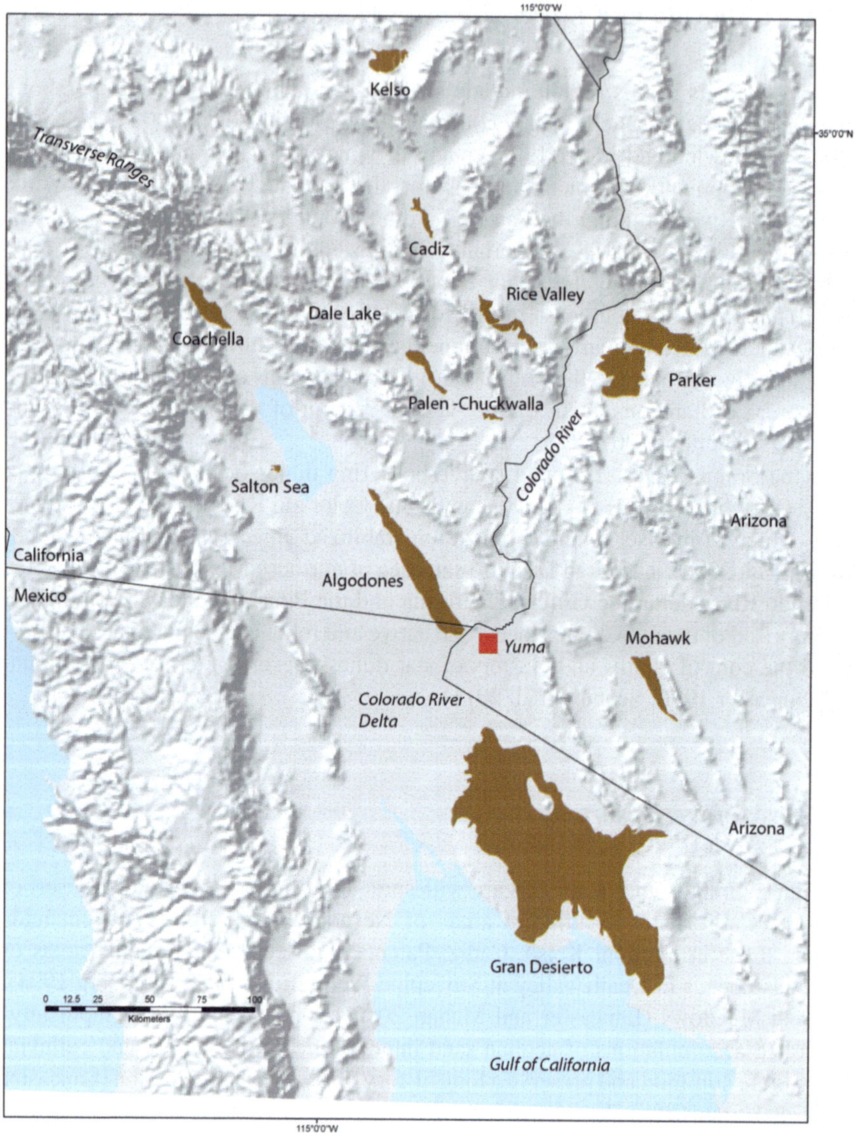


**Fig. 9.2** Dune fields of the Mojave, Colorado, and Sonoran Deserts

### 9.3.3 *Great Basin and Adjacent Areas*

Dune areas in the Great Basin Desert have not been studied extensively but include many small dune fields adjacent to playas. Some of these dunes are rich in silt and clay derived from the adjacent playa surfaces (Munroe et al. 2017). Small areas of parabolic and crescentic dunes also occur on the Great Salt Lake playa near Knolls,

Utah (Jewell and Nicoll 2011); and adjacent to Owens Lake, California (Lancaster et al. 2015; Lancaster and McCarley-Holder 2013).

Larger dune fields in central and western Nevada are comprised of star dunes surrounded by low crescentic and reversing dunes (e.g. Silver Peak, Crescent Dunes); or areas of vegetated parabolic dunes (e.g. Skull Creek dunes, Mehringer and Wigand 1986). The largest dune fields are associated with the basins of paleo-lakes Lahontan and Bonneville and are located adjacent to, or overlie, paleo-deltas of influent streams. They include the Fallon-Carson Sink dunefield, comprised largely of vegetation-stabilized parabolic and linear dunes on the paleo-delta of the Carson River (Morrison 1964), as well as the linear megadune of Sand Mountain (Snyder 1984). The Lynndyll or Little Sahara dunes of west-central Utah occupy a similar setting, located on the paleo-delta of the Sevier River as it entered Lake Bonneville. The dunefield is comprised of extensive areas of SW-NE trending elongate parabolic dunes, with smaller areas of active crescentic and barchan dunes (Sack 1987). The Winnemucca Dunes comprises a complex mosaic of different dune types that include active crescentic and transitional barchan to parabolic dunes, as well as vegetation stabilized parabolic and linear dunes and sand sheets, extending for some 60 km west to east (Pepe 2014). The dunefield occupies valleys that were inundated as part of Lake Lahontan.

## 9.4 Dunefield Examples

In this section, dune fields that have been extensively studied are discussed as examples of the variety of dune field morphology, wind regimes, sediment sources, and history in the region.

### *9.4.1 Kelso Dunes*

The Kelso Dunes form the depositional sink for the aeolian sediment transport system that extends for 60 km eastwards from the fan delta of the Mojave River as it exits Afton Canyon to the Kelso dune field via the area known as the Devils Playground.

The dunefield has been the subject of studies by a variety of workers, beginning with the seminal work of Sharp (1966). Lancaster (1993a) provided an overview of the geomorphology and sediments of the dunes, a precursor to studies of sand composition and potential sources from remote sensing (Ramsey et al. 1999) as well as mineralogy and geochemistry (Muhs et al. 2017; Yeend et al. 1984). The Kelso dunes was one of the early targets of luminescence dating in the region, providing constraints on dune ages (Clarke 1994; Sweeney et al. 2020; Wintle et al. 1994).

Although the dunefield is small (~ 122 km$^2$), it exhibits a range of dune types and as many as 14 distinct dune geomorphic units can be identified (Fig. 9.3). Four dune

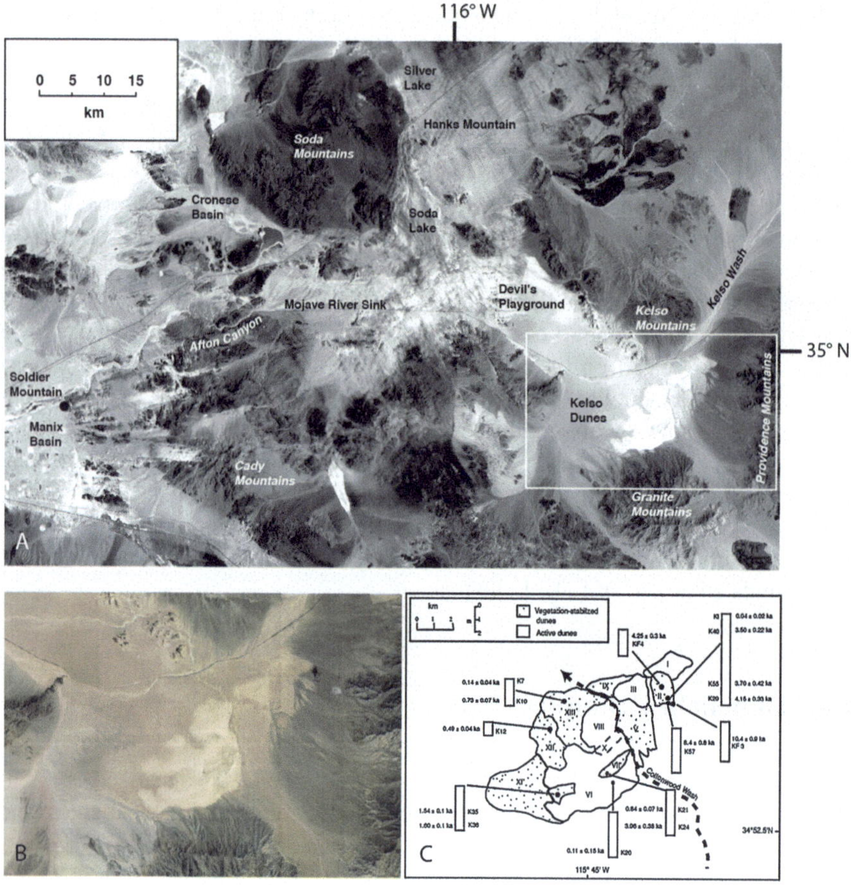


Fig. 9.3 Kelso Dunes: (**a**) Location and geomorphic context. Box indicates location of part **b** and **c**; (**b**) Google Earth Image of Kelso Dunes; (**c**) Geomorphic units at Kelso dunes and luminescence dating sites and ages. (From Lancaster and Tchakerian 2003)

units (III, VI, VIII, and X) are sparsely vegetated, while the others are all stabilized to a varying degree by vegetation and experience little or no active sand transport at present.

The wind regime of the Kelso area is dominated by winds from the W and WSW, with a peak in wind speed in spring; SE winds occur at all seasons, with a peak in the summer months; occasional winds from the N occur mainly during winter months.

As described by Lancaster and Tchakerian (2003), Kelso Dune field consists of a "core" of 3 large WSW-ENE-trending complex linear ridges up to 160 m high and 1900–2000 m apart (units VI, VIII, and X). On the west and northwest sides of the core of the dune field lie degraded, vegetated, straight-crested and barchanoid crescentic ridges up to 15 m high and linear dunes up to 5 m high (units XII – XIV). The eastern section of the dune field consists of a 1–5 m thickness of sand formed into areas of vegetation stabilized crescentic, parabolic and linear dunes (units I, II, IV, V), each cut by washes and separated from the "core" area by Cottonwood Wash, which is incised into aeolian and alluvial deposits by as much as 20 m. To the north is a smaller version of the main area of active dunes (unit III) that consists of three linear ridges up to 50 m high with superimposed 2–4- m-high crescentic dunes. Areas of low, partially active, linear and crescentic dunes and sand sheets occur on the northeast margins of the dunefield (units I, IX).

Sand streaks, sand sheets, and areas of active crescentic dunes connect the fan delta of the Mojave River (the Mojave River Sink) to the Kelso dunes via the area known as the Devils Playground. This connection indicates the importance of the Mojave River as a source of sand for the dunefield, as noted by many workers (see review in Muhs et al. 2017). The significance of local sources of sand from adjoining alluvial fans and the Kelso Wash has been much debated. Ramsey et al. (1999) were the first to clearly demonstrate the importance of local sources using a combination of mineralogical analyses and remote sensing data to show that dunes east of the Cottonwood Wash were largely derived from local sources in adjacent alluvial fans. Muhs et al. (2017) used a combination of bulk mineralogy and geochemical analysis of K feldspars to show the complexity of sand sources for the dunefield. Kelso dunes sand averages around 33% quartz, 25% K-feldspar, and 41% plagioclase (in contrast to the > 80% quartz inferred by some earlier workers (Paisley et al. 1991; Yeend et al. 1984) but similar to the results of Ramsey et al. (1999)). The sand composition is thus very similar to that of many granitic rocks, as well as sands from other parts of the Mojave Desert and adjacent areas (e.g. Lancaster et al. 2015; Muhs et al. 2003; Zimbelman and Williams 2002). The abundance of feldspar indicates a mineralogically immature sand composition and therefore a short history of weathering, erosion, and transport by wind and water. Geochemical analyses using the ratios of K/Rb and K/Ba by Muhs et al. (2017) provide valuable constraints on sand sources for the dunes. Sand from the western parts of the dunefield has a composition similar to that in the Devils Playground and the Mojave River, but also alluvium derived from the Granite Mountains to the south of the dunefield. In contrast, sand from the eastern areas of Kelso Dunes is distinctly different from the western dunes and has K/Rb and K/Ba values that indicate significant inputs from the deposits of alluvial fans that are derived from the Providence Mountains east of the dunefield.

Estimates of the age of Kelso Dunes have varied widely between "several thousand years and possibly 10,000–20,000 [years]" (Sharp 1966) to "very likely greater than 100,000 years, and quite possibly more than a million years (Yeend et al. 1984). Kelso Dunes was an early target of luminescence dating studies (Clarke 1994; Wintle et al. 1994). IRSL ages were obtained from several dune units (Fig. 9.3c), including the core of active dunes (unit VI), vegetation stabilized dunes and sand sheets on the eastern margin of the dune field (Unit II), vegetation-stabilized crescentic ridges on the northern margin of the dune field (units IX and XII), and vegetated crescentic ridges on the southwest side of the dunes (Unit XIV). IRSL dates from dune Unit II indicate that it accumulated between 10.4 and 3.5 ka. Unit VI was apparently in place by 4 ka and the crescentic dunes of unit XIV were formed around 1.5 ka and dunes on the north side of the dune field were formed, or reworked, between 0.8 and 0.5 ka. The oldest sands known are those which were deposited as sand sheets on alluvial fan surfaces as much as 5 km southeast of the present dune margins between 16 and 18 ka (Clarke 1994; McDonald and McFadden 1994) and as dunes and sand ramps northwest of the main dunefield (Sweeney et al. 2020). The pattern of luminescence ages suggests that there is an increase in the minimum ages of dune and sand sheet units from northwest to southeast, supporting the hypothesis that the dune field accumulated by stacking or shingling of successive generations of dune units on the piedmont of the Providence and Granite Mountains (Lancaster and Tchakerian 2003).

Together, the available luminescence ages and sand composition suggest that the accumulation of the Kelso dunefield likely postdates the integration of the Mojave River drainage with the Soda Lake basin via Afton Canyon, which occurred sometime after 25 ka (Reheis et al. 2007, 2012), and provided the source of much of the sand in the dunefield. Fluctuations in the hydrology of the Soda Lake basin, including the formation and desiccation of Lake Mojave, likely affected sediment supply and availability to the Kelso dunes (Kocurek and Lancaster 1999), resulting in episodic accumulation of sand, particularly between 25 and 9 ka (Sweeney et al. 2020). Likewise, periods of fan aggradation and incision affected sediment supply to the eastern part of the dunefield (McDonald et al. 2002), as documented by interstratified distal alluvial fan and aeolian sand sequences (Clarke 1994; Sweeney et al 2020) on the Providence Mountains piedmont.

### 9.4.2 Gran Desierto

The Gran Desierto sand sea of northern Mexico (Fig. 9.4) is the largest active dunefield in North America, with an area of 4816 km$^2$. It lies adjacent to the Gulf of California and the Colorado River delta in the southeastern extension of the tectonically active Salton Trough and is underlain by as much as 3200 m of fluvial-deltaic sediments that accumulated in the fault-bounded Altar Basin (Pacheco et al. 2006). The dunefield is bounded by the uplifted block of Pleistocene fluvial and deltaic sediments comprising the Sonoran Mesa and Mesa Arenosa to the west and southwest; to the east and north lie the Pinacate volcanic center and alluvial fans derived from granitic ranges of the southern part of the Basin and Range Province. The dunefield has been studied extensively (Beveridge et al. 2006; Blount et al. 1990; Kasper-Zubillaga et al. 2007; Lancaster 1992, 1993b, 1995; Lancaster et al. 1987; Scheidt et al. 2011), and the reader is referred to these detailed studies for further information.

The Gran Desierto dunefield lies in the most arid part of the Sonoran Desert and experiences a wind regime characterized by winter winds from the NNW- NNE, spring winds from the W-WNW, and summer monsoonal winds from the S-SE. The

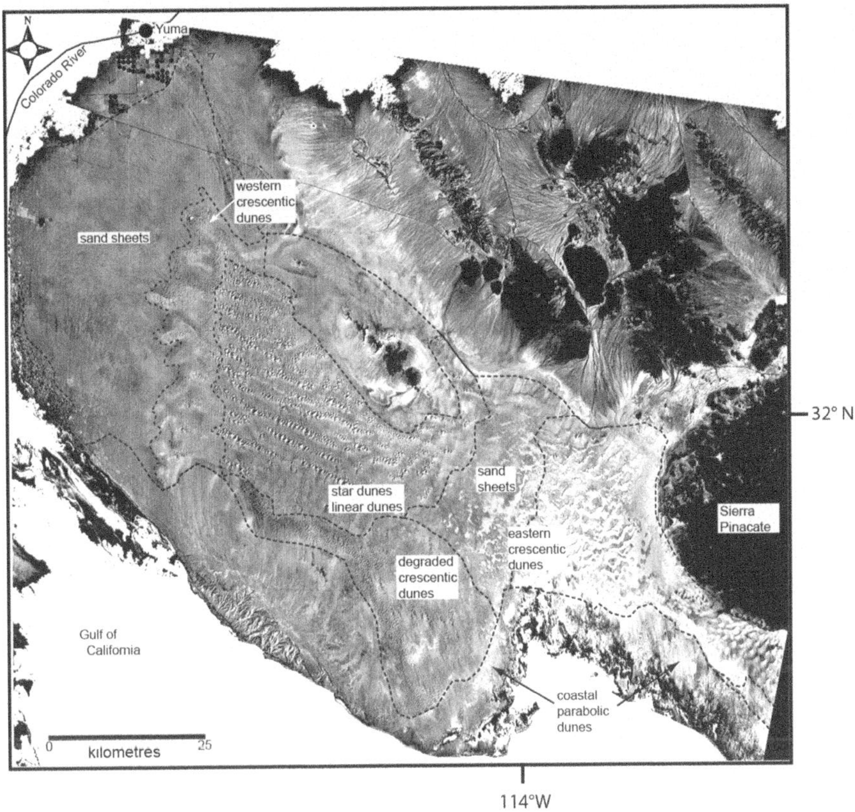


**Fig. 9.4** The Gran Desierto Sand Sea. (After Beveridge et al. 2006)

S-SE winds become more prominent towards the east of the dunefield (Lancaster 1989).

The Gran Desierto contains a variety of dune types (Fig. 9.4) forming a mosaic of different morphological types and generations of dunes, all of which overlie Colorado fluvial deposits and/or distal alluvial fan deposits. Major dune geomorphic units identified by Beveridge et al. (2006) include relict linear dunes, star dunes, degraded crescentic dunes, and active crescentic dunes. Sand sheets characterize the northwestern portion of the Gran Desierto and also separate the eastern area of crescentic dunes from the remainder of the dune field. The central part of the dune field consists of 80–100 m-high star dunes that occur in distinct chains and clusters (Lancaster 1989), with adjacent, lower-elevation reversing dunes. To the south of the star dunes and also underlying them are partly vegetated, degraded 15–20 m high crescentic dunes. These dunes are capped by a pedogenic horizon with abundant rhizoconcretions, incipient cementation, and weak caliche development (Lancaster 1992), indicating a relatively long period of stability. West and northwest of the star dunes are the western area of crescentic dunes, which are typically 5–10 m high, and partially vegetation stabilized (Lancaster 1995). The surface of the vegetation-stabilized dunes exhibits a surface similar to that on crescentic dunes to the south of the star dune area. Towards the southeast and the core area of star dunes, many of the crescentic dunes have a near-symmetrical reversing dune morphology. The eastern area of crescentic dunes is separated from the central part of the dune field by vegetated sand sheets. The eastern crescentic dunes form a clearly defined area of 5–80 m-high simple and compound crescentic dunes adjacent to the Pinacate volcanic complex. Vegetated sand sheets extend north and west of the main dune area to the Colorado River valley and south to the Bahia Del Adair. In the coastal areas of the Bahia del Adair and further east (including the Sonoyta River valley and areas of the coastal plain), there are areas of vegetation-stabilized parabolic and linear dunes.

Sediment sources for the Gran Desierto sand sea have been investigated using a combination of mineralogical and geochemical analyses and remote sensing data (Blount and Lancaster 1990; Blount et al. 1990; Kasper-Zubillaga et al. 2007; Scheidt et al. 2011). Sources include quartz-rich sand from the Colorado River, derived directly via reworking of fluvial sands or indirectly via relict Colorado River sediments such as those exposed in the Mesa Arenosa and/or coastal sand reworked by longshore drift from the Colorado River delta; feldspar-rich sand sourced from distal alluvial fan or fluvial sediments derived from Basin and Range mountains north of the sand sea; and carbonate-rich sands derived from the Bahia del Adair. Detailed analyses of remote sensing data using spectral unmixing of thermal infrared wavelengths as well as mineralogy and geochemistry show that sand composed of > 70% quartz makes up the bulk of the western crescentic dunes and the western part of the star dune area. These dunes have a composition similar to dunes sourced from Colorado River sediments (e.g. Algodones, Parker dunes). The percentage of quartz declines towards the east, so that the eastern group of crescentic dunes are comprised of ~ 70% quartz. There is a parallel trend in sand grain size and sorting, such that the eastern crescentic dunes are finer and better sorted compared to the western crescentic dunes and the star dunes. Dunes on the northern margins of the sand sea, as well as those adjacent to the Sonoyta River, are feldspar-rich and have a composition similar to dunes in the Mojave Desert derived from granitic terrains in the Basin and Range Province. Dunes in coastal areas have a mixed composition, and are either quartz rich (> 80%), or carbonate rich (7–53%), reflecting sources in both tidal flats and reworked Colorado River sediments.

The age of the sand sea is constrained by stratigraphic and geomorphic relationships. Dune sands overlie fluvial deposits of the ancestral Colorado River that formerly flowed from NNW to SSE through the area of the western star dunes to the Bahia del Adair, prior to its avulsion to the west following the uplift of the Mesa Arenosa (Blount and Lancaster 1990). OSL ages (Beveridge et al. 2006) indicate that cross-bedded gravelly sands of fluvial origin accumulated around 61.5 ka. Sand containing caliche and pebbles and underlying the western star dunes has an OSL age of 26.7 ± 4.7 ka. The dune pattern is therefore inferred to be younger than 25 ka (Beveridge et al. 2006). Sampling of specific dune morphological units provides a minimum age of 11–12 ka for the degraded crescentic dunes south of the star dune area; while the eastern crescentic dunes likely formed 7 ka ago, sourced from an influx of quartz-rich sand from reworked Colorado River sediments.

### *9.4.3 Algodones Dunes*

The Algodones Dunes (Fig. 9.5) are a compact, elongate dunefield located along the southeast margin of the Salton Trough, between the alluvial fans of the Chocolate Mountains to the east, and East Mesa to the west. The dunefield terminates abruptly at the Colorado River valley. The dunefield has received extensive study over the years starting with Norris and Norris (1961). Dunefield characteristics are summarized by Derickson et al. (2008), Sweet et al. (1988), while Muhs et al. (1995), Winspear and Pye (1995) and Muhs (2017) provide information on the source of the sand for the dunes.

A characteristic feature of the Algodones Dunefield is the juxtaposition and in some cases superposition of dunes with different morphologies (Norris and Norris 1961; Sweet et al. 1988). The core of the dune field consists of large (up to 80 m-high) compound mega-crescentic dunes that are migrating at 2 – 5 m per year to the southeast (Havholm and Kocurek 1988; Sharp 1979). To the east lie small crescentic dunes, nebkas, and sand sheets, the latter intercalated with distal alluvial deposits. To the west is an area of zibars and linear dunes. Between the linear dunes and the main compound crescentic dunes is an area of smaller crescentic dunes, which appear to be migrating over the flanks of the mega-dunes, and merging with their superimposed crescentic dunes (Derickson et al. 2008). In addition to the main dunefield, vegetation-stabilized linear dunes and sand sheets occupy the East Mesa area.

The dune field has developed under a wind regime dominated by winter and spring winds from the N and W, with NW winds dominating. Summer winds are from the SE, but are very low-energy (Sweet et al. 1988).

The origins of the Algodones dunefield have been much debated. Although the dunefield lies northwest (upwind) of the Colorado River, the quartz-rich mineralogy of the dune sands clearly indicates that they are derived from this source (Muhs 2017; Muhs et al. 1995). Prior to its flood-related avulsion in 1905–1906, the Colorado River has avulsed into the Salton Trough several times in the past 2000 years, filling paleolake Cahuilla to the 12 m asl shoreline (Waters 1983), most recently 400–550 years ago. Sand from the beaches of Lake Cahuilla is quartz-rich and deflation of sand from these and lake plain sands exposed during periods of dessication of the lake is proposed as the main source for the dunefield (Muhs et al. 1995; Stokes et al. 1997; Winspear and Pye 1995). Derickson et al. (2008) suggest that the dunefield originated as a series of inland-migrating lake-margin dune ridges that subsequently were reworked by NW winds into the crescentic dunes that dominate the dunefield today.

Luminescence and radiocarbon ages for beach ridges and aeolian sands indicate that the dunefield is likely younger than 31 ka, and has experienced significant

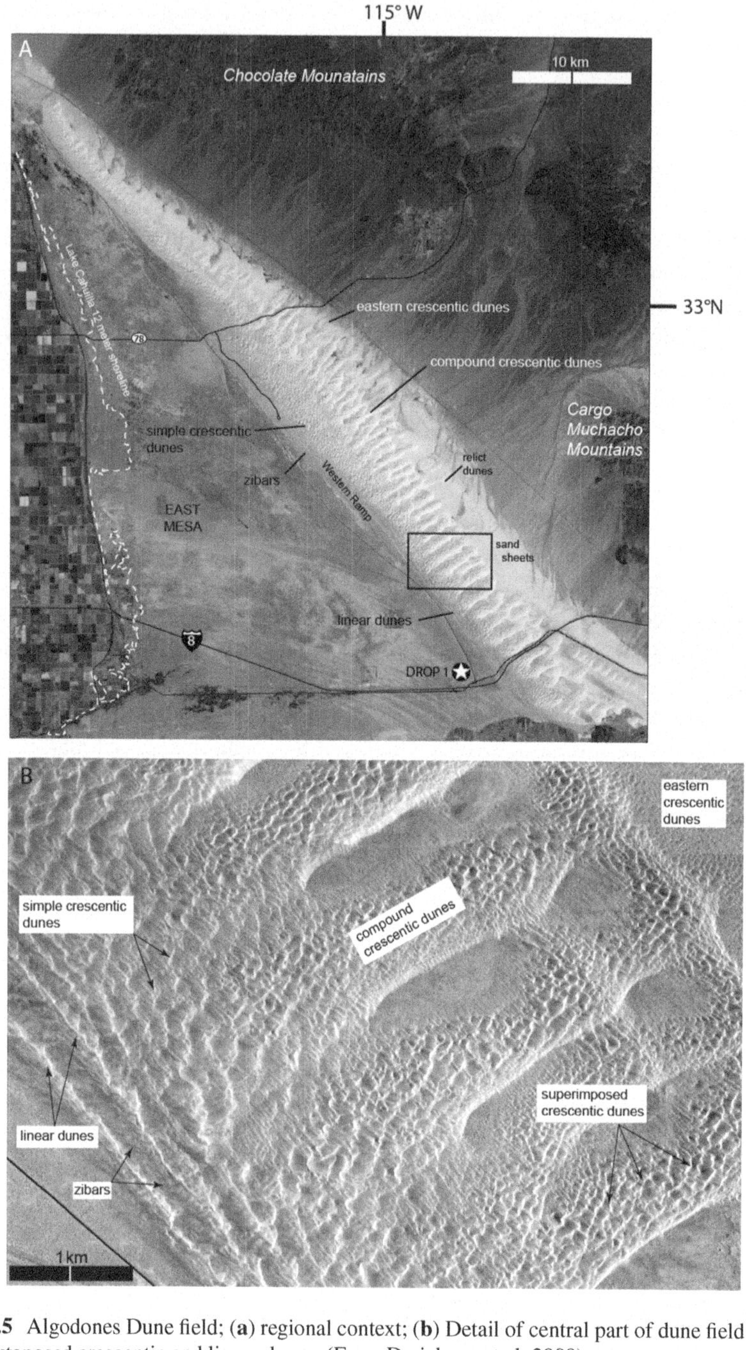


**Fig. 9.5** Algodones Dune field; (**a**) regional context; (**b**) Detail of central part of dune field showing juxtaposed crescentic and linear dunes. (From Derickson et al. 2008)

aeolian accumulation following the last major infilling of Lake Cahuilla 400–550 years ago (Derickson et al. 2008; Stokes et al. 1997). The age of the crescentic mega dunes is estimated from pattern analysis as 15–18.4 ka, a similar age to the linear megadunes of the adjacent Gran Desierto dunefield (Derickson et al. 2008). In contrast, the simple crescentic dunes on the western margins of the dunefield are estimated to be approximately 3 ka old, based on pattern analysis and an OSL age of 3.1 ± 1.1 ka for cross-bedded sands underlying the zibar (Stokes et al. 1997).

### 9.4.4 *Winnemucca Dunes*

The Winnemucca dunefield (Fig. 9.6) in north-central Nevada extends west to east for 60 km from the Desert Valley through Silver State Valley to Paradise Valley and crosses two mountain ranges- the Slumbering Hills and Bloody Run Hills. The dunefield occupies an area of 472 km$^2$ and is comprised of a mosaic of areas of active crescentic dunes together with extensive areas of active and vegetation-stabilized parabolic dunes, vegetated linear dunes, and sand sheets.

The dunefield was poorly known until recently. It was described first by Israel Russell (1885) and is briefly described in Smith (1982) and Eissmann (1990). Pepe (2014) provides a detailed mapping of the dunefield and an assessment of its sand sources and dynamics.

**Fig. 9.6** Winnemucca dunes, Nevada

The wind regime of the Winnemucca dunes is characterized by seasonally (April – July) strong SW-W winds. Winter winds are light southerly. Drift potential (DP) and resultant drift potential (RDP) at the Winnemucca airport from the years 1950–2000 shows a DP total of ~143 vector units, a RDP/DP ratio of ~0.5, and a resultant drift direction towards ~78–79$^0$ (Jewell and Nicoll 2011).

Although much of the dunefield consists of vegetation-stabilized sand sheets and linear and parabolic dunes, there are extensive areas of active dunes, which are migrating to the east at rates of up to 6.9 m/year. The majority of active dunes are dominated by various configurations of barchanoid, crescentic, and parabolic dunes, covering about 34% of the dunefield area (Fig. 9.6b).

A characteristic feature of this dunefield is the occurrence of dunes that are transitional between parabolic and barchan morphologies (termed barchanbolic by Pepe 2014). Barchanbolic dunes are typically 6–17 m-high and have a concave lee face with the shape of a barchan but also have moderately long parabolic arms. In some areas barchanbolic dunes are composed of three tails; two lateral parabolic-like tails and one medial tail that is oriented orthogonal to the crest of the dune at its midsection. Vegetation of various types stabilizes the tails on barchanbolic-parabolic dunes; these effects have produced some especially long tails that can exceed 3.2 km in length. Barchan dunes in this dunefield are unvegetated with brink heights of 3.7–25 m and are located within or near the boundaries of the crescentic complexes (Fig. 9.6c).

Areas of active crescentic dunes are developed where barchan, barchanbolic, and parabolic dunes merge and coalesce. Dunes in these complexes are typically 4.5–28 m high and consist mainly of crescentic ridges, with occasional superimposed star dune peaks. These complexes cover some 6% of the dunefield area. Isolated barchanbolic, parabolic, and undifferentiated dunes mixed with sand sheets mantle ~18% of the dunefield. Climbing dunes with transverse ridges, falling dunes and sand ramps with cliff-top shadow (echo dunes) occur where the dunefield crosses N-S oriented mountain ranges, especially at the eastern end of the dunefield.

Extensive vegetation-stabilized sand sheets dominate the western (upwind) parts of the dune field in Desert Valley, comprising around 12% of the area. Adjacent to and surrounding the active dune areas are vegetation-stabilized, semi-active (partially vegetated), and degraded dunes (dunes with a subdued topography relative to the original morphology), which cover as much as 43% of the dunefield area. These areas include vegetation stabilized parabolic dunes, 1- 3 m-high linear dunes, which may represent the trailing arms of parabolic dunes; widespread braided linear dunes; and small areas of semi-active to stabilized transverse dunes, and highly degraded dunes on lower slopes of adjacent mountain ranges.

The complexity of the patterns of dune morphology is mirrored by the complex patterns of sand grain size and sorting and bulk mineralogy. There is no consistent trend in grain size and sorting from west to east, although areas of stabilized dunes tend to be finer, but less sorted than adjacent active dunes. Bulk mineralogy determined by XRD indicates that sands in this dunefield vary widely in composition from 20–68% quartz and 80–32% feldspar (mostly albite). Lithic fragments are an important component, comprising 45% or more and vary between the different basins crossed by the dunefield. These observations suggest that sand is derived from multiple local sources which include metamorphic, plutonic, and volcanic bedrock in a similar manner that that observed in the Mojave Desert.

## 9.5 Controls on Sand Mobility and Dune Activity

Dune fields in the southwest deserts exhibit a wide range of states of sand and dune mobility, from fully active to stabilized by vegetation and degraded. In some cases, e.g. at Kelso dunes, Winnemucca dunes, and the Lynndyll dunefield, areas of active, largely unvegetated, dunes occur adjacent to vegetation-stabilized dunes. Dune field activity can be assessed using various climatic indices of aeolian mobility, such as the dune mobility index (Lancaster 1988). These indices relate sand movement and dune mobility to the ratio between wind speed above transport threshold and effective precipitation and have been shown to be a good predictor of annual measured sand movement and dune mobility (Hoover et al. 2018; Lancaster and Helm 2000).

Bach (1995) assessed dune mobility in the Mojave and Colorado deserts using the Lancaster (1988) index. He found that, although calculated values of the mobility index exceeded 200 units across many parts of the region, indicating fully active dunes, observational evidence showed that dunes were generally less active than predicted. This may be because the seasonal distribution of wind speed and vegetation cover is unfavorable for sand transport. Although wind speed peaks during the late-winter and early spring, seasonal and ephemeral vegetation cover is at a maximum at this time (Lancaster 1994).

Given the sparse distribution of observations of wind speed and direction in most desert regions, Lancaster and Hesse (2016) have developed a new approach in which digital mapping of dune field and sand sea extent has been combined with systematic observations of dune activity at 0.2° intervals from high resolution satellite image data, resulting in four classes of activity. Publically available global 1 km resolution gridded datasets for the aridity index (AI); precipitation, satellite-derived percent vegetation cover; and estimates of sand transport potential (DP) were re-sampled for each 0.2° grid cell, and dune activity was compared to vegetation cover, sand transport potential, precipitation, and the aridity index. Application of this approach to dune fields in the SW deserts requires some modification because of the small extent of most dune areas, but provides new insights into regional patterns of dune field activity in relation to climate and vegetation cover. Revised climatic indices of dune mobility (e.g. the ratio between sand transport potential and the aridity index – DP/AI) tend to decrease with latitude, but bear no clear relationship to the activity classes (Fig 9.7a). The strongest relationships to emerge are between DP/Vegetation Cover ratio and dune activity class (Fig 9.7b). Similar relationships are observed in other mid-latitude dune areas (e.g. China and Central Asia).

Preliminary data indicate that fully active (class 1) and vegetation stabilized dune fields (class 3 and 4) occur at all latitudes throughout the SW USA and northern Mexico (Fig. 9.8). Fully active dunes, however, are concentrated in a zone that

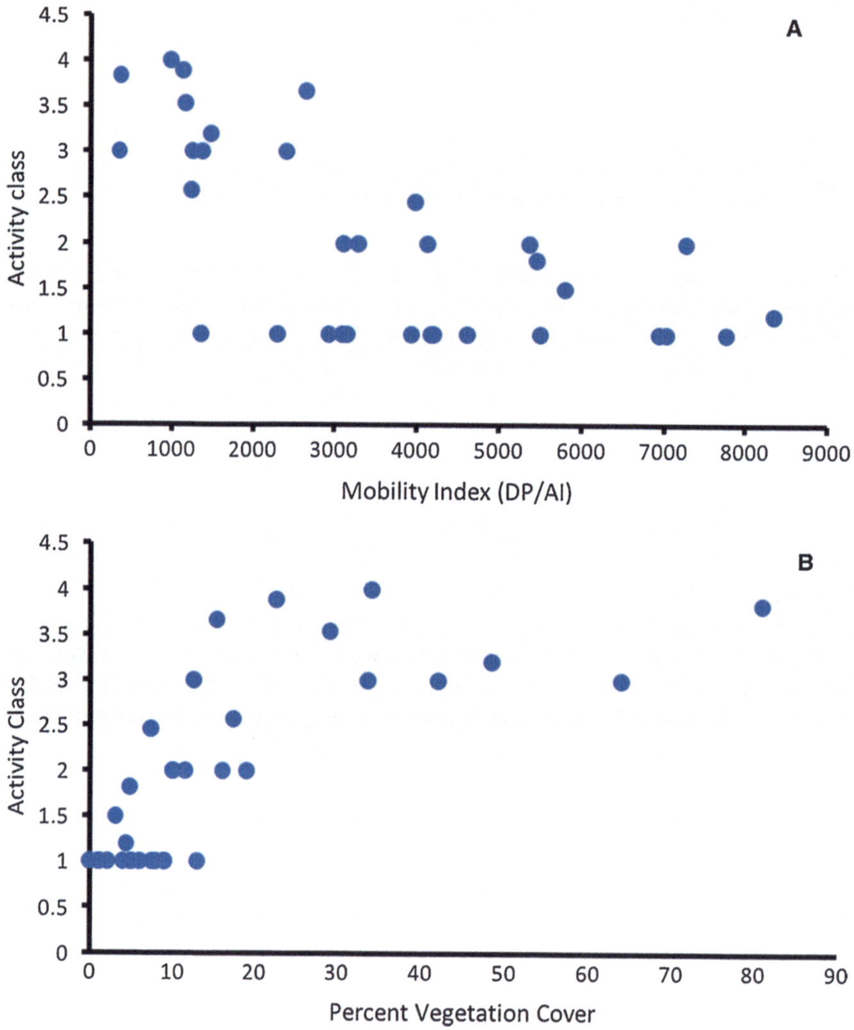


**Fig. 9.7** Dune activity classification of Basin and Range dune fields. (**a**) relationship between dunefield activity class and modified mobility index (DP/AI); (**b**) relationship between dunefield activity class and vegetation cover

extends from the Salton Trough northwards to west-central Nevada at latitude 37.66°N. This is an area that experiences generally high wind energy and DP, based on observations (Jewell and Nicoll 2011), and reanalysis data sets. Although wind energy can be locally high in the northern Great Basin, effective precipitation (increased values of the aridity index (AI)) is also higher, leading to a denser vegetation cover that reduces sand mobility in these areas. North of latitude 38°N, dune

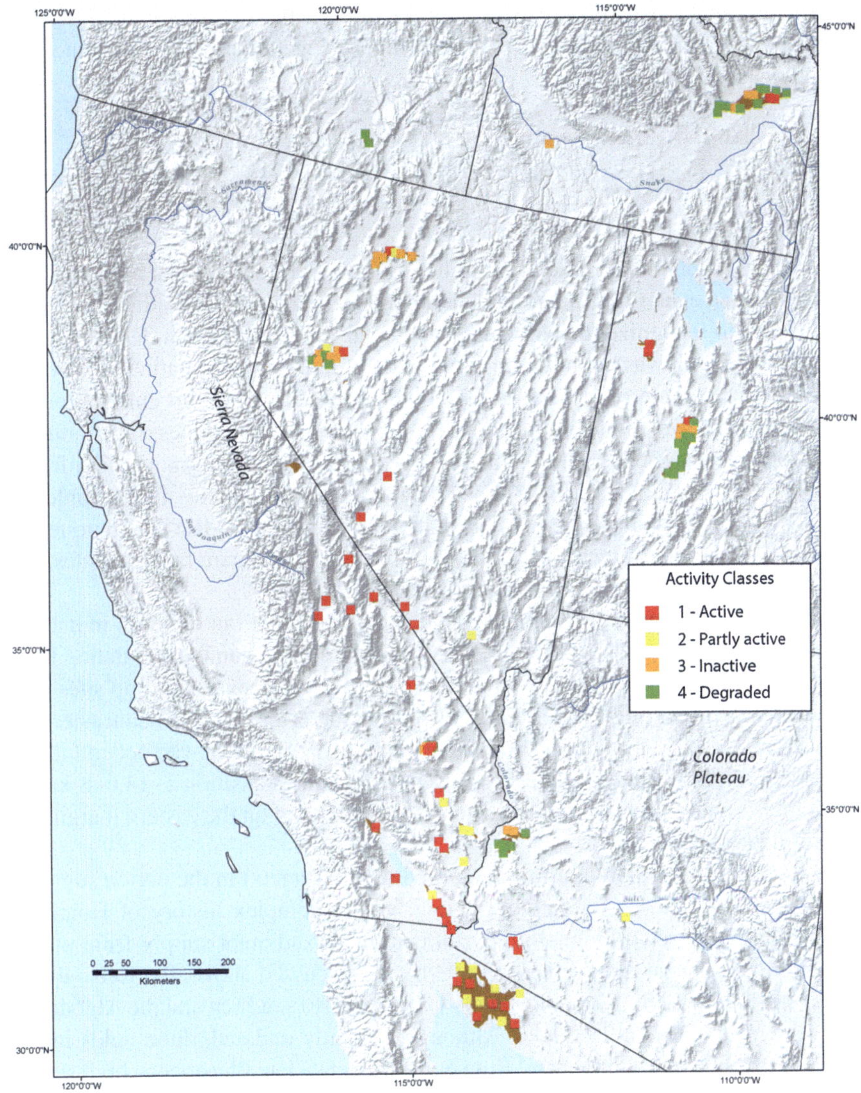


**Fig. 9.8** Spatial distribution of dune field activity classes in Basin and Range dune fields

fields are mostly vegetation stabilized (class 3 and 4), but small areas of active crescentic and parabolic dunes do occur within a generally vegetated dune landscape. The mechanism for this spatial juxtaposition of active and stabilized dunes is not clear, but it may be related to areas of disturbance by fire or human activities. Locally increased wind speed as a result of topography may also be a factor (e.g. eastern Winnemucca dunes). Once established, active dunes tend to persist, as vegetation colonization is impeded by active sand movement (Yizhaq et al. 2007). Towards the south, in the eastern Mojave and Sonoran deserts, wind energy and DP are low, as is effective precipitation (low AI values), but low wind energy favors establishment and persistence of relatively higher vegetation cover and dunes are partially vegetated.

## 9.6 Aeolian History

The history of aeolian accumulation in the southwestern deserts has been summarized by Halfen et al. (2015), building on the pioneering work of Smith (1967), who was the first to recognize the paleo-environmental record of aeolian deposits in the region. However, it was not until the application of luminescence dating in the early 1990s that a chronology of periods of aeolian deposition began to emerge for the region. Studies have employed a variety of luminescence techniques: initial studies used thermoluminescence (TL) techniques on quartz or feldspar grains and infrared stimulated luminescence (IRSL) of feldspar grains; later studies have employed optically stimulated luminescence (OSL) of quartz and post-IR-IRSL techniques on feldspar. The variety of techniques makes it difficult to compare between sites, but general patterns in space and time can be discerned (Fig. 9.9).

Sand ramps and preservation of aeolian sand in alluvial fan deposits indicate a long history of aeolian deposition in the Mojave Desert, in some cases dating back to the mid-Pleistocene (Whitney et al. 2004). Data from many sand ramp sites provide evidence for significant periods of accumulation prior to 20 ka (Lancaster and Tchakerian 2003; Mahan et al. 2007; Rendell et al. 1994). Renewed accumulation of sand occurred during the late Pleistocene-Holocene transition (~14.6–8 ka) at sand ramps in the eastern Mojave and central Mojave, and on the eastern margins of the Kelso Dunes.

In contrast, the available information on dune field age(s) in the region suggests that many dune fields are relatively young, with a complex history of Holocene accumulation and reworking, strongly influenced by sediment supply from proximal fluvial and lacustrine sources (Fig 9.9b). As discussed above, the Kelso dunefield is likely less than 25 ka old, and the Gran Desierto sand sea and the Algodones dunes are younger than 26–31 ka. Although currently undated, dune fields in the basins of paleolakes Lahontan and Bonneville post-date late Pleistocene high stands and are therefore younger than about 13–11 ka.

Short periods of accumulation in the intervals 7.0–4.5. 3–2, 1.7–1.4, 0.8–0.7, 0.5–0.3 and 0.22–0.14 ka (Clarke and Rendell 1998) occurred in source-proximal areas of the Mojave Desert, including the Cronese Basin and Kelso Dunes, where they can be related to periods of enhanced flow in the Mojave River at around 3.8, 0.7–0.8, and 0.4 ka (Enzel et al. 1992; Mahan et al. 2007). Similar relations between increased flooding of fluvial systems and aeolian depositional episodes are also evident at Ash Meadows (Lancaster and Mahan 2012) and in the Gila River valley, where crescentic dunes and sand sheets provide evidence for depositional events at 3.15; 1.95–1.36; 0.8; and 0.7–0.31 ka (Wright et al. 2011). In some locations, desiccation of lake basins following short-lived episodes of late Holocene flooding

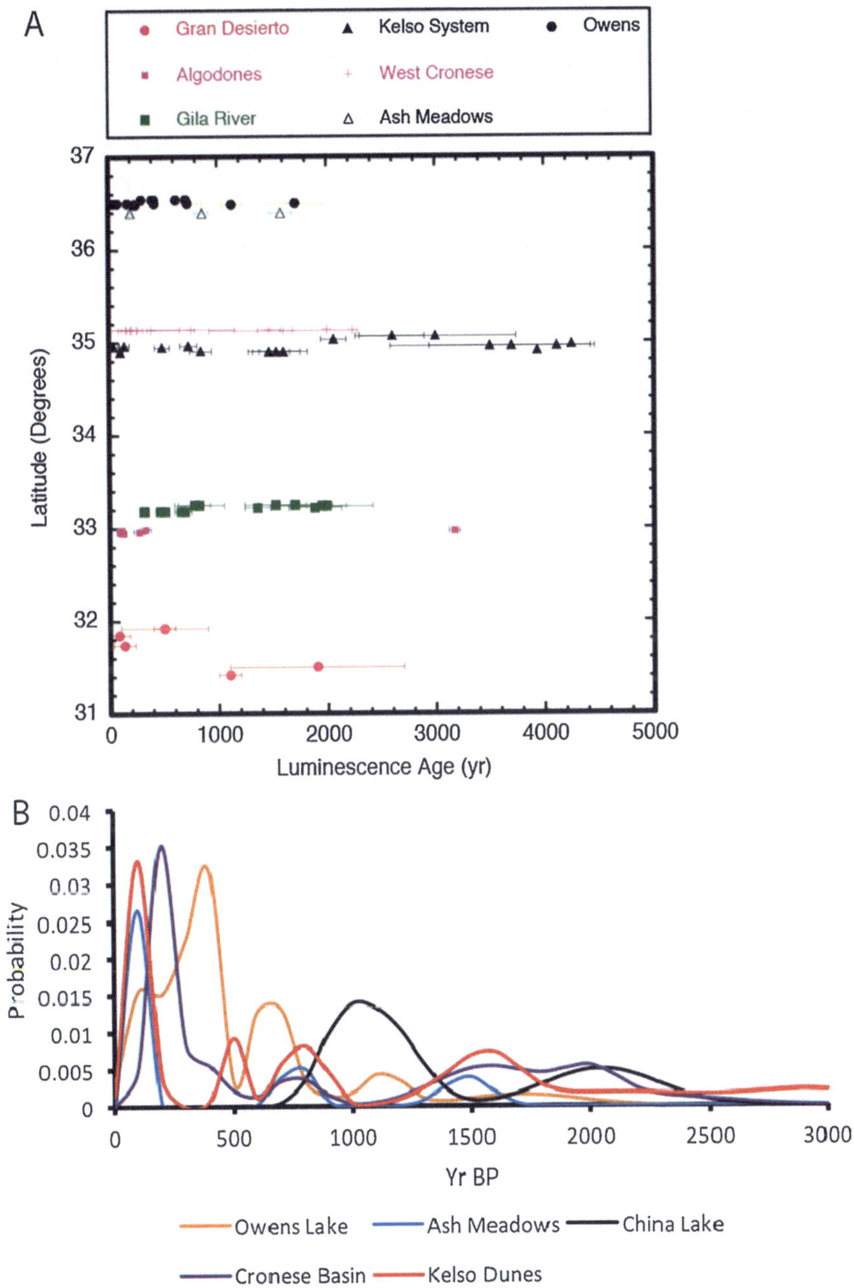


**Fig. 9.9** (**a**) Luminescence-dated Late Holocene periods of dune accumulation in the Basin and Range; (**b**) PDF plot of luminescence dates for selected dune systems in the last 3 ka

provided lagged input of sediment for nearby dune systems at Owens Lake around 0.7, 0.4 and 0.2 ka (Bacon et al. 2018); in the Salton Sea basin (Lake Cahuilla) for the Algodones dunes (Stokes et al. 1997); and at Snow Water Lake in northern Nevada at 0.2–0.3 ka (Munroe et al. 2017).

Widespread Holocene aeolian activity in the southern and western parts of the region is indicative that it is close to sand mobility thresholds, despite low wind energy and sand transport potential in many areas. Sand supply however appears to limit dune building, but where anthropogenic disturbance has occurred, significant increases in sand mobility have resulted (Laity 2003).

## References


Bach AJ (1995) Climatic controls on aeolian activity in the Mojave and Colorado Deserts, California, Department of Geography, Arizona State University, Tempe, 308p

Bach AJ, Brazel AJ, Lancaster N (1996) Temporal and spatial aspects of blowing dust in the Mojave and Colorado Deserts of Southern California. Phys Geogr 17:329–353

Bacon SN, Lancaster N, Stine S, Rhodes EJ, McCarley Holder GA (2018) A continuous 4000-year lake-level record of Owens Lake, south-Central Sierra Nevada, California, USA. Quat Res 90:276–302

Beheiry SA (1967) Sand forms in the Coachella Valley, Southern California. Ann Assoc Am Geogr 57:25–48

Beveridge C, Kocurek G, Ewing R, Lancaster N, Morthekai P, Singhvi A, Mahan S (2006) Development of spatially diverse and complex dune-field patterns: Gran Desierto Dune Field, Sonora, Mexico. Sedimentology 53:1391–1409

Blount G, Lancaster N (1990) Development of the Gran Desierto Sand Sea. Geology 18:724–728

Blount G, Smith MO, Adams JB, Greeley R, Christensen PR (1990) Regional aeolian dynamics and sand mixing in the Gran Desierto: evidence from Landsat thematic mapper images. J Geophys Res 95:15463–15482

Clarke ML (1994) Infra-red stimulated luminescence ages from aeolian sand and alluvial fan deposits from the Eastern Mojave Desert, California. Quat Sci Rev 13:533–538

Clarke ML, Rendell HM (1998) Climatic change impacts on sand supply and the formation of desert sand dunes in the southwest U.S.A. J Arid Environ 39:517–532

Dean LE (1978) The California Desert sand dunes. University of California, Riverside, Riverside, 72p

Derickson D, Kocurek G, Ewing RC, Bristow CS (2008) Origin of a complex and spatially diverse dune-field pattern, Algodones, Southeastern California. Geomorphology 99:186–204

Edgett KS, Blumberg DG (1994) Star and Linear Dunes on Mars. Icarus 112:448–464

Eissmann LJ (1990) Eolian sand transport in Western Nevada. MS Thesis University of Nevada, Reno

Enzel Y, Brown WJ, Anderson RY, McFadden LD, Wells SG (1992) Short-duration Holocene lakes in the Mojave River drainage basin, Southern California. Quat Res 38:60–73

Epps TM, Britten HB, Rust RW (1998) Historical biogeography of *Eusattis muricatus* (Coleoptera: Tenebrionidea) within the Great Basin, Western North America. J Biogeogr 25:947–968

Fryberger SG, Dean G (1979) Dune forms and wind regimes. In: McKee ED (ed) A Study of Global Sand Seas: United States Geological Survey, Professional Paper 1052, pp 137–140

Griffiths PG, Webb RH, Lancaster N, Kaehler CA, Lundstrom SC (2002) Long-term sand supply to Coachella Valley Fringe-toed Lizard (Uma inornata) habitat in the northern Coachella Valley, California. United States Geological Survey, Water-Resources Investigations Report 02-4013, Washington DC

Haff PK, Presti DE (1995) Barchan Dunes of the Salton Sea region, California. In: Tchakerian VP (ed) Desert Aeolian Processes. Chapman and Hall, New York, pp 153–178

Halfen AF, Lancaster N, Wolfe SA (2015) Interpretations and common challenges of aeolian records from North American dune fields. Quat Int 410 (Part B):75–95

Havholm KG, Kocurek G (1988) A preliminary study of the dynamics of a modern draa, Algodones, Southeastern California, USA. Sedimentology 35:649–669

Hoover RH, Gaylord DR, Cooper CM (2018) Dune mobility in the St. Anthony Dune Field, Idaho, USA: effects of meteorological variables and lag time. Geomorphology 309:29–37

Jewell PW, Nicoll K (2011) Wind regimes and aeolian transport in the Great Basin, U.S.A. Geomorphology 129:1–13

Kasper-Zubillaga JJ, Zolezzi-Ruiz H, Carranza-Edwards A, Giron-Garcia P, Ortiz-Zamora G, Palma M (2007) Sedimentological, modal analysis and geochemical studies of desert and coastal dunes, Altar Desert NW Mexico. Earth Surf Process Landf 32:489–508

Katra I, Scheidt S, Lancaster N (2009) Changes in active eolian sand at Northern Coachella Valley, California. Geomorphology 105:277–290

Kocurek G, Lancaster N (1999) Aeolian system sediment state: theory and Mojave Desert Kelso dune field example. Sedimentology 46:505–515

Laity JE (1987) Topographic effects on ventifact development, Mojave Desert, California. Phys Geogr 8:113–132

Laity J (2003) Aeolian destabilization along the Mojave River, Mojave Desert, California: linkages among fluvial, groundwater, and aeolian systems. Phys Geogr 24:196–221

Lancaster N (1988) Development of linear dunes in the southwestern Kalahari, Southern Africa. J Arid Environ 14:233–244

Lancaster N (1989) The dynamics of Star dunes: an example from the Gran Desierto, Mexico. Sedimentology 36:273–289

Lancaster N (1992) Relations between dune generations in the Gran Desierto, Mexico. Sedimentology 39:631–644

Lancaster N (1993a) Development of Kelso Dunes, Mojave Desert, California. Natl Geogr Res Explor 9:444–459

Lancaster N (1993b), Origins and sedimentary features of super bounding surfaces in the north-western part of the Gran Desierto sand sea. In: Pye, K. and Lancaster, N. (eds.) Aeolian sediments: ancient and modern. International Association of Sedimentologists Special Publication 16:71–86

Lancaster N (1994) Controls on aeolian activity: new perspectives from the Kelso Dunes, Mojave Desert, California. J Arid Environ 27:113–124

Lancaster N (1995) Origin of the Gran Desierto Sand Sea: Sonora, Mexico: evidence from dune morphology and sediments. In: Tchakerian VP (ed) Desert Aeolian Processes. Chapman and Hall, New York, pp 11–36

Lancaster N, Helm P (2000) A test of a climatic index of dune mobility using measurements from the Southwestern United States. Earth Surf Process Landf 25:197–208

Lancaster N, Hesse P (2016) Geospatial analysis of climatic boundary conditions governing dune activity. Geol Soc Am Abstr Programs 48(7)

Lancaster N, Mahan SA (2012) Holocene dune formation at Ash Meadows National Wildlife Area, Nevada. Quat Res 78:266–274

Lancaster N, McCarley-Holder G (2013) Decadal-scale evolution of a small dune field: Keeler Dunes, California 1944-2010. Geomorphology 180–181:281–291

Lancaster N, Tchakerian VP (1996) Geomorphology and sediments of sand ramps in the Mojave Desert. Geomorphology 17:151–166

Lancaster N, Tchakerian VP (2003) Late Quaternary eolian dynamics, Mojave Desert, California. In: Enzel Y, Wells SG, Lancaster N (eds) Paleoenvironments and paleohydrology of the Mojave and Southern Great Basin deserts. Geological Society of America, Boulder, pp 231–249

Lancaster N, Greeley R, Christensen PR (1987) Dunes of the Gran Desierto Sand Sea, Sonora, Mexico. Earth Surf Process Landf 12:277–288

Lancaster N, Baker S, Bacon S, McCarley-Holder G (2015) Owens Lake dune fields: composition, sources of sand, and transport pathways. Catena 134:41–49

Mahan SA, Miller DM, Menges CM, Yount JC (2007) Late Quaternary stratigraphy and luminescence geochronology of the Northeastern Mojave Desert. Quat Int 166:61–78

McDonald E, McFadden LD (1994) Quaternary stratigraphy of the Providence Mountains Piedmont and preliminary age estimates and regional stratigraphic correlations of Quaternary deposits in the Eastern Mojave Desert, California. In: McGill SF, Ross TM (eds) Geological investigations of an active margin. Geological Society of America Cordilleran Section Guidebook. San Bernardino County Museum, San Bernardino, pp 205–210

McDonald EV, McFadden LD, Wells SG (2002) Regional response of alluvial fans to the Pleistocene-Holocene climatic transition, Mojave Desert, California. In: Enzel Y, Wells SG, Lancaster N (eds) Paleoenvironments and paleohydrology of the Mojave and southern Great Basin deserts. Geological Society of America, Boulder

Mehringer PJ Jr, Wigand PE (1986) Holocene history of Skull Creek dunes, Catlow Valley, Southeastern Oregon, U.S.A. J Arid Environ 11:117–138

Morrison RB (1964) Lake Lahontan: geology of southern Carson Desert, Nevada. United States Geological Survey Professional Paper 401, 156p

Muhs DR (2017) Evaluation of simple geochemical indicators of aeolian sand provenance: Late Quaternary dune fields of North America revisited. Quat Sci Rev 171:260–296

Muhs DR, Bush CA, Cowherd SD, Mahan S (1995) Source of sand for the Algodones Dunes. In: Tchakerian VP (ed) Desert Aeolian Processes. Chapman and Hall, New York, pp 37–74

Muhs DR, Reynolds RR, Been J, Skipp G (2003) Eolian sand transport pathways in the southwestern United States: importance of the Colorado River and local sources. Quat Int 104:3–18

Muhs DR, Lancaster N, Skipp GL (2017) A complex origin for the Kelso Dunes, Mojave National Preserve, California, USA: a case study using a simple geochemical method with global applications. Geomorphology 276:222–243

Munroe JS, Gorin AL, Stone NN, Amidon WH (2017) Properties, age, and significance of dunes near Snow Water Lake, Elko County, Nevada. Quat Res 87:24–36

Norris RM, Norris KS (1961) Algodones dunes of Southeastern California. Geol Soc Am Bull 72:605–620

Pacheco M, Martín-Barajas A, Elders W, Espinosa-Cardeña JM, Helenes J, Segura A (2006) Stratigraphy and structure of the Altar basin of NW Sonora: implications for the history of the Colorado River delta and the Salton trough. Revista Mexicana de Ciencias Geológicas 23:1–22

Paisley ECI, Lancaster N, Gaddis L, Greeley R (1991) Discrimination of active and inactive sands by remote sensing: Kelso Dunes, Mojave Desert, California. Remote Sens Environ 37:153–166

Pavlik BM (1989) Phytogeography of sand dunes in the Great Basin and Mojave Deserts. J Biogeogr 16:227–238

Pelletier JD (2013) Deviations from self-similarity in barchan form and flux: the case of the Salton Sea dunes, California. J Geophys Res Earth Surf 118:2013JF002867

Pepe NE (2014) The geomorphology, eolian activity, and petrology of the Winnemucca Dune Complex, Humboldt County, Nevada, MS Thesis Geological Sciences. University of Nevada, Reno

Quade J (1986) Late Quaternary environmental changes in the upper Las Vegas Valley, Nevada. Quat Res 26:340–357

Ramsey MS, Christensen PR, Lancaster N, Howard DA (1999) Identification of sand sources and transport pathways at the Kelso Dunes, California using thermal infrared remote sensing. Geol Soc Am Bull 111:646–662

Reheis MC, Miller DM, Redwine J (2007) Quaternary Stratigraphy, Drainage-Basin development, and Geomorphology of the Lake Manix Basin, Mojave Desert, USGS Open-File Report 2007–1281, Reston, VA, 31p

Reheis MC, Bright J, Lund SP, Miller DM, Skipp G, Fleck RJ (2012) A half-million-year record of paleoclimate from the Lake Manix Core, Mojave Desert, California. Palaeogeogr Palaeoclimatol Palaeoecol 365-366:11–37

Rendell H, Lancaster N, Tchakerian VP (1994) Luminescence dating of late Pleistocene Aeolian deposits at Dale Lake and Cronese Mountains, Mojave Desert, California. Quat Sci Rev 13:417–422

Sack D (1987) Geomorphology of the Lynndyl Dunes, west-central Utah, Cenozoic geology of western Utah-Sites for precious metal and hydrocarbon accumulations. Utah Geological Association, Publication 16, Salt Lake City, pp 291–299

Scheidt S, Lancaster N, Ramsey MS (2011) Eolian dynamics and sediment mixing in the Gran Desierto, Mexico determined from thermal infrared spectroscopy and remote sensing data. Geol Soc Am Bull 123:1628–1644

Sharp RP (1966) Kelso Dunes, Mohave Desert, California. Geol Soc Am Bull 77:1045–1074

Sharp RP (1979) Intradune flats of the Algodones chain, Imperial Valley. California Bull geol Soc Am 90:908–916

Shiyuan Z, Ju L, Whiteman CD, Xindi B, Wenqing Y (2008) Climatology of high wind events in the Owens Valley, California. Mon Weather Rev 136:3536–3552

Smith HTU (1967) Past versus present wind action in the Mojave Desert region, California. U.S. Army Cambridge Research Laboratory, 26p

Smith RSU (1982) Sand dunes in the North American deserts. In: Bender G (ed) Reference handbook of the deserts of North America. Greenwood Press, Westport, pp 481–554

Snyder CT (1984) Sand Mountain, in Western Geological Excursions. Geological Society of America, Department of Geological Sciences, Mackay School of Mines, University of Nevada, Reno, pp 137–139

Stokes S, Kocurek G, Pye K, Winspear NR (1997) New evidence for the timing of aeolian sand supply to the Algodones dunefield and East Mesa area, Southeastern California, USA. Palaeogeogr Palaeoeclimatol Palaeocol 128:63–75

Sweet ML, Nielson J, Havholm K, Farralley J (1988) Algodones dune field of southeastern California: case history of a migrating modern dune field. Sedimentology 35:939–952

Sweeney MR, McDonald EV, Chabela LP, Hanson PR (2020) The role of eolian-fluvial interactions and dune dams in landscape change, late Pleistocene–Holocene, Mojave Desert. USA, GSA Bulletin

Wasklewicz TA, Meek N (1995) Provenance of aeolian sediment: the upper Coachella Valley, California. Phys Geogr 16:539–556

Waters MR (1983) Late Holocene lacustrine chronology and archaeology of ancient Lake Cahuilla, California. Quat Res 19:373–387

Whitney JW, Taylor EM, Wesling JR (2004) Quaternary stratigraphy and mapping in the Yucca Mountain area. In: Keefer WR (ed) Quaternary paleoseismology and stratigraphy of the Yucca Mountain area, Nevada. US Geological Survey, Professional Paper 1689, Washington DC

Winspear NR, Pye K (1995) Sand supply to the Algodones dunefield, south-eastern California, U.S.A. Sedimentology 42:875–892

Wintle AG, Lancaster N, Edwards SR (1994) Infrared stimulated luminescence (IRSL) dating of late-Holocene aeolian sands in the Mojave Desert, California, USA. The Holocene 4:74–78

Wright DK, Forman SL, Waters MR, Ravesloot JC (2011) Holocene eolian activation as a proxy for broad-scale landscape change on the Gila River Indian Community, Arizona. Quat Res 76:10–21

Yeend W, Dohrenwend JC, Smith RSU, Goldfarb R, Simpson RWJ, Munts SR (1984) Mineral resources and mineral resource potential of the Kelso Dunes wilderness study area (CDCA-250), San Bernardino County, California. United State Geological Survey, Open File Report 84-647, 19p

Yizhaq H, Askenazy Y, Tsoar H (2007) Why do active and stabilized dunes coexist under the same climatic conditions. Phys Rev Lett 98:188001

Zimbelman JR, Williams SH (2002) Geochemical indicators of separate sources for eolian sands in the eastern Mojave Desert, California, and western Arizona. Geol Soc Am Bull 114:490–496

Zimbelman JR, Williams SH, Tchakerian VP (1995) Sand transport paths in the Mojave Desert, Southwestern United States. In: Tchakerian VP (ed) Desert Aeolian Processes. Chapman and Hall, New York, pp 101–130

The manufacturer's authorised representative in the EU is Springer Nature Customer Service Centre GmbH, Europaplatz 3, 69115 Heidelberg, Germany. If you have any concerns regarding our products, please contact ProductSafety@springernature.com

Printed and bound by CPI Group (UK) Ltd, Croydon, CR0 4YY
23/09/2026
02221546-0012